AF888130

# THE LVOV–WARSAW SCHOOL AND CONTEMPORARY PHILOSOPHY

SYNTHESE LIBRARY

STUDIES IN EPISTEMOLOGY,

LOGIC, METHODOLOGY, AND PHILOSOPHY OF SCIENCE

*Managing Editor:*

JAAKKO HINTIKKA, *Boston University*

*Editors:*

DIRK VAN DALEN, *University of Utrecht, The Netherlands*
DONALD DAVIDSON, *University of California, Berkeley*
THEO A.F. KUIPERS, *University of Groningen, The Netherlands*
PATRICK SUPPES, *Stanford University, California*
JAN WOLEŃSKI, *Jagiellonian University, Kraków, Poland*

VOLUME 273

# THE LVOV–WARSAW SCHOOL AND CONTEMPORARY PHILOSOPHY

*Edited by*

KATARZYNA KIJANIA-PLACEK

*Jagiellonian University, Kraków, Poland*

and

JAN WOLEŃSKI

*Jagiellonian University, Kraków, Poland*

KLUWER ACADEMIC PUBLISHERS

DORDRECHT / BOSTON / LONDON

A C.I.P. Catalogue record for this book is available from the Library of Congress.

ISBN 0-7923-5105-3

Published by Kluwer Academic Publishers,
P.O. Box 17, 3300 AA Dordrecht, The Netherlands.

Sold and distributed in North, Central and South America
by Kluwer Academic Publishers,
101 Philip Drive, Norwell, MA 02061, U.S.A.

In all other countries, sold and distributed
by Kluwer Academic Publishers,
P.O. Box 322, 3300 AH Dordrecht, The Netherlands.

*Printed on acid-free paper*

All Rights Reserved
©1998 Kluwer Academic Publishers
No part of the material protected by this copyright notice may be reproduced or
utilized in any form or by any means, electronic or mechanical,
including photocopying, recording or by any information storage and
retrieval system, without written permission from the copyright owner

# TABLE OF CONTENTS

ARS·LOGICA
BONOS·MORES·COLIT
IN COGITANDO ET
DICENDO
THE FIRST LECTURE OF KAZIMIERZ TWARDOWSKI AT LVOV UNIVERSITY
WARSAW PHILOSOPHICAL SCHOOL

# PREFACE

This volume consists of papers delivered at the conference 'The Lvov-Warsaw School and Contemporary Philosophy', organised in celebration of the hundredth anniversary of Kazimierz Twardowski's first lecture as Professor of Philosophy at Lvov University. This lecture can be regarded as the starting point of the development of analytic philosophy in Poland, which culminated in the Warsaw School of Logic.

The conference was held in Lvov (15–17 November) and Warsaw (19–21 November 1995). It was organised jointly by the Ukrainian Academy of Sciences and the Polish Academy of Sciences. The general organisation was undertaken by Professor Andrzej Grzegorczyk (Polish Academy of Sciences) and Professor Marat Vernikov (Ukrainian Academy of Sciences). Professors Jaroslaw Isaievich (Ukrainian Academy of Sciences) and Jan Woleński (Jagiellonian University) were responsible for the scientific programme.

Over 100 philosophers participated in the conference. Papers published in this volume are organised according to the sections of the conference and represent a selection of the papers delivered.

The editors would like to express their gratitude to Professor Andrzej Grzegorczyk, *spiritus movens* of the conference, who, by including the present volume in a programme of publications connected with the hundredth anniversary of the Lvov-Warsaw School, provided financial support for its preparation. Finally, we express our gratitude to Dr Timothy Childers of the Academy of Sciences of the Czech Republic for correcting the English of the papers.

*Jan Woleński, Katarzyna Kijania-Placek*

# INTRODUCTION

JAN WOLEŃSKI

# THE RECEPTION OF THE LVOV-WARSAW SCHOOL

Every intellectual formation, perceived as a school, has its first and second life. The latter 'exists' in its reception by the subsequent generation, in particular in its influence on the further course of intellectual life, whereas the former is constituted by the deeds of its members. Although my main concern in this paper is the second life of the Lvov-Warsaw School, I think it is reasonable to devote some time to its first life, too, because they are not fully separable.[1] Besides, I would like to note that in speaking of the reception of the Lvov-Warsaw School, I mean not only an international dimension, but also what happened in Poland.

The first life of the Lvov-Warsaw School began exactly one hundred years ago, when Kazimierz Twardowski started his inaugural lecture as a professor of philosophy at the Lvov University. Twardowski came to Poland with a very deliberate and precise plan: he wanted to introduce scientific philosophy into Poland; scientific in the sense of his teacher Franz Brentano. In his 'Self-Portrait' (written about 1926) Twardowski describes this task in the following way: "[ ... ] I felt a calling to disseminate the style of philosophizing that I had learned from Brentano among my countrymen, and in particular to initiate the academic youth into the spirit and method of this philosophy."[2] And more: "Franz Brentano became for me the model of a philosopher who is relentless in his quest for knowledge of the truth, and of a teacher of philosophy in the spirit of antiquity who gathers students around him as younger friends. From him I learned how to strive relentlessly after substantiveness, and how to handle a method of analysis and investigation that, insofar as it is possible, guarantees that substantiveness. He proved to me by example that the most difficult problems can be clearly formulated, and the attempts at their solution no less clearly presented, provided one is clear within oneself. The emphasis that he placed on sharp distinctions that did not lapse into fruitless nitpicking was an important guideline for my own works."[3] As we know, Twardowski succeeded, but the beginnings were difficult: "He found the lecture hall almost empty. Several of his acquaintances [ ... ] and several bolder strangers used to come in, partly out of courtesy, and partly of the curiosity in order to see how the young professor looked and lectured."[4]

The fact that rooms at Twardowski's first lectures were rather empty was connected with the situation of philosophy at the Lvov University at that time. The situation was not good. There was no tradition of systematic studies in philosophy and advanced seminars, there was not even a specialized philosophical library. Thus, Twardowski had to begin almost from scratch, so to speak.



*K. Kijania-Placek and J. Woleńsky (eds.),*
*The Lvov–Warsaw School and Contemporary Philosophy,* 3–19.
© 1998 *Kluwer Academic Publishers.*

However, perhaps it was helpful for his enterprise, because he could introduce his own standards without having to fight a different philosophical tradition. Probably, Twardowski's ambitious project would have been much more difficult if it had had to be undertaken in an established philosophical community. Success came very soon: "Gradually the hall became more filled and soon it could not accommodate all those willing to listen, and after a while the lectures had to be transferred outside the university because no university hall could accommodate the listeners who from the early morning hurried to secure themselves a place."[5] And another more colorful report on Twardowski's lectures (about 1905): "At 8 o'clock, almost with winter dawn, crowds of students push to the university building on St. Nikolaus Str. in order to participate in Twardowski's lecture. A big lecture hall is overfilled, a weak light of gas lamps shines on notes. Although several people are overcome with sleep after the carnival night, there is a lot of noise and animation. The master, still quite young — he is not yet forty years old — comes from the adjacent professor's room and ceremonially takes his place. His lecture is like a sermon. Everybody is listening with great attention to those clear words on problems, very often difficult problems; one can only hear squeak of pencils. And so, almost every day, without cancellations and gaps, without any possibility of delay."[6] There were years in which Twardowski had two thousand participants in his lectures; in 1904/1905 there were more than a hundred members of his seminar and he had to introduce *numerus clausus* to limit further applications.

Twardowski attracted many young people to professional work in philosophy. Jan Łukasiewicz and Władysław Witwicki belong to the first generation (in the first decade, both began to teach in Lvov: Łukasiewicz logic, Witwicki psychology). Then came, among others, Kazimierz Ajdukiewicz, Tadeusz Czeżowski, Tadeusz Kotarbiński and Zygmunt Zawirski; Stanisław Leśniewski joined Twardowski's group in 1910. Let me mention also that Roman Ingarden was Twardowski's student before going to Husserl in Göttingen, and Władysław Tatarkiewicz had close contact with Twardowski after obtaining his Ph.D. in Mahrburg (Tatarkiewicz's Habilitation was in Lvov in 1919). About 1910, one could speak of Twardowski's philosophical school in Lvov. And it was perceived as a new quality in Polish philosophical life. There is a mystery to Twardowski's success as a teacher. All recollections indicate a very deep connection between his students and him. Let me quote three examples: "The seminar was the meeting place for students of all years, beginning with the second. It was there that under the direction and the most careful attention of the professor they prepared themselves for independent scholarly work. Classic philosophical works were read and interpreted jointly (always in the original, which required knowledge of foreign languages). All participants in the seminar would work on their topics, and at the end would submit the results to the professor for appraisal. It often occurred that one's second seminar paper was an independent scholarly contribution and could serve as a doctoral thesis or a thesis for the candidacy for a teaching post in high schools. Twardowski secured his students ideal conditions for work. All members of the seminar could use the reading room, to which they had their own keys, from 7 a.m. to 10 p.m. In that reading room they had their own desks and could avail themselves of the necessary books from a large library (which in 1930 had some 8,000 items). Rigorous and unconditionally binding regulations, thought over in minutest detail, secured the smooth functioning of this unique student's work room. Everyone had the right and the opportunity to have frequent contact with the professor who, every day, invariably received students in his office between noon and 1 p.m. He used to spend some eight to nine hours daily in the seminar, often visiting the reading room and having much personal contact with the members of the seminar. Such was the outer framework of his work, which was an educational activity unique of its kind. Twardowski kept detailed files in which every student had his or her record with the assessment of papers and examinations and a description of interests and achievements. Twardowski also kept all the papers of his disciples in the seminar archive.";[7] "After having found in Poland a fallow land covered with luxuriant weeds, he started to work and began pulling out the weeds and planting nutritious vegetables. That great, wise and extraordinarily assiduous man simply just made his aim to teach the carefree Poles how to work like the Germans do. In that sphere, of course, which was accessible to him. He was mainly concerned with human character. He was to cut to the quick with his frequently-voiced opinion about the Poles: "Die Polen sind ja so unzuverlässig!" Unreliable! To make matters worse, the criticism was justified. Hence this man, dedicated to Poland, and brought up in the atmosphere of German reliability, started waging what was, so to speak, a *Kulturkampf* in Poland [ . . . ] And he started combating short-term enthusiasm, unpunctuality, breaking of contracts, lack of systematic work, dedication to momentary whims, and forced people to be assiduous, to respect organizational principles, to engage in various forms of elementary training, to write rigorous papers and objective summaries.";[8] "Now several months ago the Professor told one of our group that he was happy and he gave the reasons for this. He referred to his age: he had not expected he would live so long. He had seen Poland regain its independence, which was for him a source of permanent joy. For many years he worked with dedication as a professor of philosophy, and had he to choose his profession over again, he would choose the same once again. He particularly enjoyed the attitude of his disciples towards him, an attitude which was quite exceptional and the extraordinary character of which he fully realized [ . . . ] All this was said by a man who was gravely ill but was not broken by the fact; on the contrary, here was a man who had dominated his suffering. This statement of his really reminds one of the last letters written by Epicurus to Hermarchus. And it is not a coincidence, because the Professor deliberately shaped his attitude after the pattern of the old Greek masters of autonomous ethics. This is why during the last years of his life he showed others by his own example how to endure physical pain and turn it into one more opportunity for the victory of concentrated will."[9] When he retired, the students offered him a special medal with the inscription: *Discipulorum amor et pietas*. On the other hand, it is known that he was extraordinarily demanding, bureaucratic, and acted with military rigor. In general, such attributes are not liked by young people. In spite of this, Twardowski continually attracted students. I mentioned the most important of his students of the first period of his activity who specialized in philosophy. But he also trained specialists in other domains of the humanities: literary studies (Juliusz Kleiner, Manfred Kridl, Zygmunt Łempicki), linguistics (Jerzy Kuryłowicz), classics (Ryszard Ganszyniec), psychology (Stefan Baley, Stefan Bachowski, Ludwik Jaxa-Bychowski, Mieczysław Kreutz), and pedagogy (Bohdan Nawroczyński, Kazimierz Sonicki). These graduated before 1914. Then came the Blausteins (Eugenia and Leopold), Izydora Dąmbska, Maria Kokoszyńska, Henryk Mehlberg and Stefan Świeżawski. In sum, Twardowski trained more than thirty full university professors; this perhaps is even the world record in academic teaching. The direct students of Twardowski successfully continued his program. Just before 1939, the Lvov-Warsaw School included about eighty active philosophers, active in the sense of publishing something.

Let me come back to some points of Twardowski's project of building philosophy in Poland. He understood that a serious philosophical life required some organizational framework and a proper atmosphere of research. In 1904, he established the Polish Philosophical Society. A fragment of his inaugural address at the first meeting of the Society gives a good picture of his idea of how philosophy should be done: "The Polish Philosophical Society will not serve any trend in philosophy exclusively because its intention is to cover all trends. It wants to be free from all one-sidedness, it wants to be as versatile as possible. The society's only dogma will be that dogmatism is the greatest enemy of all scholarly work. We want all work and opinions in our society to have a single goal, the clarification of truth. The path to that goal is scholarly criticism."[10] Twardowski applied this pluralistic understanding of philosophy to his own circle. As was reported by one of his students: "The philosophers of the Lvov group were not united by any common doctrine, by a uniform world-view. Not the content of philosophy but rather the method of philosophizing and the common language were the factors which formed the foundation of the spiritual community of those people. This is why the School could produce spiritualists and materialists, nominalists and realists, logicians and psychologists, natural philosophers and art theorists."[11] In 1911, Twardowski began to edit a new journal *Ruch Filozoficzny* (Philosophical Movement), whose main aim was to inform the Polish philosophical community about philosophical life, particularly abroad. The first issue was opened by a short paper on Polish national philosophy by Henryk Struve.[12] The author, the doyen of Polish philosophers at that time, was a professor of the Tsarist university in Warsaw. Struve did a lot for Polish philosophy, in particular for its history. He published an extensive treatise *The History of Logic in Poland as Theory of Knowledge* (1911, in Polish). In general, Struve had a very high opinion of the philosophical skills of the Poles. He maintained that Polish philosophers could achieve a synthesis composed of philosophies produced by other nations, including Germany, France and England. For Struve, the history of Polish philosophy indicated that the main points of such a synthesis had already been made. Thus, the Polish philosophers should continue and develop their own national tradition in philosophy. As to the title of his paper, let me remark that it can be misleading. The label 'Polish national philosophy' usually denotes a type of philosophical thought popular in Poland in the 19th century and connected with national political problems, mainly the loss of independence at the end of the 18th century and the means of its restoration. So called Messianism, strongly influenced by romanticism and German idealism, was the most typical kind of Polish national philosophy. Struve was definitively opposed to national philosophy in this sense. He used the name 'Polish national philosophy' to refer just to that philosophical synthesis which he pointed out in his paper. Twardowski commented on Struve in one of the subsequent issues of the journal in the same year.[13] Twardowski was also radically opposed to Messianism and similar movements; as is usual in the case of a typical Brentanist. However, for Twardowski, the problem of the relation of a national philosophy that is produced by philosophers living in a given country, to philosophies of other nations is not so simple. Twardowski pointed out that there are stronger and weaker philosophical nations. He counted England, Germany and France as countries that produced strong philosophy. Now, if a weaker country begins to develop its own national philosophy, it must be conscious of the following danger: if this nation becomes oriented to one and only one 'strong' national philosophy, its own thought will be dominated by this philosophy and ultimately will disappear, or at least will not be able to produce an original philosophy. Although Twardowski did not use these words, his perspective can be described as based on the conviction that there are centers and peripheral regions in the world of philosophy. Without any hesitation, he located Polish philosophy at the periphery. Of course, his plan was to move this philosophy more towards the center. How? Twardowski's recommendation was that Polish philosophy should be able to employ ideas produced in several strong countries. This would prevent the domination of any particular national philosophy over

Polish thought. Seven years later, Twardowski illustrated his point with a concrete example,[14] namely, he discussed the question of a good textbook on the history of philosophy in Polish. He warned against translations from English, German or French, because textbooks coming from these countries placed an excessive stress on particular national philosophies. For Twardowski, Poland needed a textbook which would be, so to speak, equidistanced from strong national philosophies. Incidentally, Tatarkiewicz's *History of Philosophy* (in Polish), which was very popular in Poland, was a realization of Twardowski's idea.[15] Nevertheless, it is one thing to say how an 'equidistanced' textbook on the history of philosophy should be written, but another thing to give a program for a national philosophy, consisting of the implementation of ideas produced in different countries. We do not have much data for evaluating whether Twardowski's students followed his idea. In fact, only Ajdukiewicz once remarked that his interests in French conventionalism were partly motivated by an explicit intention to restrict German influence on Polish philosophy.[16] Anyway, it is an interesting fact that a devoted student of Brentano's, from the very beginning of his academic career, decided to be fairly self-limited in his following of German-language philosophy. It is interesting to look at the development of Twardowski's heritage in this way. And only if we look at the second life of the tradition that originated in Poland with him will we have a chance to answer to what extent Twardowski succeeded in transforming Polish philosophy from the periphery to one located closer to the world center.

It is not known what Twardowski wanted to implement, except Brentanism, in Polish philosophy. Probably all ideas which were more or less similar to that of Brentano. For example, it was probably quite deliberate that he decided to translate, together with Łukasiewicz, Hume's *An Enquiry Concerning Human Understanding* into Polish. Perhaps he was even not quite sure what should be assimilated by his students. Following his pluralistic pattern of doing philosophy, he was inclined to give them great freedom in their investigations and topics chosen for Ph.D. theses. Certainly, the general philosophical environment brought by Twardowski to Poland favored some fields, but not others. Logic in its broad sense belonged to the favorites, but metaphysics in the German style did not. However, Twardowski certainly did not predict that mathematical logic would become a philosophical guide for the most important part of the school created by him.

The career of mathematical logic in Poland is a remarkable fact for everybody interested in the history and sociology of science.[17] The first academic contact of Poles with mathematical logic took place in 1898/99 when Twardowski lectured in Lvov on reformatory tendencies in logic, including the ideas of Boole and Schröder. Thirty years later the Warsaw School of Logic was fully recognized all over the world. A period of one generation was sufficient to build a great scientific school, although the starting point was close to nought. The

Warsaw School of Logic was the commom opus of philosophers and mathematicians. Thus, two serious projects of building Polish science met: Twardowski's program of Polish national philosophy and Zygmunt Janiszewski's program of Polish national mathematics. It is perhaps interesting to note that both projects arose in Lvov, because Janiszewski was an assistant of Wacław Sierpiński and certainly both discussed the question of how to develop Polish mathematics even before 1914. Janiszewski's main idea was very simple. Since Polish mathematicians had no chance of competing with foreign mathematicians in traditional mathematical fields like algebra, analysis or geometry, they should concentrate on novelties like set theory, topology and their applications in classical mathematics. Janiszewski's program ascribed a considerable role to mathematical logic and the foundations of mathematics. When Warsaw University was being organized after 1918, Warsaw mathematicians took a very risky step: they agreed that two philosophers, Łukasiewicz and Leśniewski were to be appointed as professors of logic in a mathematical environment. The Warsaw School of Logic, with Alfred Tarski as its most distinguished representative (Stanisław Jaśkowski, Adolf Lindenbaum, Andrzej Mostowski, Moses Presburger, Jerzy Słupecki, Bolesław Sobociński and Mordechaj Wajsberg also belonged to this group), was a product of the meeting of philosophy and mathematics in Warsaw. Tarski himself clearly pointed out Twardowski's role in this development: "Almost all researchers, who pursue the philosophy of exact sciences in Poland, are indirectly or directly the disciples of Twardowski [ . . . ]."[18] The Lvov School became the Lvov-Warsaw School.

Twardowski himself was not a fan of mathematical logic. He even warned his students against overemphasising logical tools in philosophy. However, the Warsaw School of Logic is certainly the greatest reason for the glory of his tradition. On the other hand, there is something dangerous in perceiving the Lvov-Warsaw School through the lens of the Warsaw School of Logic. Obviously, there is no reason to neglect the logical achievements of Twardowski's students as contributions to philosophy, the more so as their authors admitted this. Nevertheless, it is a mistake to reduce the Lvov-Warsaw School to a group of logicians. First, it is unfair with respect to people who did not work in logic, even in its wide sense, for they achieved remarkable results, like the Blausteins in phenomenological psychology or Tatarkiewicz in ethics. Perhaps Twardowski himself is the greatest victim of this way of looking at the Lvov-Warsaw School. Too often he is regarded only as a teacher, although his own works were important and influential. Second, the transformation of the Lvov School into the Lvov-Warsaw School did not consist only in the rise of the Warsaw School of Logic. Kotarbiński organized around himself a powerful group of philosophers in Warsaw (among others, Janina Hosiasson, Maria Ossowska, Dina Sztejnbarg and Stanisław Ossowski). Moreover, Twardowski's tradition was cultivated by himself and Ajdukiewicz in Lvov after 1918, and Twardowski's students were also professors in other academic centers — Czeżowski in Vilna, and Zawirski in Poznań and later in Krakow. Third, the glory of the Warsaw School of Logic is sometimes automatically transmitted to the whole of Twardowski's tradition. Fourth, logic, as I already remarked, can be understood in the strict or wide sense. When the former understanding is adopted, logic comprises logical calculi plus formalized metalogic (metamathematics), but, on the latter meaning, semantics and methodology of science are also part of it. It is fairly obvious that logic in its wider understanding is 'more' philosophical than logic as a mathematical science. Many works done in the Lvov-Warsaw School belonged to logic, but rather to semantics or philosophy of science than to purely mathematical logic. Once more: although there is no reason to exclude the Warsaw School of Logic from the Lvov-Warsaw School, the latter was not only a logical school. It was a school in analytic philosophy. It is important for a popular identification of analytic philosophy with Anglo-Saxon and the rest with continental traditions. The Lvov-Warsaw School was analytic with continental roots, speaking geographically.

Let me come back to Twardowski. As I already remarked, he is often perceived only as a teacher and organizer of philosophical life in Poland. But this is very far from the truth. Twardowski was a very important philosopher. Reinhardt Grossman says in the introduction to his English translation of *Zur Lehre vom Inhalt und Gegenstand der Vorstellungen* that it is a link between Brentano, Meinong and Husserl.[19] Its historical role is obscured by attributing too great an importance to Husserl's criticism of Twardowski. It sometimes creates the impression that a proper distinction of content and object was achieved by Husserl himself. This is not correct. The distinction was made and further elaborated by Twardowski, and the situation was so perceived by most scholars in the first thirty years of our century, not only in philosophy. Also Meinong's debt to Twardowski was considerable, but it is more transparent than Husserl's case: Meinong admitted that his theory of objects was essentially influenced by Twardowski. However, Twardowski's *Zur Lehre* also had other important points. For example, it was the first work in which Bolzano was seriously analyzed. Hence, Twardowski essentially influenced the further interest in Bolzano. Twardowski also offered a method of philosophizing which was of importance for the development of philosophical analysis in Poland. Among other things discussed by Twardowski in his book, there was a question of objectless presentations. One of the arguments in favor of such presentations says that we have the presentation of nothing. Twardowski proposes the following analytic strategy. We have a correlation between presentations, their contents and their objects on the one side, and names, their sense (meaning) and their designata on the second. Thus, we can investigate names and their properties in order to derive something on presentations and their properties. In the case of nothing, we should analyze the word 'nothing'. Now Twardowski demonstrates that this word is not a name. Apparently it is a name obtained by so called *infinitatio*, that is putting the negation before a name, from the word 'thing'. Twardowski observes that 'thing' is the most general name, equivalent with the name 'object'. There is an old rule for *infinitatio*: to perform this operation on a name $A$ in order to obtain the name not-$A$, both $A$ and not-$A$ must be subordinated to a generic name $B$. However, the name 'thing' as the most general cannot fall under a more generic name. Then, the word 'nothing' is not a result of *infinitatio* performed on 'thing'. Twardowski concludes that 'nothing' is not a name at all. Hence it is not proper to say that nothing is denoted by 'nothing'. For Twardowski, 'nothing' is a syncategorematic word, in fact a hidden quantifier, and the sentence 'nothing is eternal' means 'there is not an $x$ such that $x$ is eternal'. Historically speaking, Twardowski anticipates a well-known criticism of Heidegger by Carnap who remarked that Heidegger's speculations on the Nothingness which nothings, are based on grammatical misunderstanding of the word 'nothing' consisting in regarding it as a noun.[20] What is more important, Twardowski in his analysis of the word 'nothing' in the context of objectless presentations introduced a certain method of philosophical analysis, namely a method of translation of philosophical problems into semantical ones. Of course, this immediately brings to mind a translation from the material to formal mode of speaking recommended by logical empiricism. However, there is an important difference between the method used by Twardowski and subsequently executed by his followers, in particular Ajdukiewicz, Kotarbiński, Leśniewski and Tarski, and the way of doing philosophy by the Vienna Circle. The transformation of the material into the formal mode of speaking was essentially a 'desemantization' of a language, a passing from its semantic to purely syntactic properties. The method initiated by Twardowski was quite different: its aim was to clarify the logical function of words in a way which preserves semantic features of language. In fact, Twardowski's distinction of content and object had not only concrete merits for descriptive psychology, but also was decisive for the subsequent development of semantics in Poland. It is true that Twardowski consciously restricted his scientific activity in favor of his duties as a teacher. However, one of his later ideas must also be mentioned, namely the distinction of actions and product.[21] Among other things, Twardowski distinguished psychophysical actions and their products. Durable psychophysical products are a special kind of product. In particular, language and everything that it is coded in it — for example, science — belong to this category. This view determined the way of treating language and science in the Lvov-Warsaw School as objects created by real human actions, but then relatively independent of them. Someone might expect that I consider this solution as a correct account of the ontology of language, art or science. No, I see the importance of Twardowski's distinction of actions and products in another point. Several commentators on Tarski's seminal works of truth are puzzled by its title *The concept of truth in formalized languages*, and his remark that the problem of truth has no intelligible meaning for languages considered as purely formal entities without any intuitive meaning attached to them.[22] The problem is that formalized languages are usually defined as purely syntactic items without any interpretation. The puzzle is that Tarski considered the problem of truth for formalized languages, but he explicitly said that they were not formalized. There is a simple solution. A person educated in Twardowski's tradition could not consider language as a product in Twardowski's sense and deny that its signs were devoid of any meaning.[23] Every (perhaps with some restrictions) language can be formalizable, but the formalization, always given in informal metalanguage, is always related in a way to meaning-intentions of actions which create a language in questions or actions creating a given formalization.

Recently there has been an attempt to show that a general conception of language underlies the development of semantics in the 20th century philosophy. Hintikka distinguishes language as a reinterpretable calculus and language as a universal medium.[24] The latter conception was accepted by Frege, Russell and Wittgenstein, and effectively blocked semantic thinking in these philosophers, because the conception of language as a universal medium does not allow the language/metalanguage distinction which is a key stone of semantics. On the other hand, Husserl, according to Hintikka, adopted the conception of language as a reinterpretable calculus which was necessary for the development of formal semantics. It is not strange that this development started in Poland under the explicit influence of Husserl, but also in the environment created by Twardowski and his distinctions of content and object and of actions and products. This remark is a contribution to the history of continental analytic philosophy, something which is *contradictio in adiecto* for most historians of contemporary philosophy.

'Would-questions' are very risky in historical reconstructions. But let us try to speculate. It is not quite clear what would have happened in Polish science in the absence of dramatic political circumstances. Perhaps Poland would not have been rebuilt as an independent country? Perhaps the Polish Mathematical school would not have risen? Then, probably the Warsaw School of Logic would not have appeared either? Assumed conterfactually, what would have happened with Twardowski's tradition? Fortunately, we have an analogy; Brentanism, which was a very powerful movement at the beginning of our century. It dominated most the universities in the Austro-Hungarian Empire and was also popular in Germany. However, this philosophy was killed in the analytic world by more logically oriented philosophies. It is, of course, not excluded that the Lvov School would have evolved toward a sort of logical empiricism. However, I think that its fate would probably have been more similar to that of the Graz or Prague schools which either did not assimilate modern logic to a sufficient degree (Graz) or were not able to do this at all (Oskar Kraus and orthodox Brentanism). If I am right, the encounter of the Lvov School and modern logic was of the utmost importance for the development of Polish analytic philosophy.

The international career of the Lvov-Warsaw School began in the early thirties. The first volume of *Erkenntnis* (1930) contains a brief notice, mainly bibliographical on 'the Warsaw Group and related circles in Poland' (perhaps it is interesting that the Vienna Circle manifesto *Wissenschaftliche Weltauffassung Der Wiener Kreis* published in 1929 mentions no Polish name). Then, visits of logical empiricists to Poland (Menger, Carnap) and visits of Poles to Vienna (Tarski) established mutual contacts and recognition. The members of the Lvov-Warsaw School took part in all congresses for the Unity of Science. They were particularly active in Prague in 1934 and Paris in 1936. The congress in 1940 was planned for Warsaw. The appearance of Poles in the world philosophical scene was quite impressive. This is documented by all recollections. Carnap's report is this: "I found that the Polish philosophers had done a great deal of thoroughgoing and fruitful work in the field of logic and its applications to foundational problems, in particular the foundations of mathematics, and the theory of knowledge and the general theory of language, the results of which were almost unknown to philosophers in other countries. I left Warsaw grateful for many stimulating suggestions and the fruitful exchange of ideas which I had enjoyed".[25] Menger gives the following account: "It was not only the formal logical work of the Polish logicians, however, that impressed me, I also noted that they were interested in philosophical problems similar to some discussed in the Vienna Circle, but they attacked them in connection with, and partly on the basis of, their exact logical studies. They always confined themselves to concrete *questions* and completely eschewed those from vague generalities which seemed to becloud some of the Vienna discussions in the late 1920's".[26] And certainly, one of Twardowski's dreams was fulfilled: others came to Poland to learn something, for example logic, as Quine did. Thus, Poles were not only customers of foreign philosophy, they were also able to offer something in exchange. The war interrupted these contacts. After 1945, it would be difficult to reestablish them for various and complex reasons. International philosophical life was very chaotic in the first after-war years. Then, political limitations became too strong. For example, fifteen Poles were invited to the International Philosophical Congress in Amsterdam in 1948, but only two could go; the rest did not obtain passports. It was the first mass restriction of this type in communist Poland. But not the last. Although the Iron Curtain became more or less transparent after 1956, it was not fully opened before 1989, whereas the situation in Poland was certainly relatively better than in other regions of the Soviet block.[27]

There is a historical problem of whether it is correct to speak of the Lvov-Warsaw School after 1945. Certainly, it ceased to be a compact group. Twardowski and Leśniewski died just before 1939. Several members of the School became victims of the war, particularly Jews. Several died in the first years after 1945. Finally, many emigrated abroad, including those so important as Tarski and Łukasiewicz. If the School still persisted, it was rather in the individual works of its representatives, but not as a common enterprise. Certainly, the political situation was not favorable to the Lvov-Warsaw School, because it was unconditionally considered as bourgeois by Marxists. However, other factors were also important. Logic became more mathematical than philosophical, psychology completely separated from philosophy. Also the postwar period was rather unsympathetic towards 'scientistic' philosophers. In sum, although the analytic tradition never ended in Poland, the Lvov-Warsaw School certainly belongs to history. But it does not mean that it is not relevant.

If we are interested in the international reception of a concrete intellectual movement, the first question we should ask is: are there ideas which considerably influenced the world course of a given activity? In the case of the Lvov-Warsaw School, certainly five points are obvious: (a) Tarski's semantic theory of truth; (b) many-valued logic; (c) Leśniewski's systems; (d) categorial grammar; (e) Polish work in the history of logic. Although all are closely connected with logic, I regard them here as logical ideas which influenced not only logic, but also philosophy. I shall now expand on these points.

Re (a): Three distinguished contemporary philosophers entirely changed their views under Tarski's direct and explicit influence. Kazimierz Ajdukiewicz abandoned his radical conventionalism for semantic epistemology. Rudolf Carnap passed from the theory of language exclusively limited to logical syntax to a richer theory based on semantics. Karl Popper decided to base his philosophy of science on the concept of truth and came to realism. Moreover, Tarski's semantic ideas stimulated many important developments in the philosophy of language and the philosophy of science, for example Donald Davidson's approach to meaning by truth-conditions or realism in formal analysis of empirical theories. On the other hand, several other important developments, for instance Jaakko Hintikka's game-theoretical semantics, appeared, partly at least, as alternatives to Tarski's proposals. In spite of frequent declarations that Tarski's theory of truth is devoid of any philosophical meaning, its philosophical life had been rich from the very beginning. In Ayer's words (on the Paris Congress in 1935): "Philosophically the highlight of the Congress was the presentation by Tarski of a paper which summarized his theory of truth."[28]

Re (b): This is also obvious. The rise of many-valued logic opened a new chapter in applications of logic to philosophy and philosophical discussions on logic itself.

Re (c): Leśniewski's ideas are certainly not standard and orthodox. Some think that they will win in logic. Independently of this forecast, they certainly consitute an established chapter of contemporary logical and philosophical investigations. It is well documented, for example, by the fact that a special session on Leśniewski could be organized for this conference.

Re (d): This is a very special idea, but it certainly determined the contemporary thinking about language.

Re (e): These investigations, initiated by Łukasiewicz, opened a new chapter in the history of logic and completely changed the earlier picture of the development of logical ideas.

The second typical question when we discuss the problem of the reception is to find views more or less known and influential but present in a way in the international circulation of scientific results, anticipations of important ideas, and, finally, views which deserve to be known but are not. There are obvious difficulties in collecting a list of such items. Borderlines are not precise. Qualifications are motivated by subjective measures. Nobody knows everything. In consequence, what is for someone merely an anticipation appears to another person as a more or less circulated idea. However, if I were to make such a list, it would certainly include: praxiology, reism, radical conventionalism, Łukasiewicz's analysis of determinism *via* many-valued logic, Hosiasson's inductive logic and her solution of the paradox of confirmation, Mehlberg's causal theory of time, Ajdukiewicz's method of elimination of context, the attempt to modernize scholastic philosophy by logic in the Krakow Circle (a group of Catholic philosophers with Józef Bocheński and Jan Salamucha as leading members; this circle was very strongly influenced by the Lvov-Warsaw School), Zawirski's attempt to combine many-valued logic and probability theory into one system (or personally, to combine views of Łukasiewicz and antiinductionism and its general Reichenbach), Ajdukiewicz's theory of language and his criticism of various forms of idealism *via* semantics, Ginsberg's analysis of Husserl's theory of parts and wholes, and Tatarkiewicz's monumental history of aesthetics.[29] For anticipations, one can include some of those mentioned in the last section, particularly Hosiasson's inductive logic, but also Łukasiewicz's criticism of induction and his general antiinductionism (in this respect Łukasiewicz anticipated all the essential ideas of Popper) and the antifoundationalism of Łukasiewicz, Poznański and Wundheiler, and Ajdukiewicz.[30] Among ideas which deserve to be known, one can include metaethical views of the Lvov-Warsaw School which combine cognitivism with antinaturalism, Czeżowski's theory of transcendentalia, Kotarbiński's radical realism as an epistemological solution, Ajdukiewicz's later theory of fallible inferences, Polish classifications of reasonings or works of the Lvov circle of analytic phenomenology (the Blausteins, Auerbach). Even if this list is not complete (it is certainly not), it is quite impressive. Remembering that we are dealing with philosophy at the peripheries, we find a lot of ideas that were at least interesting, and sometimes quite influential.

Let me now pass to the importance of the Lvov-Warsaw School in Poland. It is very difficult to estimate the real influence of this group in Poland. Generations of students learned philosophy from textbooks written by Ajdukiewicz, Czeżowski, Kotarbiński and Tatarkiewicz and participated in classes conducted by the masters of the School. Moreover, generations of those who graduated from secondary schools were taught by teachers with a background in the Twardowskian tradition. Certainly, it had to influence the thought of the Polish intelligentsia and its attitude to the world. It is still a task for historians of ideas to investigate this question more closely. It is also a problem of evaluating the influence of the Lvov-Warsaw School on contemporary Polish philosophy. Not everybody considered it to be positive. For example, Roman Ingarden and Henryk Elzenberg were inclined to think that the School pushed Polish philosophy too much into scientism. This controversy is of course unresolvable, because it obviously depends on concrete metaphilosophical preferences. Independently of that, the dominance of the Lvov-Warsaw School in Polish philosophy was a historical fact and it had historical consequences. The School equipped Polish philosophers with very high standards of professional work which successfully protected Polish philosophy against corruption in the difficult postwar time. Paradoxically, it was the Marxists in Poland who adopted professional criteria of doing philosophy rather than converting their opponents to their ideology.[31] Now when we have full freedom in philosophical choices, the tradition of the Lvov-Warsaw School is still a strong barrier for postmodernist games. As I already noted, the Lvov-Warsaw School belongs to history, but it is still relevant. My own experience is this. When I decided to write a monograph on the School (see note 1), my motivation was purely historical. My objectives consisted in reminding my own countrymen of this remarkable event in Polish intellectual development and in arguing against a very common interpretation of the Lvov-Warsaw School as a branch of logical empiricism. My view was from the beginning that the philosophy of the School was rather a successful combination of the Brentanist tradition with modern mathematical logic. However, very soon I received several letters from scholars abroad that there was a continuous interest in the Polish analytic tradition of which I had not been aware. Thus, it became rather clear that although the School belongs to history, its relevance is not restricted only to these ideas which I listed above as present in the main stream of world philosophy. In particular, this link of Brentanist tradition and logic presented itself as something that could and even should be reviewed. And development of the nominalistic and reistic views of Brentano and Kotarbiński *via* Leśniewskian logic can be considered as one good example, and a mereological formalization of Husserl as a second.[32]

There are also some general problems to be discussed. Firstly, there is a fatal and historically unjustified gap between analytic and continental philosophy.

I do not deny that Heidegger, Derrida and similar thinkers constitute a special group of philosophers who are certainly opposed to the British analytic tradition. But it is also important to see that a good amount of analytic philosophy was also produced on the continent. Logical empiricism, Scandinavian groups and the Lvov-Warsaw School belong just to this stream and this fact must be respected. Admirers of Polish analytic philosophy sometimes say that it was as precise and clear as British philosophy and as systematic as German. This is a great compliment, probably too great. But perhaps the Polish experience showed that there is not a real conflict between clarity and precision on the one hand, and a systematic character on the other. Perhaps the Polish analytic tradition really points out how to refresh analytic philosophy: its British sources seem overexploited and dry.

There are also lessons to be derived for nations not located in the centers of culture. In spite of attempts at the unification of Europe or even the world, there is still a long life of divisions of centers and peripheries. The story of the Lvov-Warsaw School shows how difficult it is to enter into the main stream of culture, how many factors cause it, and how many good ideas are lost because they are not produced in the center. How important are, for example, the matters of language? Certainly, if Polish philosophers had published more in Western languages, their ideas would have been known better. After some works were translated, we often encountered complaints that it was too late.[33] It is nice to hear such things, but 'too late' means here that the work in question has only a historical meaning. The lessons are obvious. Nobody will popularize our own culture for us. It is a popular reaction to be inclined to a national xenophobia if our achievements are not properly evaluated: Others are guilty. This is a very stupid reaction. Of course, every nation has a duty to protect its own culture against domination from the outside. However, this culture will die very quickly if separated from the world. Barry Smith, one of the friends of the Lvov-Warsaw School, once delivered a lecture 'Why there is no Polish philosophy?'. At first, I was very surprised. However, Smith's intention was to show that, excluding the episode of Polish Messianism, there is no Polish philosophy in the sense of German philosophy, that is, as a national concern. That German (French, British, and perhaps some other) philosophy is possible is probably a privilege of nations acting in the centers. We Poles or Ukrainians (I do not mention Russians, let them speak for themselves) do not belong to this circle, and I am afraid that it will remain so. Our fate is to produce good philosophy, if we want to be recognized. According to Barry Smith, there is no Polish philosophy because Poles produced good philosophy. Perhaps it is the greatest compliment that could be addressed to the Lvov-Warsaw School by someone who belongs to a nation from the cultural center of the world. To protect this tradition is our duty. It is also our duty to remember the places where it arose, independently of their historical fate and independently of whether their present inhabitants want to continue the tradition in question. I regard this conference as an occasion for reflection on these matters. If we want to preserve our self-identity and find a place in the world, we must not avoid them.

*Institute of Philosophy, Jagiellonian University*

NOTES

[1] Most matters discussed in this paper are more fully treated in my book *Logic and Philosophy in the Lvov-Warsaw School*, Kluwer Academic Publishers, Dordrecht 1989 (this book contains an extensive bibliography, including information on foreign translations of works by members of the Lvov-Warsaw School). See also Z. Jordan, *The Development of Mathematical Logic and of Logical Positivism in Poland between Two Wars*, Oxford University Press, Oxford 1945; Z. Jordan, *Philosophy and Ideology*, D. Reidel Publishing Company, Dordrecht 1963; H. Skolimowski, *Polish Analytical Philosophy*, Routledge and Kegan Paul, London 1967; and *Polish Scientific Philosophy. The Lvov-Warsaw School*, ed. by F. Coniglione, R. Poli and J. Woleński, Rodopi, Amsterdam 1993. There are also two collections of interest: *Kotarbiński: Logic, Semantics and Ontology*, ed. by J. Woleński, Kluwer Academic Publishers, Dordrecht 1990, and *Heritage of Kazimierz Ajdukiewicz*, ed. by V. Sinisi and J. Woleński, Rodopi, Amsterdam 1995.

[2] K. Twardowski, 'Selbstdarstellung', *Grazer Philosophische Studien* **39**, p. 14.

[3] Ibid., p. 6.

[4] W. Witwicki, 'Kazimierz Twardowski' (in Polish), *Przegląd Filozoficzny* **23**, 1920, p. XI.

[5] W. Witwicki, 'Kazimierz Twardowski', p. XI.

[6] S. Łempicki, 'Kazimierz Twardowski's role in university and society' (in Polish), in *Kazimierz Twardowski: Teacher, Scientist, Citizen* (in Polish), Polskie Towarzystwo Filozoficzne, Lvov 1938, p. 31.

[7] I. Dąmbska, 'Philosophy at John Casimir's University in Lvov' (in Polish), *Zeszyty Lwowskie* **2**, 1972, p. 79.

[8] T. Kotarbiński, 'Kazimierz Twardowski' (in Polish), *Pion* **21**, 1936, p. 3.

[9] T. Kotarbiński, 'Przemówienie u trumny Kazimierza Twardowskiego' (Eulogy for Kazimierz Twardowski), *Ruch Filozoficzny* **14**, 1938, pp. 12–13.

[10] K. Twardowski, 'Address on the occasion of the inauguration of the Polish Society in Lvov' (in Polish), *Przegląd Filozoficzny* **7**, 1904, pp. 241–242.

[11] I. Dąmbska, 'Forty years of philosophy in Lvov, 1898–1938' (in Polish), *Przegląd Filozoficzny* **44**, 1948, p. 17.

[12] H. Struve, 'A word on Polish national philosophy' (in Polish), *Ruch Filozoficzny* **1**, 1911, pp. 1–3.

[13] K. Twardowski, 'One more word on Polish national philosophy' (in Polish), *Ruch Filozoficzny* **1**, 1911, pp. 113–115.

[14] See K. Twardowski, 'On the needs of Polish philosophy', *Nauka Polska* **1**, 1918, pp. 453–486.

[15] First published in 1931 and then reedited several times.

[16] This information is repeated after J. Giedymin, *Science and Convention*, Pergamon Press, Oxford 1982, p. XV.

[17] See J. Woleński, 'Mathematical logic in Poland 1900–1939: people, circles, institutions, ideas', *Modern Logic* **5**, 1995, pp. 363–405.

[18] A. Tarski, '[A letter to Otto Neurath, 24 IV 1930]', trans. from German by J. Tarski, *Grazer Philosophishe Studien* **43**, 1992, p. 20.

[19] R. Grossmann, 'Introduction', in K. Twardowski, *On the Content and Object of Presentations*, Nijhoff, The Hague 1977, p. VII; the German original was published in 1894.

[20] R. Carnap, 'Überwindung der Metaphysik durch logische Analyse der Sprache', *Erkenntnis* **2**, 1932, pp. 219–241.

[21] K. Twardowski, 'On actions and products. Some remarks from the borderline of psychology, grammar and logic' (in Polish), in *Księga Pamiątkowa ku uczczeniu 250-tej rocznicy założenia Uniwersytetu Lwowskiego przez króla Jana Kazimierza* (Commemoration Book Celebrating the 250–th Anniversary of the Foundation of the Lvov University by the King Jan Kazimierz), Uniwersytet lwowski, Lvov 1912, pp. 1–33; Eng. trans. in K. Twardowski, *Selected Writings*, ed. by J. Brandl and J. Woleński, Rodopi, Amsterdam (forthcoming).

[22] A. Tarski, *The Concept of Truth in Formalized Languages*, in A. Tarski, *Logic, Semantics, Metamathematics*, 2nd ed., Hackett, Indianapolis 1983, p. 166.

[23] For Tarski's philosophical view see my papers 'Tarski as a philosopher', in *Polish Scientific Philosophy: The Lvov-Warsaw School*, ed. by F. Coniglione, R. Poli, and J. Woleński, Rodopi, Amsterdam, 1993, pp. 319-338 and 'On Tarski's background', in *From Dedekind to Gödel*, ed. by J. Hintikka, Kluwer Academic Publishers, Dordrecht 1995, pp. 331–341.

[24] J. Hintikka, 'On the development of the model-theoretic viewpoint in logical theory', *Synthese* **77**, 1988, pp. 1–36.

[25] R. Carnap, 'Intellectual Autobiography', in *The Philosophy of Rudolf Carnap*, ed. by P. Schilpp, Open Court, La Salle 1963, p. 31.

[26] K. Menger, *Reminiscences of the Vienna Circle and the Mathematical Colloquium*, Kluwer Academic Publishers, Dordrecht 1994, p. 146–147.

[27] See my papers 'Philosophy inside communism: the case of Poland', *Studies in Soviet Thought* **43**, 1992, pp. 93–100 and 'Marxism and the professionalization of philosophy', in *Philosophy and Political Change in Eastern Europe*, ed. by B. Smith, The Hegeler Institute, La Salle 1993, pp. 63–66.

[28] A. Ayer, *A Part of My Life*, Collins, London 1977, p. 116.

[29] Most of the points included in this list are reported in works quoted in note 1.

[30] See my papers 'Some Polish contributions to fallibilism', in *Living Doubt*, ed. by G. Debrock and M. Hulswitt, Kluwer Academic Publishers, Dordrecht 1994, pp. 187–195 and 'Jan Łukasiewicz on the Liar Paradox, Logical Consequence, Truth, and Induction', *Modern Logic* **4**, 1994, pp. 392–400.

[31] See papers quoted in note 27.

[32] See papers collected in *The School of Franz Brentano*, ed. by L. Albertazzi, M. Libardi and R. Poli, Kluwer Academic Publishers, Dordrecht 1996 for a general picture of the importance of Brentanism for contemporary philosophy.

[33] Let me note some new, recently published, forthcoming or prepared translations: S. Leśniewski, *Collected Works*, ed. by S.J. Surma, J. Srzednicki, D.I. Barnett and V.F. Rickey, Kluwer Academic Publishers, Dordrecht 1992; *Zygmunt Zawirski: His Life and Work with Selected Writings on Time, Logic, and the Methodology of Science*, ed. by I. Szumilewicz-Lachman, Kluwer Academic Publishers, Dordrecht 1994; T. Czeżowski, *Selected Papers*, ed. by L. Gumański, Rodopi, Amsterdam (forthcoming); K. Twardowski, *Selected Writings*, ed. by J. Brandl and J. Woleński, Rodopi, Amsterdam (forthcoming); J. Łukasiewicz, *Collected Works*, ed. by J. Srzednicki, G. Malinowski and J. Woleński, Kluwer Academic Publishers, Dordrecht (in preparation).

# PART I

# HISTORY AND COMPARISONS

JOHANNES L. BRANDL

# TWARDOWSKI'S DISTINCTION BETWEEN ACTIONS AND PRODUCTS

Twardowski's distinction between actions and products is one of his most celebrated contributions to the Lvov-Warsaw School and to philosophy in general.[1] The distinction is intended to serve several purposes: It should help to steer a middle course between psychologism and Platonism in logic and epistemology; it should support a non-relativistic conception of truth; and it should provide a theory of linguistic meaning as a special kind of psycho-physical product. There is, however, a serious problem in interpreting Twardowski's distinction, since it can be drawn either on the ontological level as a distinction between two kinds of entities, or on the conceptual level as a distinction between two ways of conceiving of just one kind of entity. Depending on how one resolves this ambiguity, Twardowski's distinction behaves quite differently in its intended applications.

## 1. THE RANGE OF THE DISTINCTION

When Twardowski speaks of 'actions' and 'products' he uses these terms in a technical sense. Usually we employ these notions only in connection with certain physical actions. For instance, when I write a letter, the letter is the product of my writing, when I fold it, the fold is the product, and when I tear it up, the torn pieces are the product of my action. Twardowski, however, has a much broader distinction in mind. He assumes that activities like walking or fighting also have specific products, namely the walk and the fight which result from these activities; and he assumes that mental acts too have products and may therefore be treated as actions in a broad sense. For instance, Twardowski would say that *thinking* of a letter and being *tempted* to tear it apart are just as much actions as the writing and the tearing apart themselves.

This broad use of the terms 'action' and 'product' needs some justification. Twardowski therefore reminds us of the grammatical relation between verbs and derived nominals. This relation holds quite generally, whether we talk of actions, of activities, of mental acts or of physical processes, as the following examples show: "to draw/the drawing", "to walk/the walk", "to think/the thought", "to die/the death".[2] Given this linguistic fact, Twardowski proposes that all of these verbs mean or denote some kind of 'action', whereas the corresponding nouns mean or denote the product of these actions.[3,4]



*K. Kijania-Placek and J. Woleński (eds.),*
*The Lvov–Warsaw School and Contemporary Philosophy*, 23–34.
© 1998 *Kluwer Academic Publishers.*

The question now arises: What is the philosophical significance of the grammatical observation on which Twardowski relies? What can we justifiably infer about the ontology of actions and products from these linguistic data?

Since one might suspect that Twardowski is too much impressed here by an accidental feature of ordinary language, it may be helpful to compare his approach with a more recent development in natural language semantics. This comparison is intended to show that Twardowski's observation which concerns the surface grammar of ordinary language can easily be transferred to the deep structure of sentences revealed by logical analysis.

## 2. TWO PATTERNS OF ANALYSIS

According to a widely accepted theory in contemporary semantics, subject-predicate sentences often implicitly quantify over events. Using the terminology of Terry Parsons,[5] we can say, for instance, that the atomic sentence

(1) John runs

has the subatomic structure

(1a) For some event $x$, $x$ is a case of running and $x$ has John as its agent
$\exists x$ (running ($x$) & Agent ($x$, John)).

This style of analysis is applicable not only in cases where a physical action is involved. Just as we talk of an *agent*, we can also talk of the *patient* or *object* of an event. Hence we can analyse the sentence

(2) John dies

along the same lines (ignoring tense), as

(2a) For some event $x$, $x$ is a case of dying and $x$ has John as its object
$\exists x$ (dying ($x$) & Object ($x$, John)).

The general pattern here runs as follows (let us call it the A-pattern): Given a verb (or verb phrase) $V$, we can derive from it the nominal $V$-ing which refers to the event that occurs whenever something or somebody $V$s. This pattern perfectly fits Twardowski's intuitions. What we get here is a domain of 'events' which may be just as comprehensive as the domain of 'actions' in Twardowski's theory. The important difference, however, is that we arrive at this domain not simply by tracing the surface grammar of our language, but by applying a certain method of logical analysis.

What about the other kind of entity which Twardowski introduces, the 'products' of actions? Here too we can make use of the apparatus of quantificational logic. Take for instance the sentence

(3) John writes a letter

As a first step we can analyse this sentence as follows:

(3a) For some event $x$, $x$ is a case of letter-writing and $x$ has John as its agent
$\exists x$ (letter-writing ($x$) & Agent ($x$, John)).

But there is a second step we can make:

(3b) For some $x$ and for some $y$, $x$ is a case of writing, $y$ is a letter, John is the agent of $x$, and $y$ is the product of $x$.
$\exists x \exists y$ (writing ($x$) & letter ($y$) & Agent ($x$, John) & Product ($x, y$)).

Now, this kind of analysis is restricted to sentences with a transitive verb. It is therefore not directly applicable to sentences like (1) and (2). But there is a trick we can use.[6] Intransitive verbs can be replaced by transitive ones. Instead of the intransitive verb 'to walk' we can use the verbal phrase 'to take a walk', or more generally, although not colloquial English, 'to perform a walk', where 'to take' and 'to perform' are transitive verbs. In the same way we can replace 'to die' by 'to suffer death', where 'to suffer' is again transitive. Sentences (1) and (2) can now be analysed on the model of sentence (3b):

(1b) For some $x$ and for some $y$, $x$ is a case of walking, $y$ is a walk, John is the agent of $x$, and $y$ is the product of $x$.
$\exists x \exists y$ (walking ($x$) & walk ($y$) & Agent ($x$, John) & Product ($x, y$)).

(2b) For some $x$ and for some $y$, $x$ is a case of dying, $y$ is a death, John is the agent of $x$, and $y$ is the product of $x$.
$\exists x \exists y$ (dying ($x$) & death ($y$) & Agent ($x$, John) & Product ($x, y$)).

As before, we have a general pattern here (let us call it the B-pattern): Given a verb (or verb phrase) $V$, we can derive two kinds of nominals from it, the nominal $V$-ing which refers to the event that occurs whenever something or somebody $V$s, and a different nominal $N_v$, which has the same stem as the verb $V$, and which refers to the result that ensues from $V$-ing.[7] This second kind of nominal is needed in phrases like 'to perform $N_v$', 'to suffer $N_v$' etc. Again, this is completely in agreement with Twardowski's approach. What we get here are two domains, a domain of 'events' and a domain of 'results', which are equally comprehensive and which are related to each other in such a way that to every event in the one domain there corresponds a result in the other domain and *vice versa*. This is just as Twardowski conceives of the relation between 'actions' and 'products'.

Thus everything seems to be fine. Twardowski's distinction is not a mere mirage created by the surface grammar of our language; it can be recovered at the quantificational level that reveals the logical deep structure of our sentences. A doubt remains, however. Even if we can perform the analysis along the B-pattern just as easily as along the A-pattern, we still need some reason for making this further move. After all, the B-pattern carries with it an extra ontological commitment; it enlarges the domain of quantification.[8] Is there any purpose of the B-pattern which is not served equally well by the A-pattern?

This question directly leads us to the ambiguity in Twardowski's distinction between actions and products.

## 3. CONCEPTUAL AND ONTOLOGICAL DISTINCTIONS

Following the pattern of (3b) above, I have assumed that the quantifiers in (1b) and (2b) also introduce two different kinds of entities, namely events and their results. This assumption may be questioned, however. Even if the writing of a letter is not identical to the written letter, the walking may be nothing else but the walk and the dying nothing else but the death. Do we really need different domains of quantification in these cases too?

Twardowski in his time could not raise the issue in this way, but he arrives at substantially the same question when he asks himself: How is the meaning of the verb 'to walk' related to the meaning of the noun 'a walk'.[9] There are three possibilities: (a) the difference concerns only the form of expression, (b) there is a difference in meaning, and (c) there is also a difference in what these expressions refer to.

According to the first possibility we could say that verbs emphasize the dynamic aspect of an action, whereas the corresponding nouns emphasize the static aspect of what an action produces (see §2). This difference in emphasis need not count as a difference in meaning. It may be like the difference between sentences in active and passive form, e.g. 'John beats Jim' and 'Jim is beaten by John'. Frege suggested that such sentences differ only in what he calls their 'colouring'.[10] The same might be true in the case of 'John walks' and 'John takes a walk'. Here too, no difference in meaning or reference need be involved.

This response is not general enough for Twardowski, however. He points out that it covers only some special cases. Sometimes, he says, "the product virtually coalesces with the action that produces it" (§9), but "the more the dynamical and the static aspect diverge, the less artificial is the formulation that asserts the emergence of what the noun refers to from the activity referred to by the verb." (§7)

Generality requires that we take into account those cases where the noun describes a product which exists longer than the action producing it and therefore must be ontologically distinct from it. Folding a letter would be an example. The action of folding takes only two seconds, but the fold which is the result of the action may last as long as the letter exists. Hence the folding and the fold must be different entities. This is most clearly established by the fact that we can say different things about them. The folding may be quick but not sharp, whereas the fold may be sharp but not quick (see §22).[11]

In cases like this, Twardowski says, "the distinction between products and actions presents no difficulties" (§27), it is "self-imposing" (§23). Difficulties arise only when the products "exist only for as long as the activities that yield them" (§23).[12] These "non-enduring products", as Twardowski calls them, admit no more than a conceptual distinction, a *distinctio rationis*:

Non-enduring products do not exist in actuality separately from the corresponding actions, but only in conjunction with them; we can only analyse them abstractly apart from these actions. On the other hand, enduring products can and do exist in actuality apart from the actions owing to which they arise (Fn. 26).[13]

This is the crucial passage for the interpretation of Twardowski's action/product distinction. A natural reading of this passage would be that non-enduring products are like the surface of a body or the beginning of a symphony: they are undetachable parts of larger wholes. We can distinguish between the body and its surface, or the symphony and its beginning, but only by 'analysing them abstractly apart'. In actuality, the surface and the beginning do not exist apart from what they are the surface and the beginning of.

On this reading the distinction between actions and products belongs to a certain mereological ontology. Just as a body and its surface, or a symphony and its beginning, are different entities according to such an ontology, the walking would be different from the walk. The distinction may not be a real one in the sense that it distinguishes between independently existing entities; still it is a distinction between real entities which do not exist only 'in the mind of the distinguisher'.

There is another possible interpretation to be considered, however. According to this interpretation Twardowski wants to make room here for the second of the three possibilities mentioned above. Granted that the difference between actions and products is not just a matter of 'colouring' in Frege's sense, this does not imply that the distinction always concerns the reference of our terms. There may be cases where two terms differ in meaning and yet refer to the same entity. This may be the case with 'walking' and 'walk'. From an ontological point of view the walk may be nothing else but the walking which 'produces' it. Still the noun and the verb do not have the same meaning; they refer to a single entity in two different ways, namely in a static and a dynamic way respectively.

Twardowski's text is not as explicit as one would like in this respect. One can read it as if Twardowski wants to say that there is always some ontological difference between actions and events, but one can also adopt an interpretation according to which this ontological distinction may sometimes give way to a mere conceptual difference. Let us turn now to the applications which Twardowski wants to make of his distinction. It is here that any interpretation must eventually prove itself.

## 4. THE MEANING OF LINGUISTIC EXPRESSIONS

As I said at the beginning, Twardowski makes use of the action/product distinction in several ways. I shall focus first on his theory of linguistic meaning and then take up the issue of psychologism which probably was Twardowski's main concern.

In a nutshell Twardowski's theory of language and meaning is this: Linguistic expressions are the best devices we have for preserving the products of our mental acts. Once the product of a mental act is preserved in this way, it becomes the meaning of a linguistic expression.

In order to understand this conception of language and meaning we need to know three things: What is the product of a mental act? What does it mean for a product to be preserved? And why are linguistic expressions the best devices we have for preserving mental products?

For the sake of brevity, rather than treat these questions in detail, let us examine an analogy which Twardowski offers as an illustration of his theory. The function of language, Twardowski suggests, is very much like the function of a phonograph record. A record is a device for preserving music (or any kind of noise). The music which it preserves must be distinguished from the actions of the musician who originally produced it. We have here a rather complicated interplay between the non-enduring products of a musical performance and the enduring products of the musical recording:

> For example, in preserving [recording] a scream on a phonographic disc, there is the action of screaming which produces the scream — thus, a non-enduring product — and which also simultaneously and indirectly produces grooves on the record, hence — enduring products. Correlatively, these enduring products become the partial cause that makes possible the emergence of a non-enduring product which is similar to the previous one, i.e. of the sort of scream that is reproduced by the gramophone, and that is how the record grooves preserve the scream. (§35)

Two points must be emphasized here: First, the process of preservation does not change the ontological status of what it preserves. The scream, being a non-enduring product of my action, remains a non-enduring product. It only "takes on the *semblance* of some sort of perdurance", as Twardowski says (§33, my emphasis). This means that although the scream is over, screams similar to it can be produced from the record.

The second point is a consequence of the first one: The process of preservation cannot turn a conceptual distinction into an ontological one. Suppose there is only a conceptual difference between the action of screaming and the scream which it produces. Then the recorded scream will also be just conceptually distinct from the screaming which produces it.

How does this carry over to linguistic expressions? The starting point is the assumption that mental acts have only non-enduring products.[14] When I make a judgement, the judgement vanishes as soon as my judging is over, when

I have a desire, the desire ends together with my desiring, etc. In the case of mental acts, as with screaming, we have a medium of preservation, namely language. It works in very much the same way as the phonograph (see §§ 30–41).

Mental acts have a specific influence on our linguistic behaviour and hence on the utterances we make, just as the act of screaming leaves its traces on the record. As a consequence of this influence, the utterances we make *express* the mental acts which influence them. Furthermore, the utterances we make also have an influence on other people, by causing mental acts in them. Given certain conditions these mental acts will be similar to those mental acts which are expressed in our utterances, just as the screams reproduced by the phonograph are similar to the scream which was originally recorded. This similarity suggests that mental acts differ from their content like the screaming from the scream. Thus the *content* turns out to be the (non-enduring) *product* of the mental act, and the *preserved content* of a mental act turns out to be the *meaning* of the linguistic sign.

It goes without saying that this is at best the beginning of a theory about the relation between mind and language. I will mention just two questions which arise immediately:

1. What is the special way in which a mental act must influence an utterance in order to get expressed in it? How does it differ from the way in which other mental acts influence an utterance without getting expressed by it? And why, for that matter, do utterances not express the physical events which influence them?

2. What exactly are the conditions under which an utterance causes mental acts similar to those which it expresses? Twardowski merely tells us that it happens due to "an association [ ... ] or by convention". (§ 31) Clearly we want to know more about what this association or these conventions consist in.

Without answers to these questions, I think, it is impossible to say precisely how the action/product distinction works in a theory of meaning. As long as we are only given the outline of such a theory, we have, so far, no reason to prefer either the ontological or the conceptual interpretation of this distinction.

Let us therefore turn to the issue of psychologism. It is here that Twardowski saw the greatest need for his distinction.

## 5. PSYCHOLOGISM OVERCOME?

According to the ontological reading of the action/product distinction, the content of a mental act is a real, though undetachable part of the act. When a mental act gets expressed by a linguistic utterance, this part of the act becomes the meaning of the linguistic sign, provided that the utterance is capable of arousing similar acts in other people.

The trouble with this interpretation is that it essentially coincides with Twardowski's theory of names in his 1894 book *Zur Lehre vom Inhalt und Gegenstand der Vorstellungen.*[15] Twardowski himself later considered his early theory as too psychologistic,[16] and introduced the action/product distinction precisely for this reason. But if mental products are identified with real parts of mental acts, it is hard to see what progress Twardowski made beyond his early theory.

Perhaps the progress lies in the details. When Twardowski speaks of the similarity between mental acts which are expressed and aroused by a linguistic expression, he adds another twist to his story. The similarity is not so much a similarity between the acts themselves, he says, as a similarity between their products. These products must share some common attributes for the linguistic sign to be intelligible. Here, it seems, Twardowski says farewell to a psychologistic theory of meaning:

[. . . ] it is precisely these common attributes, in which these individual products accord, that we ordinarily regard as the meaning of [the linguistic expression]. [. . . ] Now, a meaning conceived in this manner is no longer a concrete mental product, but something at which we arrive by way of an abstraction performed on concrete products. (§39)

To this Twardowski adds a footnote in which he refers to Husserl and his notion of 'ideal meaning'. Then he goes on to discuss other ways in which art and science create "the illusion of the independence of mental products from mental acts". (§41) He especially mentions logic as an example and refers to Bolzano and his conception of sentences-in-themselves.[17]

It is clear that Twardowski wants to join the anti-psychologistic camp here, but it seems that he wants to do this at the least possible cost. The ontological reading of the action/product has a considerable price. According to this interpretation we would have to distinguish three kinds of entities: mental acts, mental products, and the common attributes of these products which give rise to linguistic meaning. The immediate objection to such a theory is that it multiplies entities without necessity. Mental acts themselves can have common attributes and these attributes can give rise to the meaning of linguistic expressions. Why bring in mental products as intermediary entities?

There remain two alternatives to be explored. First, there is the possibility of *identifying* the products of mental acts with the common attributes which constitute the meaning of linguistic expressions. In this case the products would have to be universals without being eternally existing. They would be universal because infinitely many mental acts can have the same product, but they would still exist in time since they are dependent on the mental acts which produce them. The difficulty for this view is to explain how a product can vanish together with the act that produces it (perhaps being a real part of this act), if *the very same product* can be produced by another mental act again. Anyhow, there is no evidence that Twardowski even considered this possibility.[18]

There is, however, still another line that Twardowski might take. He could argue that in the case of mental (and linguistic) acts the action/product distinction is to be understood as purely conceptual. There are individual mental acts and there are individual linguistic utterances. When we speak of the content or meaning of such acts or utterances we are not adding new entities to our universe. We only introduce a new concept which allows us to abstract from the individual features of the acts in question. This is the concept of the 'content' or 'meaning' of acts and utterances.

Of course, this interpretation too is not without its problems. First, what does it mean to introduce a new concept? Is it an enrichment of our ontology? If so, what is gained by the conceptual interpretation of the act/content distinction given that concepts have to be recognised in the ontology? Secondly, how are concepts related to the content of mental acts. What do we mean by saying that a concept 'enters' the content of an act? How can this be explained as the relation between a mental act and the concepts 'employed' in it?

These are the most pressing questions for the conceptual interpretation of the action/product distinction. Whereas Twardowski developed no detailed theory of meaning, he did have at his disposal a theory of concepts.[19] According to this theory, a concept is a certain kind of complex presentation, where this presentation can again be understood as a mental act or a mental product. This leads us to the following situation: If we interpret the presentation/concept distinction conceptually, then concepts themselves will turn out to be no extra ontological entities. There may be a circularity involved here, but it is not clear whether the circle is a vicious one.

In view of all this it seems that Twardowski's agreement with Husserl and Bolzano is only superficial. Ideal meanings and sentences-in-themselves are mind-independent entities. By contrast, mental products are by definition mind-dependent, however much we try to give them the illusion of independence.

## 6. CONCLUSION

I have compared two ways of interpreting the action/product distinction, and I have argued that only the conceptual reading is able to support Twardowski's goal of overcoming psychologism. I think it is important to add, in closing, that this conceptual interpretation gives language a much more prominent role.

If we think of mental content as a distinct ontological category, a mental act contains a content as a part whether or not there exists a language for expressing this act. Mental content would be an essentially language-independent category. This is not so from the conceptual perspective. According to this interpretation the content of a mental act is nothing else but the mental act in so far as it can be preserved in language. From this point of view we need the linguistic expressions as reference points to which mental acts are related in order to form the concept of a mental content. If this is true, then this concept is just as much a linguistic as it is a mental product.*

*University of Salzburg*

NOTES

* I am grateful to Joe Tougas for his comments and corrections of my English.

[1] See Woleński (1989), pp. 41ff; Smith (1989), pp. 338ff., and Albertazzi (1993), pp. 19f. The distinction is introduced by Twardowski in his paper 'O czynnościach i wytworach. Kilka uwag z pogranicza psychologii, gramatyki i logiki' (Actions and Products. Several Remarks on the Borderline of Psychology, Grammar and Logic), first published in Polish in 1912 in a volume commemorating the 250th anniversary of the founding of Lvov University. Excerpts from this paper were translated in Pelc (1979), pp. 13-27. A full English translation will be available in Brandl/Woleński (in preparation). There also exists a German version of the same paper by Twardowski which deviates in certain respects from the Polish version. A publication of the German version, entitled 'Funktionen und Gebilde', is also in preparation (see bibliography at the end). References in the text (by paragraph or by footnote) are to the English translation of Twardowski's paper.

[2] It is not quite clear how far Twardowski wants to apply this distinction to the range of purely physical processes. Later in the paper he denies that the configuration of veins on the surface of a leaf are a genuine product. This suggests that the processes 'producing' this pattern are not covered by Twardowski's distinction. However, no criterion is given for distinguishing these processes from others, like the process of dying, which Twardowski does want to include.

[3] The disjunctive phrase 'mean or denote' indicates the ambiguity in Twardowski's theory. Twardowski does not commit himself in this respect when he says 'that the relation of the verb to its corresponding noun, ... expresses the relation of some action to that which emerges owing to that action, as a result of, by means of it." (§8). This can either be a relation between the meaning of the verb and the meaning of the noun or a relation between the entities to which these terms refer.

[4] It is interesting to note here that in the German version of his paper Twardowski uses the term 'function' to cover both actions (*Tätigkeiten*) and processes (*Vorgänge*).

[5] See Parsons (1990), pp. 3 and 6.

[6] This kind of trick is actually used by Twardowski in extending the notion of an 'inner object' beyond the so-called *etymological figure*. 'A walk', he says, is the inner object of 'to walk', but not because it gives rise to the etymological figure 'walking a walk'; the real reason, according to Twardowski, is that 'taking a walk' is equivalent to 'walking' (§5 and §24).

[7] Notice that the two nominals may be orthographically indistinguishable. In English and German, for instance, this is the case with the nominal 'yearning' ('Verlangen'). It can either be used for the event of desiring, or for the desire which one suffers thereby. Twardowski discusses such ambiguities in §§ 16–17.

[8] I assume for the sake of argument that the quantifiers are objectual, not substitutional.

[9] This is in fact the opening question in §1.

[10] See Frege (1892), pp. 45f.

[11] This does not rule out, of course, that there are predicates that apply both to an action and its product. Just as a walk can be quick or a scream horrific, one can walk quickly and scream horrifically.

[12] That there is no clear boundary between the two cases may be due to the fact "that enduring objects can last over spans of varying duration, [ . . . ]. Thus, a shout rebounded in an echo lasts longer than the action of shouting, although, generally speaking, it certainly ought to be regarded as a non-enduring product." (Fn. 24)

[13] Fn. 26. It is worth noting the different way Twardowski puts this point in the German version of this passage: "Und so treten denn hier [bei dauernden Gebilden] Funktion und Gebilde als etwas real Verschiedenes ganz deutlich auseinander, während dort, wo es sich um die Verschiedenheit von Funktion und vergänglichem Gebilde handelt, vielleicht nur eine *distinctio rationis* vorliegt." (§27)

[14] "Now, there is no place for mental products within the domain of enduring products."(§28) It is worth noting that on this conception it would be wrong to call a belief the 'product' of a judgement, because the belief would be an enduring state (a disposition) which continues to exist after a judgement is made.

[15] See Twardowski (1982), pp. 11f.

[16] The influence of Husserl in this respect is documented in Schuhmann (1993).

[17] "Bernard Bolzano was the first to articulate in detail this perspective on the subject-matter of logic. He called judgements Sätze-an-sich when they have been made independent from the actions of judging in the manner discussed above." (Fn. 44)

[18] A view of this kind is defended by Tomasz Placek as an improvement on Twardowski's position. See Placek (1996).

[19] Twardowski developed this theory in his paper 'Über begriffliche Vorstellungen' (1903). This was still before he drew the action/product distinction, and he did not later revise his theory in the light of that distinction. Twardowski himself mentions this fact in a footnote to the Polish translation of his (1902) paper which appeared in 1924.

## REFERENCES


Albertazzi, L.: 1993, 'Brentano, Twardowski, and Polish Scientific Philosophy', in Coniglione (1993), pp. 11-40.

Brandl, J., Woleński, J., *Kasimir Twardowski. Selected Writings*, Rodopi, Amsterdam (in preparation).

Cohen, R.S., Schnelle T. (eds.): 1986, *Cognition and Fact*, D. Reidel Publishing Company, Dordrecht.

Coniglione, F., Poli, R., Woleński, J. (eds.): 1993, *Polish Scientific Philosophy: The Lvov-Warsaw School*, Rodopi, Amsterdam.

Frege, G.: 1892, 'Über Sinn und Bedeutung', in *Funktion, Begriff, Bedeutung*, Vandenhoeck, Göttingen, 1962, pp. 40-65.

Giedymin, J.: 1986, 'Polish Philosophy in the Inter-War Period' in Cohen (1986), pp. 179-215.

Parsons, T.: 1990, *Events in the Semantics of English. A Study in Subatomic Semantics*, The MIT Press, Cambridge.

Pelc, J. (ed.): 1979, *Semiotics in Poland. 1894-1969*, D. Reidel Publishing Company, Dordrecht.

Placek, T.: 1996, 'Thought as a Product of Thinking', forthcoming in *Conceptus* **29**.

Schuhmann, K.: 1993, 'Husserl and Twardowski', in Coniglione (1993), pp. 41-58.

Smith, B.: 1989, 'Kasimir Twardowski: An Essay on the Borderlines of Ontology, Psychology and Logic' in Szaniawski (1989), pp. 313-373.

Szaniawski, K. (ed.): 1989, *The Vienna Circle and the Lvov-Warsaw School*, Kluwer Academic Publishers, Dordrecht.

Twardowski, K.: 1894/1982, *Zur Lehre vom Inhalt und Gegenstand der Vorstellungen. Eine psychologische Untersuchung*, ed. by R. Haller, Philosophia, Munich.

Twardowski, K.: 1903, 'Über begriffliche Vorstellungen', Wissenschaftliche Beilage zum 16. Jahresbericht der philosophischen Gesellschaft an der Universität Vienna, Barth, Leipzig, pp. 1-28. An English translation of this paper will be contained in Brandl/Woleński (in preparation).

Twardowski, K.: 1912, 'Actions and Products. Several Remarks on the Borderline of Psychology, Grammar and Logic.' in Brandl/Woleński (in preparation).

Twardowski, K.: 1996, 'Funktionen und Gebilde', ed. by J. Brandl, forthcoming in *Conceptus* **29**.

Woleński, J.: 1989, *Logic and Philosophy in the Lvov-Warsaw School*, D. Reidel Publishing Company, Dordrecht.

TADEUSZ CZARNECKI

# ON AJDUKIEWICZ'S EMPIRICAL MEANING-RULE AND WITTGENSTEIN'S DEFINING CRITERION

I intend to discuss some selected topics of Ajdukiewicz's 'Language and Meaning'[1] and Wittgenstein's *The Blue Book*,[2] both works dating from 1934. Though thoroughly independent and rooted in different methods of philosophising, they are worth comparing as being similar in accounting for the role of sentences in the specification of meaning. In analysing the structure of a meaning-rule and a criterion I particularly focus on the postulate of the intersubjectivity of meaning — undoubtedly a pillar of the semantic programmes Ajdukiewicz and Wittgenstein propose. First, I inquire into the security measures Ajdukiewicz takes to make his empirical meaning-rule intersubjective. Then, I compare the paradigm of the empirical meaning-rule with Wittgenstein's defining criterion to judge its intersubjectivity. I conclude that Ajdukiewicz's measures are insufficient, at least if confronted with Wittgenstein's extreme requirements for intersubjectivity.

Ajdukiewicz states clearly at the beginning of his paper that he ultimately aims at a definition of the meaning of a word which would explain the intersubjectivity of meaning and justify his 'radical conventionalism'.[3] In accordance with this, he avoids reducing the issue of meaning to psychic phenomena, i.e., relating it wholly to individual psychic acts. Instead, he is determined to speak about the meaning of a word inasmuch as the word is taken from a particular language and considered in relation to the language. The chain of steps taken to secure intersubjectivity is framed by the distinction between judgements in the psychological sense and judgements in the logical sense and the decision to rely only on the latter as they are adequately expressible in sentences of a language.[4] Thus, a language and sentences of that language are considered the fundamental warrants of intersubjectivity and — via meaning-rules — essential factors of the definition of meaning he is to propose.

To justify his commitment to sentences Ajdukiewicz discusses the relationship between a judgement and a sentence. He believes that some judgement-processes, psychic though they are, can be adequately expressed in sentences. Such processes appear in the area of articulate or verbal judging. Articulate judging is in principle a mental process, but of a particular character, namely, one of its two essential components is what Ajdukiewicz calls 'sentence-representation'. The sentence-representation, or, the appearance of sentences in the mind, is indispensable to the act of articulate judging and natural to the degree that it is not a matter of full awareness but rather of intuition. In short,



*K. Kijania-Placek and J. Woleński (eds.),*
*The Lvov–Warsaw School and Contemporary Philosophy*, 35–42.
© 1998 *Kluwer Academic Publishers.*

articulate judging is of necessity the judging through sentences in a language. To highlight the importance of sentence-representation Ajdukiewicz suggests that articulate judging, in opposition to inarticulate, is the mature form of judging. The other component of judging he finds it necessary to mention is the assertion-moment, i.e., the act of accepting or rejecting sentences, classified as a non-intuitive, conscious process. The mention of the assertion-moment betokens the role that convictions expressed by sentences, as opposed to naked sentences represented in the mind, are to play in the revealing of meaning.

The fact that he undoubtedly stays on the level of subjective, psychic processes while describing components of the inner structure of a judgement does not worry him. He thinks it is sufficient to enter subsequently the level of judgements in the logical sense, i.e., to restrict himself to considering logical operations on sentences, and the intersubjectivity of meaning will still be preserved. For example, it is enough to consider affirmation and negation, which are logical operations, and genuine components of the content of a logical judgement, instead of acceptance and rejection, which are kinds of psychic processes and do not belong to the content of logical judgements. So, Ajdukiewicz thinks he can safely speak about articulate judgements as they are adequately expressible in sentences. Significantly, he clearly sees the necessity of speaking about sentences because sentences, when used in affirmative and negative judgements, become essential tools of the specification of the meaning of words.

To explain the role of sentences in the specification of meaning he switches from the issue of meaning to the issue of understanding and introduces the notion of a meaning-rule as a criterion of understanding a word. His key claim is that the meaning of a word determines to some extent the rules of its use, but not vice versa.[5] Consequences of the claim are that meaning as such is primary to meaning-rules, meaning does not exhaust itself in meaning-rules and meaning-rules are only negative measures of understanding, i.e., acting in accordance with rules does not prove that the meaning has been grasped, whereas the violation of a rule is a definite proof of a misunderstanding of the meaning.

Some worrying things happen, however, when Ajdukiewicz progresses from this general outline of his programme to particulars connected with the introduction of the notion of an empirical meaning-rule. An empirical meaning-rule is to be a procedure which, by reference to a necessary condition of understanding an empirical term, enables us to decide whether someone attaches to the term a different meaning than someone else. The paradigmatic example illustrating this procedure, crucial — I assume — for Ajdukiewicz's general conception of the meaning of empirical terms, goes as follows: An exposed sensitive nerve in *A*'s tooth is touched. *A* cries. Another person, *B*, who observes what is going on, does not doubt that *A* is in pain. Nevertheless, *B* asks

*A* if he is in pain. Suppose now, Ajdukiewicz says, *A* rejects the sentence 'I am in pain'. Then, *B* can choose one of three possibilities to rationalise *A*'s rejection: 1) *A* lies, 2) *A* does not feel pain, or, 3) *A* does not lie and feels pain but attaches to the sentence 'I am in pain' another meaning than *B* himself.[6] This is the last possibility that suggests a general procedure for discovering misunderstandings: if an experience which is necessary for the acceptance of a sentence has been already pointed out, then he who rejects the sentence despite the occurrence of the experience attaches to it a meaning different from the established one.[7]

The example does not leave a shadow of doubt that Ajdukiewicz relates the meaning of empirical terms straight to the senses. He locates the Archimedean point for measuring understanding in the senses and assigns to someone's basic experiences the role of a warrant of the intersubjectivity of meaning. An empirical meaning-rule is always a 'sentence-experience' pair. A very close pair too, as he adds that only the occurrence of a satisfactory experience is essential for understanding, i.e., such an experience that the transition from the experience to the acceptance of the sentence occurs in one step.[8]

I think it is important to stress that the paradigmatic example includes three elements: 1) the outer fact that *A* cries when his nerve is touched, 2) *A*'s inner experience of pain and, 3) *A*'s rejection of the sentence 'I am in pain'. *A*'s rejection of the sentence, the linguistic element, is undoubtedly a necessary one for Ajdukiewicz's procedure. An open question, however, is which one of the remaining pair of extra-linguistic elements would really be better from the point of view of intersubjectivity: the outer fact that *A* cries when the nerve is touched or the inner occurrence of pain. Ajdukiewicz chooses the occurrence of pain as the complementary element in his paradigmatic example. The role of the fact of touching the nerve is taken at most as a secondary one.

Ajdukiewicz is aware of some difficulties the choice leads to. But he does not think that they are significantly disturbing. One of them is that in practice one cannot determine with complete certainty whether another person actually has a particular inner experience.[9] This is the doubt whether one can be certain of the occurrence of something which is beyond one's direct access. But this counts little for him in comparison with the fact that the procedure he has invented, as a whole, is apodictically certain whenever the experience has been previously co-ordinated with a sentence.[10] The procedure involves the appeal to experience but is still analytic since it links together two constitutive elements. That is what really counts. The brighter side of the issue encourages him to propose a general scheme of an analytic meaning-rule:

> Only that person connects with the words of a language $S$ the meanings co-ordinated with them by that language who, in situations of type $L$, is prepared to accept a sentence of the sort $Z$.[11]

Needless to say, the apparently neutral expression 'in situations of type $L$', in the case of empirical rules, should be interpreted as 'while experiencing a perception of type *P*'.

Another difficulty Ajdukiewicz discusses concerns his forced consent to compound rules. He distinguishes between simple and compound empirical rules and suggests that his paradigmatic rule is a simple one, i.e., depicts a straight co-ordination of a sentence with a perception. Compound rules, which require the acceptance of a number of sentences in the context of a series of sequential perceptions, are unavoidable whenever, as he puts it, the so-called 'statements about the external world' are concerned.[12] Compound rules, as well as 'external objects', cause some discomfort since only the simple ones, including as their components singular perceptions, can be precisely formulated and, therefore, are really desired. The price for their precision is high because only words referring to psychic entities can have simple empirical meaning. Nevertheless, Ajdukiewicz concludes that the difficulty is negligible. He sees no obstacle to believing in the possibility of building an intersubjective empirical language by establishing simple rules which define characteristic connections between sentences and perceptions. On the contrary, he takes it to be the only possibility. Never again does he turn back to the problem of 'external objects', and nowhere does he try — at least in 'Language and Meaning' — to explain how the transition from simple perceptions to compound external objects is made. In other words, Ajdukiewicz seems to accept unreservedly the traditional empiricist thesis of the priority of simple perceptions over compound external objects and does not anticipate any danger connected with it. To show that there is a danger and what the danger consists in I turn to a short presentation of Wittgenstein's views on criteria.

I have already said that there are two main points where Ajdukiewicz and Wittgenstein's views meet. They share the belief that sentences are essential elements of meaning-rules and pay special attention to the intersubjectivity of meaning. In fact, Wittgenstein is obsessed by intersubjectivity. Wittgenstein outlines his notion of a defining criterion in a passage in *The Blue Book* where he says that the question of whether something is or is not the case can be answered either by giving criteria or symptoms:

> If medical science calls angina an inflammation caused by a particular bacillus, and we ask in a particular case 'why do you say this man has got angina?' then the answer 'I have found the bacillus so-and-so in his blood' gives us the criterion, or what we may call the defining criterion of angina. If on the other hand the answer was, 'his throat is inflamed', this might give us a symptom of angina. Then to say 'A man has angina if this bacillus is found in him' is tautology or it is a loose way of stating the definition of angina.[13]

Wittgenstein's criterion can be interpreted as at least a necessary condition of understanding of the empirical term 'angina'. If so, one can put Wittgenstein's sentences in Ajdukiewicz's general scheme and get an empirical meaning-rule of the word 'angina':

> Only that person connects with the word 'angina' of a language the meaning co-ordinated with it by that language who, having found the bacillus so-and-so in a man's blood, is prepared to accept the sentence 'The man has angina'.

However, besides surface similarities, this record of the rule reveals deep differences between both philosophers. First, the phenomena Wittgenstein appeals to in the defining part of a criterion are outer phenomena. The defining sentence is always an object-sentence, i.e., it describes a situation that may be observed by many people. Next, he explicitly uses the form of a non-tautological conditional 'If A, then B' to express a criterial relationship between two different sentences. That is, in every case two different terms are involved. None of the terms is defined by itself, no sentence expresses the defining part for itself. A criterion is called 'tautology' only to underline its semantic function; the conditional 'If a man has angina, then the man has angina' would express no criterion at all. Now, the question arises whether one could use two different sentences to formulate a conditional for Ajdukiewicz's necessary condition of understanding. Though a simple empirical rule consists of a 'sentence-perception' pair, which secures the qualitative difference of levels between the co-ordinated elements, it seems that if Ajdukiewicz tried to describe precisely the perception in a sentence, he would always have to use the very sentence which has already been co-ordinated with the perception. As a result, in the case of the word 'pain' he would obtain the conditional: 'If a man is in pain, then the man is in pain'; a tautology in the logical sense. In fact, this tautology can be expressed in Ajdukiewicz's scheme to maintain the qualitative difference of levels between the two elements:

> Only that person connects with the word 'pain' of a language the meaning co-ordinated with it by that language who, being in pain, is prepared to accept the sentence 'I am in pain'.

But the consolation following from the fact is only apparent. The difference of levels is maintained because the rule is, to use Wittgenstein's phrase, a loose way of stating the ostensive definition of the word 'pain'. And one can suspect that all Ajdukiewicz's simple empirical rules are disguised ostensive definitions of empirical terms. Of course, Wittgenstein has nothing against ostensive definitions as basic semantic tools. Nevertheless, he warns that one should be careful with them. In particular, one should not think that an ostensive definition is the only tool one needs to build an empirical language or that an ostensive definition can be used in every situation. His message is twofold: For some empirical terms criteria consisting of genuine conditionals are unavoidable. In some situations ostensive definitions cannot be successful.

To realise that Ajdukiewicz's position leads to serious difficulties it is enough to reconstruct the reason why Wittgenstein uses object-sentences as defining sentences, i.e., why he is a professed supporter of the thesis of the semantic priority of external objects, or external situations, over perceptions. In the passage that precedes the introduction of a 'criterion' he discusses the problem: What is the reason for claiming that one knows that someone else is in pain?:

Now one may go on and ask: How do you know that he has got toothache when he holds his cheek? The answer to this might be, 'I say, he has toothache when he holds his cheek because I hold my cheek when I have toothache'. But what if we went on asking:– And why do you suppose that toothache corresponds to his holding his cheek just because your toothache corresponds to your holding your cheek? You will be at a loss to answer this question, and find that we strike rock bottom, that is we have come down to conventions.[14]

What Wittgenstein depreciates in this passage is the Argument from Analogy, i.e., the idea that on the basis of what we ourselves do when we have an inner experience we can know what the others experience when we see what they do. In his opinion, in the case of the word 'pain' the Argument from Analogy is useless. One must appeal directly to some established, conventional criteria in order to decide whether someone else is in pain, and, the defining parts of the criteria must consist of object-sentences describing outer situations. The ultimate reason for this strong thesis lies in his Private Language Argument. I do not intend to unfold here the argument in its entirety and I restrict myself to a basic comment. The Private Language Argument is a criticism of the possibility of a private ostensive definition of a word referring to an inner experience. Significantly, Wittgenstein's exemplary word is 'pain'. He claims that a person cannot decide about the occurrence of someone else's pain by analogy with his own pain because the person is unable to coordinate the word 'pain' with his own inner experience of pain by means of an ostensive definition. If the person after all uses the word 'pain', he owes it to other people, who themselves use the word on the ground of some outer criteria, mainly, people's natural behaviour in paradigmatic situations. Well known passages from *Philosophical Investigations* summarise the issue in the following way:

An 'inner process' stands in need of outward criteria.[15]

[...]That is to say: if we construe the grammar of the expression of sensation on the model of 'object and designation' the object drops out of consideration as irrelevant.[16]

Unlike Wittgenstein, Ajdukiewicz relies on the Argument from Analogy when he speaks about the reason for knowing someone else's pain. He says:

If ever we meet a situation in which someone experiences what for us is the normal reaction to a severely irritated nerve, and in spite of this the person is not ready to accept the sentence 'it hurts' but rather rejects this sentence or even accepts the sentence 'it feels good', then as already explained we see in this situation a proof that he does not attach to the sentences 'it hurts' and 'it feels good' the same meaning that we do.[17]

He takes the step because he is convinced that his paradigm of a simple empirical rule works and the Argument from Analogy is its necessary supplement, regardless of how strong a one. If, however, this paradigm is a disguised ostensive definition, then it is within the range of the Private Language Argument and Ajdukiewicz can be accused of violating (of course if still judged from

Wittgenstein's viewpoint) fundamental requirements for the intersubjectivity of meaning. He implicitly assumes that 1) words about inner experiences — including words about simple perceptions — initially gain their meaning by instances of private ostensive definitions and 2) some outer situations are subsequently found which correspond to the primary co-ordination and make it possible for a person to decide in the future, by analogy with his own experiences in these situations, whether someone else has the same experiences and uses words with the same meaning. For Wittgenstein, both assumptions are wrong.

The consequences of Wittgenstein's attack on privacy are far from pleasant for Ajdukiewicz's vision of language. The possibility of straight relations between empirical terms and inner experiences — via private ostensive definitions — is denied. As Wittgenstein claims that there exists no direct relation between a language and the mind, the thesis of 'sentence-representation' as well as the thesis of adequate co-ordination between judgements and sentences should be evaluated as overly optimistic. The theses do not elucidate the intersubjectivity of meaning-rules and have nothing to do with the intersubjectivity of meaning. Generally, Ajdukiewicz's empirical language is more a private than intersubjective language, or, what would be for Wittgenstein the same, an impossible language. Wittgenstein would have said that such a conception of empirical language as Ajdukiewicz's results from adopting an unfortunate formulation of the general question about meaning which is: What is meaning? The formulation should be replaced by: What is an explanation of meaning? Wittgenstein says:

> Asking first 'What's an explanation of meaning?' has two advantages. You in a sense bring the question 'what is meaning' down to earth. For, surely, to understand the meaning of 'meaning' you ought also to understand the meaning of 'explanation of meaning'. Roughly: 'let's ask what the explanation of meaning is, for whatever that explains will be the meaning'. Studying the grammar of the expression 'explanation of meaning' will teach you something about the grammar of the word 'meaning' and will cure you of the temptation to look about you for some object which you might call 'the meaning'.[18]

He believes that by sticking to the second formulation one is able to understand that no private sensation can be a proper component of an explanation of the meaning of a word. Ajdukiewicz is only partially relieved from the impact of the first one. He recognises the role of sentences in explanations of meaning and the preservation of its intersubjectivity. On the other hand, he does not manage to cure himself of the temptation to look for extra-linguistic components of meaning-rules in the sphere of private objects. Now, the Private Language Argument may be considered extravagant or just wrong. This is not a rare reaction. But Wittgenstein's message that all empirical terms, including words about inner experiences, need outer, shared criteria or proper ostensive definitions is worth the attention of anyone who values both intersubjectivity and, like Ajdukiewicz did when he wrote 'Language and Meaning', radical conventionalism.

I do not claim that I have proved that Ajdukiewicz failed to preserve the intersubjectivity of meaning. Nevertheless, I think his position as a protagonist of intersubjectivity and conventionalism would have been better if he had chosen an outer situation as the second constitutive element of his empirical rules and made the basic rules differ from private ostensive definitions.

*Cracow University of Technology*

NOTES

[1] K. Ajdukiewicz: 1978, 'Language and Meaning', in J. Giedymin (ed.): *The Scientific World-Perspective and other Essays (1938–1963)*, D. Reidel Publishing Company, Dordrecht, pp. 35–66.

[2] L. Wittgenstein: 1958, *The Blue and Brown Books*, Basil Blackwell, Oxford.

[3] For a discussion of Ajdukiewicz's semantics and conventionalism see J. Giedymin, 'Radical Conventionalism, its Background and Evolution: Poincare, Le Roy, Ajdukiewicz', in V. Sinisi, J. Woleński (eds.): *The Heritage of Kazimierz Ajdukiewicz*, Rodopi, Amsterdam 1995, pp. 120–135; J. Woleński: 1989, *Logic and Philosophy in the Lvov-Warsaw School*, Kluwer Academic Publishers, Dordrecht, pp. 199–213.

[4] Ajdukiewicz (1978), pp. 35–39.

[5] Ibid., p. 39.

[6] Ibid., p. 40.

[7] Ibid., p. 40.

[8] Ibid., p. 41.

[9] Ibid., p. 41.

[10] Ibid., p. 41.

[11] Ibid., p. 43.

[12] Ibid., pp. 46–47.

[13] Wittgenstein (1958), pp. 24–25.

[14] Ibid., p. 24.

[15] L. Wittgenstein, *Philosophical Investigations*, Basil Blackwell, Oxford, 1963, §580.

[16] Ibid., §393.

[17] Ajdukiewicz (1978), p. 46.

[18] Wittgenstein (1958), p. 1.

RYSZARD JADCZAK

# INSPIRATIONS AND CONTROVERSIES: FROM THE LETTERS BETWEEN K. TWARDOWSKI AND A. MEINONG

The correspondence between Kazimierz Twardowski (1866–1938) and Alexius Meinong (1853–1920) contains interesting material on three matters, namely: 1) their research work, 2) the creation and functioning of psychological laboratories in Graz and Lvov, 3) the atmosphere in philosophical centers in German-speaking areas (mainly in Austria-Hungary). A part of Twardowski's letters, stored in the archives of Vienna University, has been published.[1] A large collection of Meinong's letters quoted in this paper are preserved in the *Kazimierz Twardowski Archive*.[2] Meinong's letters, all of them typed, have Twardowski's notes with the dates of his replies. It enables us, despite the fact that many of Twardowski's letters are missing, to trace the records and frequency of correspondence between the two scientists.

Meinong's first short letter from the K.T.A. collection is dated 23 December 1893 (K 22), from Graz. Twardowski's note indicates that the answer was written on 18 February 1894. At that time Twardowski was employed in the Mathematics Office of the Vienna Insurance Company *Lebensversicherungs-anstalt des I. Allgemeinen Beamtenvereins des Österreichischen Monarchie*, working simultaneously on his paper for qualification as an assistant professor. In this letter Meinong, thanking Twardowski for sending him his paper, wrote: 'Even at first glance I found in the text a lot of interesting things, and I hope to make use of some of the ideas in my present work, not only in one field.' No doubt he refers to Twardowski's doctoral work, *Idee und Perzeption*, published in 1892.[3] The work was the basis of Twardowski's doctoral degree in philosophy in 1891. The paper was considerably influenced by his studies at Vienna University (1886–1889), mainly under F. Brentano, although the professor conferring his degree was R. Zimmermann. During his studies in Vienna, Twardowski met, among others, A. Höfler, Ch.V. Ehrenfelds, J.K. Kreibig and H. Schmidkunz, with whom he afterwards kept in touch.

Twardowski was qualified as assistant professor on the basis of his *Zur Lehre vom Inhalt und Gegenstand der Vorstellungen*.[4] This work, regarded as the most significant in Twardowski's oeuvre, brought him respect and was mentioned in the German philosophical literature. Twardowski wrote in his *Autobiography*: "Writing the paper I put forward a task, referring to one statement in *Logic*, published by A. Höfler with the participation of Meinong (Vienna, 1890, paragraph 6), to make a detailed distinction between the content of the



*K. Kijania-Placek and J. Woleński (eds.),*
*The Lvov–Warsaw School and Contemporary Philosophy*, 43–52.
© 1998 *Kluwer Academic Publishers.*

presentation and the object of the presentation, and to consider their mutual relationship. Realization of this task required a number of terminological explications, as well as an analysis of the structure of what, I thought, was to be defined as the content and the object of presentation, which in turn led to describing the ideas of 'feature', 'property' etc. Studying the relationship of the content and object of presentation eventually led to considering the problem of the object of a general presentation."[5]

Having obtained *veniam legendi*, Twardowski began lecturing as a private assistant professor at Vienna University. At the same time he applied for a department in one of the two Polish universities functioning at that time. Finally, by the decision of the Austrian Ministry of Education on 18 October 1895, Twardowski was appointed as associate professor at the Department of Philosophy of Lvov University, from 1 November 1895. The other associate professor of philosophy at that time was Aleksander Skórski. Thus the Lvov period in Twardowski's activity began. Meinong congratulated him on the promotion in his letter of 28 October 1895 (K.T.A., K 21).

Intensive correspondence between Meinong and Twardowski started no earlier than in 1897. In his letter of 2 July 1897 (K.T.A., K 22), Meinong informed Twardowski that he himself was born in Lvov in 1853, studied philosophy under Brentano, but also attended the lectures of Zimmermann and Vogt. He wrote his assistant professor's thesis in Vienna in 1878. He had been an associate professor in Graz since 1882, and a full professor since 1889. He also gave Twardowski the titles of several of his more significant works, but only the ones that he considered "more important than the ones with passing value." Apart from the more important dissertations he listed two reviews, because he "put in them — let's say temporarily — all that I regard as fundamental and that I intend to put into my big work on recognition, including the new study of relation theory; apart from that the second paper contains criticism of some points of Brentano's theory of judgments", to which he "paid some attention."[6] He also admitted that A. Höfler, Ch.V. Ehrenfelds, A. Oelzelt-Nevin, K. Zindler, E. Martinak and St. Witasek had similar styles of work.

In the letter of 11 July 1897 Twardowski asked Meinong for some details concerning the origin of the laboratory in Graz to, as he wrote, popularize his contribution in the field, and inquired for more information about Meinong's birth in Lvov. He also informed Meinong that he intended to work out a theory of reasoning based on the research of Brentano, Meinong and Höfler. He quoted a main idea of his general outline, which assumed the distinction of the act, content and object. He wrote: "I intend to achieve two things by this reasoning: 1) freeing all sciences of reasoning from ambiguity which emerges from the idea of existence, 2) combining the theory of Brentano, Meinong and Höfler with Sigwart's theory, where the three elements (act, content and object) are clearly distinguished (act = consciousness, content = objective merit and the object is what, according to Sigwart, "puts itself in unity" in conceptions."[7] He also asked Meinong if he intended to work on similar subjects in his theory of recognition, writing that "in that case", his work "would be useless, although it would be sad to give up this subject", which had become his "favorite one."[8] The work Twardowski mentioned was his later paper from 1889, entitled *Conceptions and Ideas*.

Meinong gave Twardowski a very detailed response to this letter (nearly 15 pages) on 22 July 1897 (K.T.A., K 22). From additional information he gave about his birth in Lvov, we learn that his father, a colonel, was in a garrison in Lvov for three years. He himself spent one year there. Afterwards, before his father was promoted to general, he lived in Galician garrisons until 1858. He added that he had an opportunity to learn a little Polish then, but at that time he did not remember any. He broadly described his activity in the field of experimental psychology. Its beginning was dated to his Vienna period. During the winter semester of 1880/81 he gave lectures on the psychology of spatial conceptions. After the first period of adaptation as a professor in Graz, he began to act to introduce propaedeutics of philosophy in Austrian grammar schools, emphasizing the meaning of demonstration in psychology. At the same time (1886) he began cooperating with Höfler on a textbook entitled *Logic*, edited after ten years, where he emphasized the necessity of psychological experiment in teaching. Without waiting for government subsidies, he decided to pay the expenses connected with arranging a psychological laboratory, from his own resources. In his letter Meinong described the problems of purchasing and assembling the equipment, searching for a location for the laboratory and classrooms (at first he gave his lectures in the Department of Law's rooms). Nevertheless, in spite of the satisfaction with the work and great interest on the part of students, Meinong decided to give up the experimental lectures after five semesters, having expected that they would have no longer been his private activity. Only the growth of the university made it possible for the lectures to continue. At that time Meinong obtained financial support from the Ministry of Education and, with his assistant Dr. Witasek, he was able to take up regular lectures and research work. He took advantage of the equipment brought from Bern by Dr. Oelzelt-Nevin. "And if at present", wrote Meinong, "Austrian universities, one by one, equipped with better means and possibilities, try to arrange experimental psychology laboratories because they do not want to stay behind compared to Graz University, it means that the Graz laboratory has contributed much to develop this branch of science in Austria."

As far as the dissertation intended by Twardowski, relating for the most part to the distinction between act, content and object he had made in his *Zur Lehre*, is concerned, Meinong himself wrote about his interests in the same subject and his intention of concentrating on the problem in his paper 'Theory of Recognition', which he hoped to publish in a year's time. He mentioned that the fact that Twardowski was still dealing with the problem of content and object was evident and promising, particularly for somebody who had found so many helpful ideas in the book, as he did. Apart from that he found some matters in the book to which he was opposed to by their nature. Meinong explained that his objections concerned particularly the idea of existence as the content of reasoning. He wanted to relate to this and other questions in his 'Theory of Recognition'. Keeping his opinion on Twardowski's future work until it would be published in German, he encouraged him to continue his theoretical research on reasoning and to find some time to look at his theory when it came out, as he might find some useful things there and agree with his opinion without undermining the point of view he had come to.

Twardowski responded to this letter on 25 July 1897. He thanked Meinong for the information on teaching propaedeutics of philosophy in Austria and his efforts in establishing an experimental psychology laboratory in Graz. Referring to Meinong's experience in making the professors and students understand the necessity of continuing philosophical education, he mentioned that he faced the same problems in Lvov. The letter also contains interesting information on Twardowski's first lectures in experimental psychology in Lvov, starting with the academic year 1896/97. "My colleague, a lecturer in experimental physics, was so kind", we read in the letter, "as to find a room for me in the newly opened (since October) building of the institute. The room will constitute the beginning of the psychological laboratory. All of these actions are arranged, of course, in secret, without informing the authorities. I also want to collect secretly all of the necessary equipment to start classes in experimental psychology after a few semesters."[9]

The letter may be significant in the discussion of the priority in experimental research which took place between the Krakow and Lvov centers. Władysław Heinrich lectured at Jagiellonian University in 1897/98, on 'The development of ideas and methods of psychological studies'. In the reports of Akademia Umiejętności (Academy of Sciences and Arts, Krakow 1898, issue 9) he published his work 'About fluctuation in minimal intensity of optical and acoustic impressions', which was the first work on experimental psychology in a Polish-speaking area. Heinrich carried out his first experiments at the Institute of Physics owing to the kindness of August Witkowski. On the list of lectures of the Jagiellonian University for 1903/04 there is the "experimental psychology laboratory under the supervision of Dr. Władysław Heinrich." The results of Heinrich's work were reported in his *Psychology of Feelings*, published in Krakow in 1907.

The list of lectures for the spring semester 1897/98 at Lvov university shows that Twardowski had lectures entitled 'General ideas of psychophysics', and in the winter semester 1898/99 a lecture entitled 'About optical illusions'. Twardowski led the classes in experimental psychology in 1901/1902. All the activity was possible due to the kindness of Ignacy Zakrzewski, in the lecture hall of the Institute of Physics and the rooms of the vacant Department of Mathematical Physics. The situation became much worse with the appointment of personnel for the latter. Another possibility of starting the classes again emerged when the philosophical seminar obtained a separate room in 1905. However, the application to the Ministry in Vienna asking for a separate psychological laboratory with proper funding was rejected. Twardowski started regular two-hour classes in experimental psychology no earlier than in 1907/1908.

Referring to his dissertation *Zur Lehre* that he sent to Meinong, Twardowski admitted in the same letter of 25 July 1897 that the problems he considered there were the most difficult ones and even then he would "write some of them in a different way, or not write at all". But at the same time he defended his theory of reasoning, although he treated it as "an attempt to form as simple and homogeneous description of the phenomenon of reasoning as possible, which would not amount to a complete theory", he wanted to "put a question for discussion whether, either by exposing existence as a constitutional element coordinated with the act and the object of reasoning, it would be possible to solve various difficulties related to it, or at least bring them nearer to the solution." He also wrote "At any rate I will do everything to obtain your approval and I do wish to avoid the mistake of forcing my own concept even if it delays a better solution."[10]

Meinong's next letter comes from 22 November 1897 (K.T.A., K 22). He declared in it his eagerness to read Twardowski's text on reasoning, and also promised to send him additional information about the equipment of the Graz laboratory and ways of obtaining it. He mentioned that since September 1897 he had had a philosophical seminar in Graz.

The next letter from Meinong, from 6 May 1898 (K.T.A., K 22) tells us that Twardowski sent him his dissertation *Conceptions and Ideas*. Although it dates from the second period of Twardowski's scientific work, it still belongs to the first Vienna period as far as the content is concerned and it is regarded as a continuation of his assistant-professor thesis. Written under new conditions, it had different character, paying much attention to methodical consistency and clarity. Twardowski was broadly critically of Meinong's opinion of the idea of abstraction and relation *inter alia*. Meinong was not able to react to those remarks because the work was published in Polish. Twardowski presented his theory of idea explained in *Conceptions and Ideas* on 18 November 1902 at a general meeting of the Philosophical Society at Vienna University. The paper was later published in German.[11] Meinong's letter of 6 May 1898 shows that Twardowski informed him about the application of the Department of Philosophy of Lvov University for the promotion to full professors of himself and his colleague Skórski. That is why Meinong wrote: "I wish I had known about that when I was in Vienna (applying to the University for the improvement of the working conditions of his philosophical seminar and psychological laboratory — R. J.). I could have helped a little then. Your recent publications establish your position evidently. I do hope I will soon be able to welcome you as a full professor." The applications mentioned above as well as Twardowski's and Skórski's profiles were sent from Lvov on 20 September 1897.

Twardowski's letter of 13 May 1898, refers to Meinong's opinion of supporting the application and presents the situation in Lvov. He wrote that the Institute of Philosophy decided to recommend both candidates, that is, Twardowski and Skórski. But it turned out that the Ministry decided to establish only one department with a full professor. Therefore, the problem of competition between Twardowski and Skórski arose. In the light of this, he asked Meinong to support his case. Twardowski understood that his request, as that of a person involved in the matter, might meet with objections, but suggested asking professors S. Pawlicki, M. Straszewski or H. Struve for their opinion.[12]

Meinong answered in the letter of 17 May 1898 (K.T.A., K 1–37). We read that there were three important obstacles to supporting one of the two candidates. First, there was no data concerning Skórski, and relying only upon the opinions of the philosophers mentioned by Twardowski would not help much, as Meinong would not like to base his personal opinions on the knowledge of others rather than his own, especially since 'the others' were not specialists in the field. Secondly, Twardowski's recent publication, which might have been the reason for his promotion, was not available to Meinong because it was printed in Polish only. Thirdly, he could not give his opinion on the matters of other universities unless asked, therefore his advice for Twardowski was to make the authorities, *viz.* the Ministry of Galicia, direct a formal letter to Meinong asking for his opinion: "In that case I would be able to say as many good things as I know."

As we can suppose from the remarks in Meinong's letter, Twardowski had earlier declared his willingness to visit Graz and to see the work of the seminar and psychological laboratory. Meinong was really interested in this declaration.

It is difficult to determine what effect, if any, Meinong's support had in the case of Twardowski's promotion. The fact is that the Austrian Ministry decided on 15 September 1898 to appoint Twardowski as a full professor from 1 October 1898. Twardowski was sure of Meinong's support.

Twardowski informed Meinong about his plans to visit, for scientific purposes, Graz and other psychological institutes in Western Europe, as we can infer from Meinong's letter of 31 December 1898 (K.T.A., K 1–39). Meinong suggested visiting Graz at the end of his journey in order to be able to compare what he had seen before with Meinong's institute and to be able to stay longer and talk with him.

Meinong sent his next letter on 11 November 1899 (K.T.A., K 1–39), in which he anxiously inquired about the reason for Twardowski's delay in writing (he received no reply to his letter of December and no response to his publication *Über Gegenstände höherer Ordnung und deren Verhältnis zur inneren Wahrnehmung*).

Meinong received the reply no earlier than 14 November 1899. Apologizing for the delay, Twardowski wrote about Meinong's recent dissertation, which woke him up from "please forgive me and do not treat it as being conceited — dogmatic doze, in which I have been so far in the company of all of Brentano's disciples, concerning inner perception. I really start doubting whether we are at all able to perceive anything else inside us other than feelings. As you see, this is Schopenhauer's view. Nevertheless, these doubts involve the necessity of answering the entire chain of questions and dealing with various pseudo-facts. I am just trying to clear this matter out, and if you do not find the idea senseless, I am eager to give more details."[13]

The part of the letter quoted above is interesting because of Twardowski's declaration of his viewpoint on the subject of psychology and the role of so called inner experience: Twardowski adopted the position of psychologism. In the paper 'Psychology towards Physiology and Philosophy' from 1897, he regarded "spiritual phenomena" as a subject of psychology, among which he included ideas and wishes, convictions and affections. The knowledge of these "psychic symptoms" is possible owing to inner experience, because of which: "We possess the awareness of direct action of all spiritual phenomena within us, the awareness that constitutes the basis of inner experience." "Psychology then", we read in the next passage, "provides philosophy not only with a method, but also an object. If we were not equipped with our inner experience, and thus did not know the symptoms of intellectual life, psychology could not exist at all, as well as logic, ethics, aesthetics, the theory of recognition, and even metaphysics."[14]

As we can suppose, on the basis of his remarks put down in the letter of 14 November 1899 suggesting the possibility of limiting inner perception only to the sphere of feelings, Twardowski decided then to revise some of his statements. Finally, it resulted in giving up psychologism entirely, which was related to working out a new theory of reasoning, and of the act and the creation.[15]

In the same letter Twardowski confirmed his willingness to obtain funds in order to visit some centers in Western Europe, including Graz, to meet Meinong in person and to visit his laboratory. He also wrote that he intended to apply to the Austrian ministry for subsidies to equip his psychological laboratory. The documents kept at Lvov University confirm his applications.

Twardowski asked Meinong to join the action to support H. Schmidkunz (his old colleague from his Vienna studies, from 1890 to 1894 a private assistant professor of philosophy in Munich) in his attempts to obtain a scientific degree.[16]

Later correspondence between Meinong and Twardowski comes from 1903. In the letter of 3 May 1903 (K.T.A., K 3–80) Meinong turned to Twardowski asking him for his support for A. Höfler, who applied for a department in Prague after O. Willmann, for which there were many candidates (mainly Willmann's students). He suggested "private" talks with Ćwikliński. Ludwik Ćwikliński was a professor of classical philology at Lvov University at that time, and since 1899 a deputy to the Vienna Parliament: therefore his political influence was significant.

The fact that Twardowski took up the idea is confirmed in Meinong's letter of 22 August 1903 (K.T.A., K 3–81), in which he thanked Twardowski for the "optimistic news in Höfler matter". The latter obtained the post in Prague in 1903. The letter informs us that Twardowski sent Meinong his work 'Über sogenannte relative Wahrheiten' printed in 1902 and translated from Polish by Mścisław Wartenberg.[17] Wartenberg was the head of the second Department of Philosophy at Lvov University.

In April 1904 Twardowski participated in an Experimental Psychology Seminar in Giessen, and then, being on vacation and having obtained a government grant, he set out to visit psychological laboratories. He came to Graz in the middle of May 1904 and met Meinong. He visited Prague, Halle, Würzburg, Wroclaw, Leipzig, Göttingen and Paris. He returned to Lvov at the beginning of July. Neither the further correspondence nor Twardowski's work give direct remarks concerning the trip and the visit to Graz. The visible fact is that just after the journey his attempts to obtain financial aid for purchasing equipment for the psychological laboratory at Lvov University intensified. He took advantage of his being elected dean of the Department of Philosophy.

Twardowski may have expressed his opinions on his staying in Graz, as Meinong wrote him a postcard on 2 May 1905 (K.T.A., K 3–84), in which he thanked Twardowski for his letter.

Meinong's next letter to Twardowski comes from no sooner than 10 February 1907 (K.T.A., K 3–85). Its main content was Meinong's request for support for A. Höfler, who intended to move from Prague to a full professorship of pedagogics in Vienna. Höfler obtained the post in 1907. Meinong also wrote that "The main novelty in Graz since your visit here is Kreibig, of course, who — I suppose — seems to feel quite well here and has already earned our gratitude. And what is going on in my native town, particularly in the street, whose name I have forgotten again?" As we see, during his staying in Graz, Twardowski had some time for private conversations with Meinong (among others on the name of St. Sophie Street). Twardowski's answer to this letter has the date of 13 February 1907.

There are no more traces available of the correspondence between these two scientists. The parts of the letters presented above, however, make it possible to understand better some parts of the scientific biographies of both Kazimierz Twardowski and Alexius Meinong.

*Nicholas Copernicus University, Toruń*

## NOTES

[1] *Philosophenbriefe. Aus der Wissenschaftlichen Korrespondenz von Alexius Meinong mit Franz Brentano . . .* (Philosophical Letters. From the Scientific Correspondence between Alexius Meinong and Franz Brentano) hrsg. u. mit Ann. vers. von Rudolf Kindinger — Akad. Druck — u. Verl. — Anst., Graz, 1965.

[2] *Archiwum Kazimierza Twardowskiego.* Biblioteka Instytutu Filozofii i Socjologii UW (Kazimierz Twardowski Archive, The Library of the Institute of Philosophy and Sociology, Warsaw University) (further referred to as K.T.A.). Meinong's letters call numbers: K 1. 34–40; K 3. 80–85: K 22. Referring to the collection in the text I provide call number and date.

[3] *Idee und Perzeption. Eine erkenntnis-theoretische Untersuchung aus Descartes*, Konegen, Vienna, 1892. Polish ed. trans. by E. Paczkowska-Łagowska in *Archiwum Historii Filozofii i Myśli Społecznej* **22**, 1976, pp. 321–343.

[4] *Zur Lehre vom Inhalt und Gegenstand der Vorstellungen. Eine psychologische Untersuchung*, Hoerder, Vienna, 1894. Polish edition trans. by I. Dąmbska in K. Twardowski, *Wybrane pisma filozoficzne*, (Selected Philosophical Writings, Kazimierz Twardowski), PWN, Warsaw, 1965, pp. 3-91.

[5] Kazimierz Twardowski: *Selbstdarstellung*, Herausgegeben und mit einem Nachwort versehen von Jan Woleński (Krakau), redigiert von Thomas Binder (Graz), *Grazer Philosophische Studien* **39-1991**, p. 11.

[6] The two articles mentioned by Meinong were published in *Göttinger Gelegenheits Anzeiger* as 'Anzeige von Kries's 'Die Principien der Wahrscheinlichkeits-Rechnung' ', 1890, and 'Anzeige von Hillebrand's 'Die neuen Theorien der kategorischen Schlüsse' ', 1892.

[7] *Philosophenbriefe . . .* , op. cit., no 7203 O. UB.

[8] ibid.

[9] *Philosophenbriefe . . .* , op. cit., no 7204 O. UB.

[10] ibid.

[11] K. Twardowski, *Wyobrażenia i pojęcia* (Conceptions and Ideas), H. Altenberg, Lvov, 1898, *Über begriffliche Vorstellungen*, Lvov, 1903, Polish edition: *O istocie pojęć*, PTF, Lvov, 1924.

[12] *Philosophenbriefe . . .* , op. cit., no. 7206 O. UB.

[13] *Philosophenbriefe . . .* , op. cit., no 7209 O. UB.

[14] K. Twardowski, 'Psychologia wobec fizjologii i filozofii', in *Selected Philosophical Writings, Kazimierz Twardowski* (in Polish), PWN, Warsaw, 1965, p. 95 and 109. Eng. trans. as 'Psychology vs. Physiology and Philosophy' in Brandl, J., Woleński, J., *Kasimir Twardowski. Selected Writings*, Rodopi, Amsterdam (forthcoming).

[15] K. Twardowski notes, that his abandoning of psychologism was partly caused by the reading of the first volume of E. Husserl's *Logische Untersuchungen* of 1900, see *Philosophenbriefe*, op. cit., pp. 19-20.

[16] *Philosophenbriefe . . .* , op. cit., no. 7209 O. UB.

[17] K. Twardowski, 'O tak zwanych prawdach względnych' (On So-called Relative Truths), in *Księga Pamiątkowa Uniwersytetu Lwowskiego ku uczczeniu pięćsetnej rocznicy fundacji Jagiellońskiej Uniwersytetu Krakowskiego* (Commemoration Book of the Lvov University Celebrating the 500–th Anniversary of the Foundation of the Jagiellonian University in Krakow), Nakładem Senatu Uniwersytetu lwowskiego, Lvov, 1900. Trans. into German by M. Wartenberg as 'Über sogenannte relative Wahrheiten', *Archiv für systematische Philosophie* **8**, no. 4, Berlin, 1902, (also in reprint).

WŁADYSŁAW KRAJEWSKI

# THE LVOV-WARSAW SCHOOL — THE FIRST SCHOOL OF NON-POSITIVIST SCIENTIFIC AND ANALYTIC PHILOSOPHY

## 1. THE SCIENTIFIC PHILOSOPHY

Scientific philosophy is based on science in various ways. It exploits and generalises the results of the sciences, from mathematics to psychology and sociology. It uses scientific methods. It tries to be as exact as science.

Scientific philosophy has existed at least since the 19th century (Comte, Mill, Whewell, Engels, Spencer etc.). In the 20th century we have many kinds of scientific philosophy: conventionalism (Poincaré, Duhem), scientific rationalism (Brunschvicg, Meyerson, Bachelard), dialectical idoneism (Gonseth), dialectical materialism (in spite of speculations which often separated Marxist philosophy from scientific philosophy), and last but not least, logical analysis (Russell, Vienna Circle) which began analytic and scientific philosophy.

## 2. ANALYTIC PHILOSOPHY

Analytical philosophy arose in our century. It tries to be as clear and exact as possible. It analyses meanings of words, formulates definitions and distinguishes analytic and synthetic statements; in general, it uses formal logic. In this respect it goes even further than science which (excepting mathematics) rarely analyses its concepts with the tools of formal logic. There are two main kinds of analytic philosophy. The first one analyses ordinary language (and the language of philosophers). It is not scientific. The second kind analyses the language and methods of science. It is scientific.

We may also divide up analytic philosophy in another way: into narrow and broad. Narrow analytic philosophy confines itself to the analysis of language. Broad analytic philosophy starts with such analyses but does not confine itself to it: it tries to solve philosophical problems, often to give a synthesis.

G.E. Moore created not scientific but broad analytic philosophy (he used the analysis of language of philosophy to solve epistemological and ethical problems); B. Russell a broad and scientific one (analysis of language and solution of philosophical problems of mathematics); the late Wittgenstein and the Oxford school of linguistic philosophy a narrow and non-scientific one (only the analysis of common language); the Vienna Circle (in any case, in its early period) a scientific but narrow (positivist) analytic philosophy (only



*K. Kijania-Placek and J. Woleński (eds.),*
*The Lvov–Warsaw School and Contemporary Philosophy*, 53–54.
© 1998 *Kluwer Academic Publishers.*

the analysis of the language of science), K.R. Popper a broad and scientific one (he used the analysis of science to solve epistemological and ontological problems).

### 3. THE LVOV-WARSAW SCHOOL

The Lvov-Warsaw School developed a broad and scientific analytic philosophy. It began to do it a little later than Russell, who, however, did not create a school, and earlier than Popper. Kazimierz Twardowski and all of his followers tried to be clear and exact, to analyse the language of science and of philosophy and to solve, on the basis of such an analysis, philosophical problems. They used, to a lesser or greater degree, the tools of formal logic.

The members of the school often said that the philosophy which they wanted to develop should be rational, use scientific methods and be rather science than literature, analyse concepts, use logic, etc. All these postulates were not merely declarations. Many of members of the school were eminent mathematical logicians and at the same time philosophers (Stanisław Leśniewski, Jan Łukasiewicz, Alfred Tarski). Others seriously studied empirical sciences – physics, biology, psychology, sociology. None of the members of the Lvov-Warsaw School confined himself solely to the analysis of language. All of them tried to solve various philosophical problems, in the domain of epistemology (Kazimierz Twardowski, Kazimierz Ajdukiewicz, Maria Kokoszyńska-Lutmanowa, Izydora Dąmbska), of ontology (Tadeusz Kotarbiński, Jan Łukasiewicz, Henryk Mehlberg), and of ethics (Tadeusz Kotarbiński, Tadeusz Czeżowski, Władysław Tatarkiewicz, Maria Ossowska).

The Lvov-Warsaw School was the first philosophical school which systematically elaborated a scientific and non-positivist analytic philosophy.

*Institute of Philosophy, University of Warsaw*

ELŻBIETA PAKSZYS

# WOMEN'S CONTRIBUTIONS TO THE ACHIEVEMENTS OF THE LVOV-WARSAW SCHOOL: A SURVEY*

## 1. AN INTRODUCTION

According to conventional wisdom philosophy is a bastion of patriarchy: masculine bias determines both the object of study (quite often discrediting femininity) and the dominant gender of the 'agents/subjects' representing the field and carrying on work in it. Rarely in history has a woman been included in the list of great thinkers.[1] Nevertheless, within the relatively short span of the last two centuries, when women gradually began obtaining education comparable to that of men, women philosophers have appeared and today we shall look at some interesting examples of the contributions they have made to this area of the humanities.

In the case of Poland, quite a number of women made contributions to the analytical philosophy of the Lvov-Warsaw School, something which seems to have been a general result of Polish women's access to higher education at the turn of the century (1884–1918). However, nothing in the 19th century prepares us for this, since at that time Polish women were only occasionally present in philosophy.[2]

The women of the Lvov-Warsaw School represent the generation of our grandmothers, in some cases even great grandmothers, since they were born at the end of the 19th or beginning of the 20th century. Two of them, as Polish Jews, were killed by the Nazis; the rest, however, passed away during the 1970s and 80s. Only one of them, Janina Kotarbińska (1901–1996), was still active, recently engaged mostly in preserving her late husband's philosophy.

The total number of members of the Lvov-Warsaw School includes ten women engaged in research on contemporary philosophical issues who developed their own careers during the 1930-70s. However, it should be stressed that in many respects these women have only become visible recently, now that our focusing on gender issues in philosophy has been justified by the feminist inquiry of the last two decades. Here it is important to observe that the majority of the Lvov-Warsaw philosophers did not display any kind of feminine or feminist identity, hence one needs to be cautious when considering their achievements as feminist thinkers.

The choice of the humanities — and in particular of philosophy — as a subject of study was conditioned by diverse factors and motivations at work in the



*K. Kijania-Placek and J. Woleński (eds.),*
*The Lvov–Warsaw School and Contemporary Philosophy*, 55–71.
© 1998 *Kluwer Academic Publishers.*

women of those times. Within a Poland torn by divisions and within a rigidly conservative society, it was certainly easier for women to expect approval for choosing to work within 'service' professions, like teaching, even in higher education (Żarnowska, Szwarc, 1992). In the case of the Lvov-Warsaw School women, however, the choice of philosophy or logic happened to be more a not-yet fully conscious decision than a youthful seeking for the self (in case of Janina Kotarbińska or Izydora Dąmbska), or an effect of an early budding of intellectual capacities, which found its way to analytical philosophy (in Maria Ossowska's case).

Yet the ability of female academics to meet the standards and to pursue their profession in a field so dominated by men seems to have been part of the complex effect of the profeminist attitude of the School's founder Kazimierz Twardowski (Pachucka, 1958) as well as the general ambiance of support for higher education for women within Lvov University. The Master grafted onto his pupils his own understanding for women's emancipatory endeavours, which later bloomed during the Warsaw stage of the School. The liberal stance of the Lvov-Warsaw School towards gender, nationality and religion was well known.

Certainly, among the other important factors that should be mentioned here were the positive, long-lasting teacher-student relationships and partnership marriages which seem to happened quite often among representatives of the Lvov-Warsaw School. Also, in the reborn Poland of the time we encounter truly non-sexist educational policies, which resulted in many female scholars of philosophy receiving grants from 'The National Culture Fund' for temporary studies abroad.

## 2. THE PRESENCE OF WOMEN IN THE LVOV-WARSAW SCHOOL: AN ATTEMPT AT RANKING THEM IN TERMS OF THEIR ACHIEVEMENTS

I will now try to provide a broad general outline of the contributions of the women of the Lvov-Warsaw School, setting them within philosophy in terms of certain individual levels of 'prominence' or their representation in similar subdisciplines within philosophy. An integral part of this analysis involves the compiling of (a still incomplete) bibliography of these women's publications. With the help of this bibliography it is possible to rank to the scope and diversity of the women's interests as well as to rank the originality of their achievements. All this involves trying to 'measure' — how immeasurable! — the amount and quality of the publications. It will also be necessary to refer to generally established assessments and judgements of this output, which undoubtedly should help in the formulation of a tentative ranking. Also vital will be the tracing of particular scholar's styles of argument, suggesting — possibly — the Lvov-Warsaw School women's agreement or disagreement with the main programme of the School.

The philosophical attitude established within the School seems to be expressed most concisely by K. Ajdukiewicz's (1934, 1935) phrase — and can be understood by comparison to the Vienna Circle programme — as 'logistic antiirrationalism'.[3] It can be seen, however, that formulated in this way the School's programme does not include its founder's passion for scientific psychology. It should be also stressed that the majority of women representing the School pursued K. Twardowski's interest in descriptive psychology (*Słownik Psychologów Polskich*, 1992). For the purposes of the present analysis, Ajdukiewicz's list of the postulates should be extended to include psychological studies, as well as historico-philosophical investigations. However, among the most conspicuous fields of interest within the School were logic, philosophy of science and semantics. All these subjects appear in different combinations in the works of the women philosophers considered below. It is possible to differentiate the logicians and semanticians among them; however, the labels are usually quite arbitrary and do not thoroughly embrace a particular writer's interests.

Two very prominent figures among the women of this philosophical group are Izydora Dąmbska and Maria Ossowska. Izydora Dąmbska (1904–1983) seems to have been the most versatile and most prolific of all the women philosophers of the School. Introduced to philosophy in Lvov by K. Twardowski and K. Ajdukiewicz, she also attended M. Schlick's seminar in Vienna. Her early publication (1933) systematises conditions of general laws and is considered to be one of the first devoted to this topic in the world.[4] In (1937) she tackled the problem of scientific knowledge and irrationalism and in (1937a) presented the general attitude of the Lvov-Warsaw School towards the ideas of one of the most outstanding writers representing the opposition — Ludwig Fleck.[5] Before the war, Dąmbska expressed her ideas on semantics when examining conditional sentences (1938a) and she developed this issue later in (1949) when reflecting on the philosophy of proper names and in (1962) in her study on analogy. She tried to define some notions of grammar in the light of logic (1964). Later she gathered all her semantic studies into a book 'On Instruments and Objects of Cognition' (1967), considered by Jan Woleński the most important work of all her oeuvre. Dąmbska's interest in problems of convention (1938) also began early on, but only later did she present a fuller account in a small book devoted to this topic (1975). She was also active in the history of philosophy, and always interpreted it in a very contemporary way as in her well-known book on 17th century French scepticism (1958), in her study of the concept of language in K. Ajdukiewicz's philosophy (1965), in two studies on Plato (1972a), in her discussion of semiotic concepts in Twardowski's philosophy (1975a) and in evaluating the influence of F. Brentano on Polish philosophy (1978). Semiotics and epistemology seem to have preoccupied her equally in her study of the semiotic functions of silence (1971) and in her last work, an introduction to the semiotics of the ancient Greeks (1984). She also translated many philosophical works into Polish. Dąmbska's influence has been acknowledged: among others Jan Woleński dedicated to her his broad studies of Polish analytic philosophy (Woleński, 1985, 1989).

Semantics was also the first philosophical area in which Maria Ossowska (née Niedźwiecka) (1896–1974) began her career. Educated at Warsaw University in 1915–19, where she was a pupil of Łukasiewicz and Kotarbiński, she also studied in Paris (Sorbonne, 1921–23) and later on she left for England to do research in G. E. Moore's (1925–27?) and Bronisław Malinowski's (1933–35) seminars. After the war, in the period 1949–56, she, together with her husband and many other representatives of the School, was denied the right to lecture.[6] Later they were both given chairs at Warsaw University. In her early works (1929) and (193), Ossowska studied issues ranging from logic to grammar and the relations of words to thoughts. Together with Stanisław Ossowski she published one of the first books in the world devoted to the science of science (1935). After the Second World War, Ossowska clearly switched to the humanities when developing her reflections on the sociology of ethics. She developed the foundation for her descriptive ethics (1947) that marks both a continuation and an overcoming of traditionally developed normative ethics, an approach introduced by Twardowski. In her later books (1949) and (1956) Ossowska reflects on the psychology of morals and analyses bourgeois morality. Her approach is still typically analytic, however: in Skolimowski's words "To emphasise a departure from speculation, Ossowska called her inquiry a theory of morality or the science of morals" (Skolimowski, 1967, p. 199).

When trying to evaluate generally the position of the leaders among women of the School just named we can say this: The oeuvre of Dąmbska is particularly characterised by its abundance and universality, oscillating between the School's programme, i.e. rationalism and its opposite, irrationalism.[7] It would seem that Ossowska's achievements display perhaps even greater originality in defining a new discipline, 'her own' discipline developing along the way of her individual evolution and moving from agreement with the School's spirit of semantics to the thoroughly new 'science of morals'.[8]

The largest group of women in the Lvov-Warsaw School were theorists of formal logic and the methodology of science, working as well in semantics and semiotics, and occasionally on psychology and the history of philosophy and logic.

A career tragically broken by the Holocaust is that of the highly promising logician Janina Hosiasson-Lindenbaum (1899–1942), a pupil of Kotarbiński married to Adolf Lindenbaum (1904–1941?).[9] Polish Jews, the Lindenbaums were both killed by the Nazis during WW II. According to Woleński's evaluation of Hosiasson's works (1928), (1934), and (1941), her theory of induction was one of the most interesting ever created within the School. Woleński also maintains that if Hosiasson's ideas had been known earlier, they "would have certainly played a vital role in the contemporary development of the theory of induction". (Woleński 1987, p. 115) Her solution to the so-called paradox of confirmation in (1940) and (1941) is famous.[10] Hosiasson's posthumous publications (1946) and (1948–49) concern epistemological problems of progress in cognition. However, this scholar of great promise passed away too early to carry through her ideas.[11]

Another outstanding philosopher and methodologist of the period is Janina Kotarbińska (1901–1996).[12] A pupil and co-worker of Tadeusz Kotarbiński, she later become his wife. Among Kotarbińska's studies, the most important ones seem to address epistemological problems. She starts from the necessity of elemental relations (Sztejnbarg, 1931) and indeterminism in the natural sciences (physics, biology) (1932) (1932a) and in the humanities (1933); her work is crowned by her study of chance (1934). In her post war publications Kotarbińska is mostly oriented towards semantics in the studies on definitions (1955), (1959), (1960) and on the notion of sign (1956). However, like the others in the School she was engaged in defending induction as a common method of cognition and became a well-known critic of Popper's falsificationism (1962). Since the 1980s, after Kotarbiński's death, she has been mostly occupied with explaining her husband's philosophy.[13] Kotarbińska's bibliography, still in preparation, comprises about thirty publications (Kotarbińska, 1990).

Maria Kokoszyńska-Lutman (1905–1981) represents Lvov as a pupil of Twardowski and Ajdukiewicz. She also came under the influence of the Vienna Circle, having had the opportunity to attend Vienna Circle seminars. After WW II, like many other representatives of the School, she was only allowed to deliver lectures on logic. In her early work (1938) Kokoszyńska dealt with epistemological ideas when proposing to battle metaphysics. After the war she developed theories concerning the relativity of truth in (1948) and (1951), and took part in a discussion with the Marxists on the relationship between logic and dialectics in (1955) and (1957a). Her main field of interest, however, seems to have been methodology and especially the relationship of induction-deduction. In (1957) she tried to define 'good', distinguished from 'bad', induction, and in (1962), (1967) reflects broadly on the axiomatic method. In semantics, Kokoszyńska follows Ajdukiewicz when differentiating 'semantic' from 'syntactic' meanings (1969). Woleński ascribes to her the "Polish concept of analytic sentences" developed early in (1947), because this subject seems to be present in almost all her works. Considering only the scope of

Kokoszyńska's works it is apparent that she is still not sufficiently well recognised.[14]

Another woman principally employed in logic was Seweryna Łuszczewska-Romahn (1904–1978), a student of the well-known Lvov School of mathematics,[15] and in philosophy a pupil of Twardowski and Ajdukiewicz as well as of Roman Ingarden (1893–1970). After WW II she was associated with the Adam Mickiewicz University in Poznań. Among her interests were the study of the equivocation of the language of science in her (1948), and the concept of classification as a function of a distance (1961). Her philosophical horizon seems not to have surpassed that of her teachers' (1962) and her last publication (1977) concerns Twardowski's theory of knowledge, so it can be considered epistemological and historico-philosophical. Woleński (1985) ascribes to Łuszczewska-Romahn a "nonformal analysis of notions."[16]

Still other philosophical disciplines explored by women of the Lvov-Warsaw School, ethics and the history of philosophy, were the subjects of Daniela Gromska's (nèe Tennerówna) (1889–1973) reflection. While Twardowski's disciple in Lvov, she was one of the first women in the School to publish. In (1914) Tennerówna reflects on existence as a 'content' of judgement-making and judgement on the margin of Twardowski's theory of judgement. After the war in (1948) she assessed the losses Polish philosophy had suffered.[17] Except for her didactic work at the Jagiellonian University in Krakow, Gromska devoted herself mainly to numerous translations and editorial jobs as well as to the popularisation of her teachers' ethical ideas (1959).[18]

There were three women in the Lvov-Warsaw School engaged in work on psychology, at that time still considered a part of philosophy. One was Eugenia (née Ginsberg) Blaustein (1905–1942) — killed together with her husband Leopold as Polish Jews by the Nazis. She is presumed to have had only one publication (1931). The second psychologist was Helena Słoniewska (1897–1982) a gifted scholar, active in research and the methodology of psychological practice (Słoniewska, 1934, 1959). Also as a witness to the times, she occasionally took part in the historico-philosophical debates (1968, 1978). Her list of publications, although not yet compiled, should total more than ten titles.

The last of the ten female names ascribed to the Lvov-Warsaw School is Irena Filozofówna (1906–1967). Incomplete data in her case makes it difficult to assess her place within the School. She is listed among Kotarbiński's students in *Fragmenty Filozoficzne* (1934).[19] Jadacki (1987) mentions her involvement in semiotics but Kotarbińska considered her a "psychologist of theater".[20] Indeed, Filozofówna left two articles concerning the semantics of imaginative images (1931) and the Stanisławski method (1932). But her own name is a linguistic puzzle in itself, being of the form typically used in a Polish (inflection) for maiden names which expresses the clear meaning of 'being a philosopher's daughter'. However, the question remains: where was she in the ranks of Kotarbiński's 'daughters'?

## 3. DISCUSSION AND CONCLUSIONS

Let us now make a tentative analysis of the best-recognised professional careers among the women of the Lvov-Warsaw School. It can be said that they usually worked on their own account, although their sometimes quite long apprenticeships could not have been possible without their teachers' full approval and support. One example is Dina Sztejnbarg's ten years as Kotarbiński's assistant (1929–39). She recalled her "very close" co-operation with her mentor on *his book* only (Kotarbiński, 1929).[21] Another instance is the four year apprenticeship of Izydora Dąmbska to Twardowski (1926–30).

We shall presume that while the doctoral theses, the majority of them completed in the 1920s, were most often a direct result of the teachers' and the supervisors' influence, the subjects of the habilitation theses reflected more independent choices, conditioned by individual discoveries and previous international contacts. It looks as if this indeed happened in the case of Dąmbska, who in her habilitation thesis (1937) dealt with the problem of irrationalism. However impressive the number of influences reflected in her publications (epistemological, semantic, semiotic or historico-philosophical), the very variety of her work causes certain difficulties in establishing her reputation.[22] In contrast, however, Ossowska's works up to her habilitation thesis (1931) reflect a continuation of the Lvov-Warsaw School semantics. Her development from the School's main programme — if it ever took place — was much slower and happened during the period of her full independence, that is, after the war.

The possibility of overcoming the quite orthodox rationalist and logical programme of the School can be ascribed to Dąmbska in her early appreciation of irrationalism as a constant philosophical adversary of the general superrationalist attitude of the School. While Ossowska's eventual branching out could have been accomplished only indirectly, during her mature period she adopted the attitude of a fully objective observer of moral phenomena. In this way she proposed to describe morality within psychology and sociology, or — as she called it — "the science of morals" (Skolimowski, 1967). In this case, however, the peculiar choice of subject for study, together with a lack of the consequences in the realisation of a declared ideal of an objective spectator would only serve for some possible departures from the most orthodox elements of the Lvov-Warsaw School programme.[23]

It seems that other female representatives of the School thoroughly accepted its rationalist-logical 'paradigm'. However, it should be possible (in future) to evaluate their real position through investigating their own understanding of philosophy and comparing them to already worked out interpretations of the subject. This would make it possible to evaluate the extent of agreement with, or departure from, the paradigm of both leaders — Dąmbska and Ossowska — as well as of the rest of the outstanding women logicians: Kotarbińska, Kokoszyńska-Lutman, Łuszczewska-Romahn and others.

I would now like to consider the creativity of the women of the Lvov--Warsaw School against the broader background of Polish philosophy between the wars. As regards the proportion of women to men: 10 of the 80 members of the School, the women making up 13%, reflecting only "an average feminisation" of Polish philosophy in pre-war period.[24] The group of ten, however, can be treated rather conventionally when we consider how many other women were supposedly more loosely connected with the School during its Lvov period. Among them were Zofia Pasławska-Drakslerowa, ? Jakubowska, Maria Franklówna and Józefina Mehlberg. Later, in the Warsaw stage of the School, there were even more, since the list of Kotarbiński's disciples includes 20 women out of a total of 51 persons.[25] Jadacki (1987) names other women in connection with semiotics in the Lvov-Warsaw School: ? Jawicówna, Zofia Lissa (1908–1980), Irena Raczyńska (1900–1981), ? Radlińska-Ostaszewska, ? Radomska-Strzemecka, Krystyna Zwolińska (not all names and dates have yet been found). We can suppose that within the School and around it, connected more 'socially' than professionally, there were many more women, whose presence was not documented other than in pictures, lists of pupils and memoires.[26]

The Second World War losses among female members of the School have been indicated earlier. Here it should be pointed out that two prisoners of the Nazi camps were Janina Kamińska (Kotarbińska) and Seweryna Łuszczewska-Romahn. Izydora Dąmbska was active in the Underground Movement (AK), while Kamińska and Ossowska worked unintimidated in so-called 'secret education'.

Certainly, the post-war period seems to have been difficult and painful for all members of the Lvov-Warsaw School. Until 1956 their activity was restricted as alleged "representatives of bourgeois philosophy", in a Poland dominated by Marxist ideology. Later on, however, the most prominent women of the Lvov-Warsaw School were granted the highest honours in science and education. By the 1960s and 70s they had retired.

The phenomenon of productive women in Polish analytical philosophy is not an easy one to interpret, especially when one attempts to consider it from a contemporary, relatively egalitarian perspective toward sex/gender, a perspective now shaped by the relatively wide representation of women in this profession. Certainly, with an average of 20–25% of women active in Polish philosophy today, the percentage of women of Lvov-Warsaw School has doubled. However, it is also clear that there is still quite a long way to go to achieve gender equality in this domain. However, the proper perspective from which to evaluate the presence of women in the Lvov-Warsaw School, however, is noting the virtual absence of women in world philosophy at this time. Only in that context can one realise how strong was the taboo challenged by the profeminist attitudes of K. Twardowski and his pupils.

Because other contemporary philosophical currents, like German Phenomenology and American Pragmatism (Seigfried, 1991) had a very small number of active and creative women, we can that in the case of the Lvov-Warsaw School we are dealing with a unique cultural phenomenon (Pakszys, 1996). Unfortunately, the lack of materials for comparison makes such generalisations difficult or risky, even within the current of analytical philosophy today. I hope, however, that such a comparison will be possible in the future. As a beginning, we can suggest the comparison of individual scholars who display some similarities, like — for instance — Suzanne Langer, representing American pragmatism, and Izydora Dąmbska. They both display a kind of versatility, in spite of their analytical orientation.[27] It seems further that such comparative studies would be really informative for the broader issue of female presence in analytic philosophy.

This type of philosophy, whose logical, seemingly 'dry' and sober reflective character seems to resist feminist interpretation, may lie at the core of any account of the extent to which women have been suppressed in philosophy. In other words: in the abundantly developing wave of contemporary feminist insights into philosophy this example of not a passive, but a creative presence of women analytical philosophers seems a paradox which cannot easily be subordinated to the common (and comfortable) charges of victimological character, which views women as mere victims of male domination in this domain (Lloyd, 1984). Moreover, the Lvov-Warsaw School also seems a challenge to so-called essentialist tendencies which seek to find an explanation for male bias (androcentrism) within analytical philosophy, by identifying maleness with rationality, objectivity and logic. These components of a general philosophical attitude, so typical of the analytical viewpoint, are currently ascribed as being synonymous with the masculine intellect and further complicate the relation of human and intellect.[28]

This argument has been prepared, perhaps, without a sufficient understanding of that exaggeration or even of female bias. However, as a point of departure for all discovery and study of neglected women and their place within the Lvov-Warsaw School, we can follow a feminist agenda when, for instance, sharing the opinion of indeed often neglected female achievements in the past and in today's philosophy. Hence, from the outset an evaluation of the patriarchal character of philosophy is not to be treated only as empty rhetoric.

My discussion of the Lvov-Warsaw School sets aside the question of the masculinity or maleness of philosophy,[29] while reinforcing (male)gender representatives as the discipline's predominant agents. However, both conditions seem to be reciprocal. But one must ask whether separating women from the broader human context of the Lvov-Warsaw School is justified. Or to put it in a different way: how far can the presence or absence of an element of feminist consciousness in these women justify such separation? Now, consensus is impossible. We found among the women of the Lvov-Warsaw School notable examples of such a possible attitude, in Maria Ossowska (when dwelling on the problems of human dignity) and Janina Hosiasson (the translator of Russell's sociological and pedagogical works could be suspected of some feminist affiliation). But there are also clearly negative examples in the case of Kotarbińska, who shows no understanding of feminist issues.[30] This material is quite scant and too diffuse even for speculation about whether or not women from the Lvov-Warsaw School had pro-feminist ideas. It seems, then, that justification for my proposal of a new reshuffling of the 'cards' in philosophy can be sought in the new contexts of today's feminist thought. My discussion is an example of women's studies in the history of philosophy; that is, a new discipline which focuses on certain neglected anthropological determinants of the agents in philosophical cognition. For it seems that now is the time to study the impact of sex/gender and race (not white only) as the constant 'currents' or contextual — natural/cultural — conditions of philosophical reflection (Pakszys, 1994, 1995).

*Institute of Philosophy, Adam Mickiewicz University, Poznań*

## NOTES

* In preparation of this paper I feel deeply indebted to my American friends: Sandra Harding and Bonnie Scott (University of Delaware, Newark), Meg Newcomer (Lublin) and Claire Hill (Paris), who commented on its content as well as polished its language.

[1] See the discussion in feminist literature: Bordo (1988), Gatens (1991), Grimshaw (1986), Lloyd (1984), Mahowald (1986/7), Pakszys (1994) and (1995), Scheibinger (1989), Seigfried (1991).

[2] Such was the position of the very first Polish woman writing on philosophical (theistic) topics — Eleonora Ziemięcka (1819–1869). Among the philosophers of that earlier epoch we find some unique examples of female thinkers and writers representing contemporary currents: Positivism by Eliza Orzeszkowa (1841–1910), Marxism by Rosa Luxemburg (1871–1919) and Empiriocriticism by Józefa Kodisowa (1865–1940).The last were the first Polish women who received Ph.Ds from foreign universities: Kodisowa in 1892 and Luxemburg in 1897. See Hulewicz (1939), Mazurczak (1995), Pachucka (1958), Pakszys (1996), Petrażycka-Tomicka (1914), Walicki (1986).

[3] This stance can be summarised in some five main postulates: 1. antiirrationalism (verifiability of statements), 2. clarity of notions and precision of language, 3. logical (symbolic) apparatus of notions, 4. scientific knowledge as a main subject of study (in particular metatheoretical and semantic research), 5. intra-theoretical inquiry, i.e. concerning the foundations of science, especially the deductive disciplines.

[4] See the opinion of Polish authority in this subject, W. Krajewski (1982).

[5] Fleck was bacteriologist and philosopher, active in Lvov before WW II but his orientation towards science was not logical but sociological ('collectivist'). See his main book: *Entwicklung einer wissenschaftlicher Tatsche: Denkstil und Denkkollektiv*, (Basel, 1935). English translation is forwarded by T. Kuhn: *Genesis and Development of Scientific Fact*, (Univ. of Chicago Press, 1981). First Polish translation of the book: L. Fleck, *Powstanie i rozwój faktu naukowego*, was edited even later (Lublin, 1986).

[6] This was a kind of persecution directed by Marxist ideologists towards most representatives of pre-war philosophy.

[7] Dąmbska's full bibliography contains about 300 publications, see Bober (1992).

[8] Ossowska's posthumous bibliography numbers about 200 titles, see Ossowska (1983).

[9] A. Lindenbaum was considered to be second in rank, after Alfred Tarski (1902–1983), as a logician of the Lvov-Warsaw School.

[10] T. Kotarbiński (1967, p. 12) commemorated Janina Hosiasson, Popper (1990) also maintained a high opinion of her ideas over the years.

[11] J. Hosiasson's entire oeuvre totals nine articles and three translations into Polis Russell's books. See bibliography.

[12] In her youth she published under her maiden name, Dina Sztejnbarg, and after the Holocaust was also known as Janina Kamińska (a pseudonym used during the Nazi occupation). See bibliography and *Wywiady z Prof. J. Kotarbińską*.

[13] She was still busy with the task of introducing Poland's traditionally Catholic society to Kotarbiński's independent ethics, that is an ethics without simple commitments to any religious faith containing the highest humanist ideas. Janina Kotarbińska died in 1996.

[14] The bibliography of M. Kokoszyńska-Lutman numbers about 70 works, see Kokoszyńska (1976).

[15] Łuszczewska-Romahn studied mathematics under the direction of Hugo D. Steinhaus (1887–1972) and Stefan Banach (1892–1945).

[16] This scholar's oeuvre comprises some 25 publications. See Batóg (1979).

[17] According to later estimates about 25 members of the School were victims of the war, see Woleński (1985, p. 28).

[18] Gromska's full bibliography comprises some 25 titles. See her obituary and bibliography in *Studia Filozoficzne* **4(101)**, 1974, pp. 207–209.

[19] This pupil's initiative begins a famous series of three publications honouring Kotarbiński's 15th and 40th anniversary of teaching and his 80th birthday. See bibliography.

[20] See *Wywiady* /The Interviews with Prof. J. Kotarbińska, in bibliography.

[21] Ibid.

[22] The best strategy for studying these works is to take a broad, cultural approach, which makes the necessary generalisations possible. See Bober (1992).

[23] This is a conclusion from one of Ossowska's pupil's retrospective reflection, see Lazari-Pawłowska (1975).

[24] According to the *List of the Third Polish Philosophical Assembly*, see bibliography: *Księga Pamiątkowa* (1936).

[25] See *Fragmenty Filozoficzne* (1934).

[26] In my later study I recognise two generations of women of the Lvov-Warsaw School, see E. Pakszys (1997) (in print).

[27] See bibliography for Dąmbska and Langer.

[28] There is a discussion on this matter in American feminist philosophy. See for instance Bordo (1988) and Mahowald (1986/87).

[29] See especially: Grimshaw (1986), Lloyd (1984) and Schiebinger (1989).

[30] She expressed this in *Wywiady*/The Interviews, see bibliography.

## REFERENCES

Ajdukiewicz, K.: 1934, 'Logistyczny antyirracjonalizm w Polsce' (Logistic Antiirrationalism in Poland), *Przegląd Filozoficzny* **37/4**, pp. 325–327.

Ajdukiewicz, K.: 1935, 'Der Logistische Antiirrationalismus in Polen', *Erkenntnis* **5**, pp. 151–164.

Batóg, T.: 1979, 'Seweryna Łuszczewska-Romahnowa (1904–1978)', *Studia Filozoficzne* **1**, pp. 189–194.

Blaustein (Ginsberg) E.: 1931, 'W sprawie pojęć samoistności i niesamoistności' (On the Concepts of Existential Dependence and Independence), in *Księga Pamiątkowa Polskiego Towarzystwa Filozoficznego we Lwowie* (A Memorial Book of the Polish Philosophical Society in Lvov), pp. 143–168. Eng. trans. in Smith (1982).

Bober, I.A.: 1992, *Filozofia kultury Izydory Dąmbskiej* (I. Dąmbska's Philosophy of Culture), WSP, Kielce.

Bordo, S.: 1988, 'Feminist Scepticism and the 'Maleness' of Philosophy', *The Journal of Philosophy* **11**, pp. 619–29.

Dąmbska, I.: 1933, *O prawach w nauce* (On Laws in Science), Gubrynowicz i Syn, Lvov.

Dąmbska, I.: 1937, 'Irracjonalizm a poznanie naukowe' (Irrationalism and Scientific Knowledge), *Kwartalnik Filozoficzny* **14(2–3)**, pp. 83–118, 185–212.

Dąmbska, I.: 1937a, 'Czy intersubiektywne podobieństwo wrażeń zmysłowych jest niezbędnym założeniem nauk przyrodniczych?' (Is Intersubjective Similarity of Sense Data a Necessary Condition of Natural Science?), *Przegląd Filozoficzny* **40/3**, pp. 83–118.

Dąmbska, I.: 1938, 'Konwencjonalizm a relatywizm' (Conventionalism and Relativism), *Kwartalnik Filozoficzny* **15/4**, pp. 328–37.

Dąmbska, I.: 1938a, 'Z semantyki zdań warunkowych' (Semantic Issue of Conditional Sentences), *Przegląd Filozoficzny* **41(3)**, pp. 241–267.

Dąmbska, I.: 1949, 'Z filozofii imion własnych' (Issues in the Philosophy of Proper Names), *Kwartalnik Filozoficzny* **18**, pp. 241–261. Eng. trans. in Pelc (1979), pp. 131–143.

Dąmbska, I.: 1958, *Sceptycyzm francuski XVI i XVII w.* (French Scepticism in the 16th and 17th c.), Towarzystwo Naukowe w Toruniu, Toruń.

Dąmbska, I.: 1962, *Dwa studia z teorii naukowego poznania* (Two Studies on the Theory of Scientific Knowledge), Towarzystwo Naukowe w Toruniu, Toruń.

Dąmbska, I.: 1964, 'Niektóre pojęcia gramatyki w świetle logiki' (Some Notions of Grammar in the Light of Logic), in *Szkice filozoficzne — Romanowi Ingardenowi w darze* (Philosophical Essays dedicated to Roman Ingarden), PWN, Warsaw, pp. 221–238.

Dąmbska, I.: 1965, 'Koncepcja języka w filozofii Kazimierza Ajdukiewicza' (The Concept of Language in K. Ajdukiewicz's Philosophy), *Ruch Filozoficzny* **24/1–2**, pp. 3–9.

Dąmbska, I.: 1967, *O narzędziach i przedmiotach poznania* (On Instruments and Objects of Cognition), PWN, Warsaw.

Dąmbska, I.: 1971, 'O funkcjach semiotycznych milczenia' (On Semiotic Functions of Silence), *Studia Semiotyczne* **2**, pp. 77–88.

Dąmbska, I.: 1972, 'Filozofia na Uniwersytecie Jana Kazimierza we Lwowie w latach 1918–1939' (Philosophy at Jan Kazimierz University in Lvov 1918–1939), *Zeszyty Lwowskie* **2**, London, pp. 76–90.

Dąmbska, I.: 1972a, *Dwa studia o Platonie* (Two Studies on Plato), Ossolineum, Wroclaw.

Dąmbska, I.: 1975, *O konwencjach i konwencjonalizmie* (On Conventions and Conventionalism), Ossolineum, Wroclaw.

Dąmbska, I.: 1975a, *Znaki i myśli. Wybór pism z semiotyki, teorii nauki i historii filozofii* (Signs and Thoughts. Selected Writings on Semiotics, Theory of Science and History of Philosophy), PWN, Warsaw.

Dąmbska, I.: 1978, 'Franz Brentano et la pensée philosophique en Pologne: Casimir Twardowski et son ecole', *Grazer Philosophishe Studien* **5**, pp. 118–129.
Dąmbska, I.: 1984, *Wprowadzenie do starożytnej semiotyki greckiej. Studia i teksty* (An Introduction to Ancient Greek Semiotics. Studies and Texts), Ossolineum, Wroclaw.
Filozofówna, I.: 1931, 'W sprawie wyobrażeń imaginatywnych' (Dyskusja z L. Blausteinem) (In the Matter of Imaginative Images (Discussion with L. Blaustein)), *Przegląd Filozoficzny* **34**, pp. 39–50, 122–141.
Filozofówna, I.: 1932, 'Uwagi o tak zwanym "Systemie Stanisławskiego" ' (Remarks on the so-called "Stanisławski System"), *Przegląd Filozoficzny* **35**, pp. 339–346.
*Fragmenty Filozoficzne*, 1934, Księga pamiątkowa dedykowana Prof. T. Kotarbińskiemu przez jego uczniów w 15–lecie pracy nauczycielskiej — seria pierwsza (Essays in Honour of Prof. T. Kotarbiński on the Occasion of His 15th Year of Teaching), PWN, Warsaw, 1934.
*Fragmenty Filozoficzne*, 1959, Księga pamiątkowa ku uczczeniu 40–lecia pracy nauczycielskiej w Uniwersytecie Warszawskim Prof. T. Kotarbińskiego — seria druga (Essays in Honour of Prof. Kotarbiński on the Occasion of His 40th Year of Teaching), PWN, Warsaw, 1959.
*Fragmenty Filozoficzne*, 1967, Księga pamiątkowa ku czci Prof. T. Kotarbińskiego w 80. rocznicę urodzin — seria trzecia (Essays in Honour of Prof. T. Kotarbiński On the Occasion of His 80th Birthday), PWN, Warsaw, 1967.
Gatens, M.: 1991, *Feminism and Philosophy, Perspectives on Difference and Equality*, Indiana Univ. Press, Bloomington and Indianapolis.
Grimshaw, J.: 1986, *Philosophy and Feminist Thinking*, Univ. of Minnesota Press, Minneapolis.
Gromska, D.: 1948, 'Philosophes Polonaise mortes entre 1938 et 1945', *Studia Philosophica* **3**, pp. 31–97.
Gromska, D.: 1959, 'Poglądy etyczne Kazimierza Twardowskiego' (Ethical Ideas of K. Twardowski), *Ruch Filozoficzny* **17**, pp. 9–17.
Hempoliński, M. (ed.): 1987, *Polska filozofia analityczna, analiza logiczna i semiotyczna w szkole lwowsko-warszawskiej* (Polish Analytical Philosophy, Logical and Semiotic Analysis in the Lvov-Warsaw School), Ossolineum, Wroclaw.
Hosiasson, J.: 1928, 'Definicje rozumowania indukcyjnego' (Definitions of Inductive Reasoning), *Przegląd Filozoficzny* **31**, pp. 352–67.
Hosiasson, J.: 1931, 'Why Do We Prefer Probabilities Relative to Many Data?', *Mind* **40/157**, pp. 23–36.
Hosiasson, J.: 1934, 'O prawomocności indukcji hipotetycznej' (On the Validity of Hypothetical Induction), in *Fragmenty Filozoficzne*, 1st series, pp. 11–34.
Hosiasson-Lindenbaum, J.: 1935, 'Uwagi w sprawie wnioskowań indukcyjnych' (Remarks on Inductive Reasoning), *Kwartalnik Psychologiczny* **7**, pp. 275–300.
Hosiasson-Lindenbaum, J.: 1936, 'La theorie des probabilites est-elle une loguique generalisée?', in *Actes du Congres Internationale de Philosophie Scientifique* **4**, Paris, pp. 58–64.
Hosiasson-Lindenbaum, J.: 1940, 'On Confirmation', *Journal of Symbolic Logic* **6**, pp. 133–148.
Hosiasson-Lindenbaum, J.: 1941, 'Induction et analogie, comparison de leur fondament', *Mind* **50/201**, pp. 351–365.
Hosiasson-Lindenbaum, J.: 1946, 'Postęp wiedzy z punktu widzenia poznawczego' (Advancement of Knowledge from a Cognitive Point of View), *Przegląd Filozoficzny* **44**, pp. 59–65.
Hosiasson-Lindenbaum, J.: 1948/49, 'Theoretical Aspects of the Advancement of Knowledge', *Synthese* **7/4–5**, pp. 253–261.
Hulewicz, J.: 1939, *Sprawa wyższego wykształcenia kobiet w Polsce w wieku XIX* (The Issue of Higher Education of Women in 19th cent. Poland), PAU, Krakow.
Jadacki, J.J.: 1987, 'Semiotyka szkoły lwowsko-warszawskiej: główne pojęcia' (Semiotics of the Lvov-Warsaw School: Main Notions), in Hempoliński (1987), pp. 131–218.

Kamińska, J.: 1946, 'Idealizm', *Wiedza i Życie* **15/6**, pp. 518–528.

Kamińska, J.: 1947, 'Ewolucja Koła Wiedeńskiego' (The Evolution of the Vienna Circle), *Myśl Współczesna* **2**, pp. 145–160.

Kokoszyńska-Lutman, M.: 1938, 'W sprawie walki z metafizyką' (On the Battle Against Metaphysics), *Przegląd Filozoficzny* **41**, pp. 9–24.

Kokoszyńska-Lutman, M.: 1947, 'O różnych rodzajach zdań' (On Various Kinds of Sentences), *Przegląd Filozoficzny* **43**, pp. 22–51.

Kokoszyńska-Lutman, M.: 1948, 'What does 'relativity of truth' mean?', *Studia Philosophica* **3**, pp. 167–176.

Kokoszyńska-Lutman, M.: 1948a, 'O pewnym warunku semantycznej koncepcji wiedzy' (On a Certain Condition of the Semantic Concept of Knowledge), *Przegląd Filozoficzny* **44**, pp. 372–381.

Kokoszyńska-Lutman, M.: 1951, 'A Refutation of the Relativism of Truth', *Studia Philosophica* **4**, pp. 93–149.

Kokoszyńska-Lutman, M.: 1955, 'W sprawie stosunku logiki do dialektyki' (Logic and Dialectics), *Studia Logica* **3**, pp. 184–207.

Kokoszyńska-Lutman, M.: 1957, 'O 'dobrej' i 'złej' indukcji' (On 'Good' and 'Bad' Induction), *Studia Logica* **5**, pp. 43–70.

Kokoszyńska-Lutman, M.: 1957a, 'Krytyka niektórych poglądów na stosunek logiki i dialektyki' (Critique of Some Ideas Concerning the Relationship Between Logic and Dialectics), *Studia Filozoficzne* **3**, pp. 118–151.

Kokoszyńska-Lutman, M.: 1962, 'O dwojakim rozumieniu uzasadniania dedukcyjnego' (Two Concepts of Deductive Justification), *Studia Logica* **13**, pp. 177–196.

Kokoszyńska-Lutman, M.: 1965, 'O niektórych osiągnięciach Kazimierza Ajdukiewicza w zakresie logiki, metodologii nauk i filozofii' (On Some of K. Ajdukiewicz's Achievements in Logic, Methodology of Science and Philosophy), *Studia Filozoficzne* **1(40)**, pp. 5–16.

Kokoszyńska-Lutman, M.: 1967, 'W sprawie różnicy pomiędzy naukami dedukcyjnymi i niededukcyjnymi' (On the Difference Between Deductive and Non-deductive Sciences), in *Fragmenty Filozoficzne*, 3rd series, pp. 43–74.

Kokoszyńska-Lutman, M.: 1969, 'Znaczenie syntaktyczne i semantyczne wyrażeń językowych' (Syntactic vs. Semantic Meaning of Linguistic Expressions), in *Rozprawy Filozoficzne* (Philosophical Dissertations), ed. by L. Gumański, Towarzystwo Naukowe w Toruniu, Toruń, pp. 143–175.

Kokoszyńska-Lutman, M.: 1970, 'Złudzenia aprioryzmu' (The Illusions of Apriorism), *Ruch Filozoficzny* **28**, pp. 72–78.

Kokoszyńska-Lutman, M.: 1973, 'Niektóre metodologiczne problemy integracji nauk', (On Some Methodological Problems of the Integration of Sciences), *Studia Filozoficzne* **4**, pp. 47–56.

Kokoszyńska-Lutman, M.: 1976, 'Bibliografia prac Marii Kokoszyńskiej-Lutman' (Bibliography of M. Kokoszyńska-Lutman's Works), *Acta Univ. Wratislaviensis* **290**, Prace Filozoficzne (Logika 5), Wroclaw, pp. 3–6.

Kotarbińska, J.: 1955, 'Definicja' (On Definition), *Studia Logica* **2**, pp. 301–327.

Kotarbińska, J.: 1956, 'Pojęcie znaku' (The Concept of Sign), *Studia Logica* **6**, pp. 57–134.

Kotarbińska, J.: 1959, 'Tak zwana definicja dejktyczna' (The So-called Deictic Definition), in *Fragmenty Filozoficzne*, 2nd series, pp. 44–74.

Kotarbińska, J.: 1960, 'On Ostensive Definitions', *Philosophy of Science* **27/1**, pp. 1–22.

Kotarbińska, J.: 1962, 'Kontrowersja: dedukcjonizm-indukcjonizm', *Studia Filozoficzne* **1**, pp. 25–42. Eng. trans. as 'The Controversy: Deductivism vs. Inductivism' in Nagel, E., Suppes, P., Tarski, A. (eds.), *Logic, Methodology and Philosophy of Science*, Stanford UP, Stanford, 1962, pp. 265–274.

Kotarbińska, J.: 1965, 'Controversy on the Applicability Limits of Logical Method', *Logique et Analyse* **29**, pp. 3–29.
Kotarbińska, J.: 1967, 'Kłopoty z istnieniem', in *Fragmenty Filozoficzne*, 3rd series, pp. 129–146. Eng. trans. as 'Puzzles of Existence' in Pelc (1979), pp. 208–226.
Kotarbińska, J.: 1990, *Z zagadnień teorii nauki i teorii języka* (Some Problems of the Theory of Science and the Theory of Language), PWN, Warsaw.
Kotarbiński, T.: 1929, *Elementy teorii poznania, logiki formalnej i metodologii nauk* (Elements of Epistemology, Formal Logic and Methodology of Science), Ossolineum, Lvov. Eng. trans. as Kotarbiński (1966).
Kotarbiński, T.: 1966, *Gnosiology. The Scientific Approach to the Theory of Knowledge*, Pergamon Press-Ossolineum, Oxford-Wroclaw. Eng. trans. of 2nd ed. of Kotarbiński (1929), with appendices.
Kotarbiński, T.: 1967, Introduction: *Polish Logic 1920–1939*, ed. by S. McCall, Oxford Clarendon Press, Oxford, pp. 1–14.
Krajewski, W.: 1982, Introduction: *Polish Essays in the Philosophy of Natural Sciences*, D. Reidel Publishing Company, Dordrecht, pp. xi–xxviii.
*Księga Pamiątkowa Trzeciego Polskiego Zjazdu Filozoficznego: 1936*, (A Memorial Book of the Third Polish Philosophical Assembly), Krakow, *Przegląd Filozoficzny* **39**.
Langer, S.K.: 1930, *The Practice of Philosophy* (with a preface by A.N. Whitehead), H. Holt & Comp., New York.
Langer, S.K.: 1937, *An Introduction to Symbolic Logic*, Doser Publ. Inc., New York.
Langer, S.K.: 1957, *Philosophy in a New Key, A study in the Symbolism of Reason, Rite and Art*, Harvard College. Polish trans. as *Nowy sens filozofii. Rozważania o symbolach myśli, obrzędu i sztuki*, PWN, Warsaw 1975.
Langer, S.K.: 1962, *Philosophical Sketches*, J. Hopkins Press, Baltimore.
Lazari-Pawłowska, I.: 1975, 'Maria Ossowska jako badacz moralności' (M. Ossowska as a Researcher of Morality), *Studia Filozoficzne* **12**, pp. 91–107.
Lloyd, G.: 1984, *The Man of Reason, 'Male' and 'Female' in Western Philosophy*, Univ. of Minnesota Press, Minneapolis.
Łuszczewska-Romahn, S.: 1948, 'Wieloznaczność a język nauki' , *Kwartalnik Filozoficzny* **17/1**, pp. 47–58. Eng. trans. as 'Ambiguity and the Language of Science' in Pelc (1979), pp. 148–159.
Łuszczewska-Romahn, S.: 1961, 'Classification as a Kind of Distance Function', *Studia Logica* **12**, pp. 41–81.
Łuszczewska-Romahn, S.: 1962, 'Czy filozofia obumiera?' (Is Philosophy Dying Away?) *Studia Filozoficzne* **1**, pp. 187–196.
Łuszczewska-Romahn, S.: 1977, 'Teoria wiedzy Kazimierza Twardowskiego' (Kazimierz Twardowski's Theory of Knowledge), in Skarga B. (ed.): *Polska myśl filozoficzna i społeczna* (Polish Philosophical and Social Thought), vol. 3, Książka i Wiedza, Warsaw, pp. 86–125.
Mahowald, M.B.: 1986/87, 'A Majority Perspective: Feminine and Feminist Elements in American Philosophy', *Cross Currents* winter, pp. 410–417.
Mazurczak, D.: 1995, 'Dążenie kobiet polskich do wyższego wykształcenia na przełomie XIX-XX wieku' (Polish Women's Endeavours to Higher Education in the Turn of 19th and 20th cent), in Miluska, J., Pakszys, E. (1995), pp. 179–192.
Miluska, J., Pakszys, E. (eds.): 1995, *Humanistyka i Płeć (I), Studia Kobiece z Psychologii, Filozofii i Historii* (The Humanities and Gender (I), Women's Studies in Psychology, Philosophy and History), Wyd. Nauk. UAM, Poznań.
Ossowska, M.: 1929, 'Stosunek logiki i gramatyki' (Relationship Between Logic and Grammar), *Kwartalnik Filozoficzny* **7/2**, pp. 231–261.

Ossowska, M.: 1931, 'Słowa i myśli' (Words and Thoughts), *Przegląd Filozoficzny* **34/4**, pp. 203–258.

Ossowska, M., Ossowski, S.: 1935, 'Nauka o nauce' (The Science of Science), *Nauka Polska* **20**, pp. 1–12.

Ossowska, M.: 1947, *Podstawy nauki o moralności* (The Foundations of the Science of Morality), Czytelnik, Warsaw.

Ossowska, M.: 1949, *Motywy postępowania — Z zagadnień psychologii moralności* (Human Motivation: An Inquiry into the Psychology of Moral Life), Książka i Wiedza, Warsaw.

Ossowska, M.: 1956, *Moralność mieszczańska* (Middle Class Morality), Ossolineum-Łódzkie Towarzystwo Naukowe, Łódź.

Ossowska, M.: 1966, *Myśl moralna oświecenia angielskiego* (Moral Thought of the English Enlightenment), PWN, Warsaw.

Ossowska, M.: 1983, *O człowieku, moralności i nauce. Miscellanea* (On Human Beings, Morality and Science, Miscellanea), PWN, Warsaw.

Pachucka, R.: 1958, *Pamiętniki z lat 1886–1914* (Memoirs from the Years 1886–1914), Ossolineum, Wroclaw.

Pakszys, E.: 1994, 'Epistemologia feministyczna albo kobieca kontestacja poznania i co z tego wynika dla nauczania filozofii dzisiaj' (Feminist Epistemology or Women's Contestation of Cognition and its Consequences for Teaching Philosophy Today), in Buksiński, T. (ed.), *Filozofia w dobie przemian* (Philosophy in a Time of Transformation), Wyd. Nauk. Instytutu Filozofii UAM, Poznań, pp. 103–111.

Pakszys, E.: 1995, 'Płeć a rozwój nauki: Problemy epistemologii feministycznej' (Gender and the Development of Science: The Problems of Feminist Epistemology), in Miluska, J., Pakszys, E. (1995), pp. 85–98.

Pakszys, E.: 1996, 'Kobiety w polskiej filozofii analitycznej. W stulecie szkoły lwowsko-warszawskiej (1895–1995)' (Women in Polish Analytical Philosophy. The Centenary of the Lvov-Warsaw School (1895–1995)), in Żarnowska, A., Szwarc, A., (eds.), *Kobieta i Kultura* (Women and Culture), DiG, Warsaw, pp. 89–98.

Pakszys, E.: 1997, 'Kobiety w filozofii polskiej, dwa pokolenia szkoły lwowsko-warszawskiej' (Women in Polish Philosophy, Two Generations of The Lvov-Warsaw School), in Pakszys, E., Sobczyńska, D., *Humanistyka i płeć (II). Kobiety w poznaniu naukowym, wczoraj i dziś* (The Humanities and Gender (II). Women in Scientific Cognition, Yesterday and Today), Wyd. Nauk. UAM, Poznań (in print).

Pelc, J.: 1979, *Semiotics in Poland 1894–1969*, D. Reidel Publishing Company, Dordrecht, PWN, Warsaw.

Petrażycka-Tomicka, J.: 1914, *Z dziejów kobiety polskiej, cztery odczyty* (From the History of Polish Woman, Four Lectures), A.W. Szyjkowski, Lvov.

Popper, K.R.: 1990, *A World of Propensities*, Thoemmes, Bristol.

Russell, B.: 1932, *O wychowaniu, ze specjalnym uwzględnieniem wczesnego dzieciństwa* (On Education, 1926), Polish trans. by J. Hosiasson, Nasza Księgarnia, Warsaw.

Russell, B.: 1933, *Wychowanie a ustrój społeczny* (Education and Social Order, 1932), Polish trans. by J. Hosiasson, Tow. Wyd. "Rój", Warsaw.

Russell, B.: 1939, *Zarys Filozofii* (An Outline of Philosophy, 1927), Polish trans. by J. Hosiasson, Wyd. Przeworski, Warsaw.

Schiebinger, L.: 1989, *The Mind Has No Sex? Women in the Origins of Modern Science*, Harvard Univ. Press, Cambridge-London.

Seigfried, C.H., 1991, 'Where Are All the Pragmatist Feminists?', *Hypatia* **6/2**, pp. 1–20.

Skolimowski, H.: 1967, *Polish Analytical Philosophy, A Survey and Comparison with British Analytical Philosophy*, The Humanities Press, New York.

Słoniewska, H.: 1934, 'W sprawie interpretacji objawów przeżyć psychicznych' (On the Interpretation of Symptoms of Psychic Phenomena), *Kwartalnik Psychologiczny* **5**, pp. 5–49.

Słoniewska, H.: 1959, 'Psychologiczna analiza zainteresowania' (Psychological Analysis of Interests), Poznańskie Towarzystwo Przyjaciół Nauk, vol. 10/1, Poznań, pp. 1–111.

Słoniewska, H.: 1968, 'Poglądy psychologiczne Kazimierza Twardowskiego' (Psychological Ideas of Kazimierz Twardowski), *Przegląd Psychologiczny* **15**, pp. 87–105.

Słoniewska, H.: 1978, 'Kazimierz Twardowski nauczyciel i wychowawca' (Kazimierz Twardowski as a Teacher and Pedagogue), in Stróżewski, W. (ed.), *Studia z teorii poznania i filozofii wartości* (dedykowane I. Dąmbskiej) (Studies in the Theory of Knowledge and the Philosophy of Values, dedicated to I. Dąmbska), Ossolineum, Wroclaw-Gdańsk, pp. 9–22.

*Słownik Psychologów Polskich*, 1992, (Dictionary of Polish Psychologists), Wyd. Instytut Psychologii UAM, Poznań.

Smith, B. (ed.): 1982, *Parts and Moments. Studies in Logic and Formal Ontology*, Philosophia Verlag, Munich.

Sztejnbarg, D.: 1931, 'O tzw. konieczności związków przyrodzonych', (On So-called Necessity of Elemental Bonds), *Sprawozdania z Posiedzeń Tow. Nauk. Warszawskiego Wydz. Nauk Społ.* **24**, pp. 67–108.

Sztejnbarg, D.: 1932, 'Zagadnienie indeterminizmu na terenie fizyki' (The Problem of Indeterminism in Physics), *Przegląd Filozoficzny* **35**, pp. 34–69.

Sztejnbarg, D.: 1932a, 'Zagadnienie indeterminizmu na terenie biologii', (The Problem of Indeterminism in Biology), *Przegląd Filozoficzny* **35**, pp. 245–272.

Sztejnbarg, D.: 1933, 'Zagadnienie indeterminizmu na terenie nauk humanistycznych' (The Problem of Indeterminism in the Humanities), *Przegląd Filozoficzny* **36**, pp. 77–106.

Sztejnbarg, D.: 1934, 'Analiza pojęcia przypadku' (An Analysis of the Concept of Chance), in *Fragmenty Filozoficzne*, 1st series, pp. 161–179.

Tennerówna, D.: 1913, 'W sprawie zgodności w filozofii' (On the Issue of Agreement in Philosophy), *Ruch Filozoficzny* **3**, pp. 49–50.

Tennerówna, D.: 1914, 'Istnienie jako 'treść' sądzenia i sądu' (Existence as a 'Content' of Judging and Judgment), *Przegląd Filozoficzny* **17/4**, pp. 465–483.

Walicki, A. et al (ed.): 1986, *Zarys dziejów filozofii polskiej 1815–1918*, (An Outline of the History of Polish Philosophy 1815–1918), PWN, Warsaw.

Woleński, J.: 1981, 'Janina Hosiasson-Lindenbaumowa o konfirmacji' (Janina Hosiasson-Lindenbaum on Confirmation), *Ruch Filozoficzny* **39**, pp. 32–33.

Woleński, J.: 1985, *Filozoficzna szkoła lwowsko-warszawska*, PWN, Warsaw. Eng. trans. as Woleński (1989).

Woleński, J.: 1987, 'Teorie i analizy w szkole lwowsko-warszawskiej', (Theories and Analyses in the Lvov-Warsaw School), in Hempoliński (1987), pp. 69–130.

Woleński J.: 1989, *Logic and Philosophy in the Lvov-Warsaw School*, Kluwer Academic Publishers, Dordrecht.

Woleński, J.: 1990, *Kotarbiński: Logic, Semantics and Ontology*, Kluwer Academic Publishers, Dordrecht.

*Wywiady z Prof. Janiną Kotarbińską 1993–94* (Interviews with Prof. J. Kotarbińska 1993–94), ed. by E. Pakszys (unpublished manuscript).

Żarnowska, A., Szwarc, A. (eds.): 1992, *Kobieta i edukacja na ziemiach polskich w XIX i XX w.* (Woman and Education in Polish Territories in 19th and 20th centuries), Instytut Historyczny Uniwersytetu Warszawskiego, Warsaw.

ARTUR ROJSZCZAK

# TRUTH-BEARERS FROM TWARDOWSKI TO TARSKI

## 1. THE INTRODUCTORY QUESTIONS

It seems to be clear what the truth-bearers are in Tarski's semantic definition of truth given in (1933).[1] They are sentences. The predicate 'is true' is affirmed in the metalanguage of the sentences in the object language, i.e. the language in which we talk about things. Then the first question is: do the metalinguistic expressions which tell us about sentences univocally define the notion of a sentence? The second question is: why should sentences and not, for example, judgements, play the role of truth-bearers?

The answer to the first question is a negative one. Tarski uses the notion of the sentence in at least four senses. In the first 'sentence' means: pure syntactically defined expression of a language.[2] The second meaning of 'sentence' is: a psycho-physical product of a psychological event.[3] In the third, the sentence is a physical object (something like an inscription).[4] And in the fourth, sentences are defined as functions without free variables.[5] Furthermore, if we assume that the sentence as psycho-physical or as physical object can be understood as an utterance or an inscription and if we add that each of them can be understood either as tokens or as types or as general names then we have about eleven more possible understandings of the term 'sentence' in Tarski's work.[6] To understand the second question about why we should choose sentences as truth-bearers needs some more historical and systematical remarks.

## 2. THE HISTORICAL CONTEXT FOR TARSKI'S CHOICE OF TRUTH-BEARERS: THE BRENTANIAN TRADITION

Alfred Tarski was a pupil of Jan Łukasiewicz, Tadeusz Kotarbiński and Stanisław Leśniewski. And Tarski was deeply influenced philosophically by his teachers. For example: Tarski wrote his dissertation under Leśniewski, who, as well as Łukasiewicz, was Tarski's teacher in logic. As Tarski points out in (1933), the first paragraph of his work about truth was influenced very strongly by Leśniewski. It is unclear which parts of it are from Tarski and which come from Leśniewski, but the topic of the first paragraph — the notion of truth in natural language — contains a lot of information about Tarski's view of language, about his understanding of a sentence and especially of his views on the function of quotation marks. Kotarbiński was Tarski's main teacher in philosophy. Tarski's famous collection of papers was dedicated to Kotarbiński.[7] The



*K. Kijania-Placek and J. Woleński (eds.),*
*The Lvov–Warsaw School and Contemporary Philosophy,* 73–84.
© 1998 *Kluwer Academic Publishers.*

so-called classical formulation of the truth-definition comes from Kotarbiński. And the original version of Tarski (1933) refers to Kotarbiński's *Elements*.[8] Kotarbiński and Leśniewski discussed together a problem of the objectivity of truth.[9] But Kazimierz Ajdukiewicz should also be mentioned in the context of Tarski's (1933) truth definition. Ajdukiewicz's papers are included in the short bibliography at the beginning of Tarski's book. At this time Ajdukiewicz had prepared his two very important papers about language and meaning.[10] I think that these papers were influenced by the same philosophical problems as Tarski's work on the problem of truth-bearers. This can be shown if we look at Ajdukiewicz's lectures on epistemology in the academic year 1930–31.[11]

Łukasiewicz, Leśniewski and Kotarbiński belong to the first generation of Twardowski's students. Considering that Twardowski was one of the six principal pupils of Brentano one can inquire as to the influence of Brentano and his students on the philosophical discussions about truth and truth-bearers in the Lvov-Warsaw School. The Brentanian heritage in the philosophy of Twardowski is well known. But the influences of the Brentanian tradition are also quite clear in the philosophy of Twardowski's students. Jan Łukasiewicz spent some time in Graz working with Meinong.[12] Łukasiewicz's appendix to the book about Aristotle,[13] one of the first manuals of mathematical logic in Poland, refers to the 'Algebre de la logique' of Couturat, who was seen as a logician connected with the phenomenology of Husserl.[14] As Kotarbiński points out, before Leśniewski came to the circle of Twardowski, he was under the influence of Cornelius and J. S. Mill on the one hand, and of A. Marty, one of Brentano's students, on the other.[15] Kazimierz Ajdukiewicz spent some time in Göttingen, where he heard the lectures of Hilbert and of Husserl. One of his first papers was written and constructed in the mode of Husserl's first Logical Investigation.[16] Several times the mentioned authors refer to Husserl's 'Logical Investigations' and to Marty's 'Untersuchungen zur Grundlegung der allgemeinen Grammatik und Sprachphilosophie'.[17] Summing up, there are enough grounds to assume that the theory of meaning and language and the theory of truth in the Brentanian tradition was influential for the philosophical background of Tarski's thought.[18] Tarski himself was conscious of the Twardowskian heritage in Polish philosophy:

> Almost all researchers, who pursue the philosophy of exact sciences in Poland, are indirectly or directly the disciples of Twardowski, although his own works could hardly be counted within this domain.[19]

## 3. THE PHILOSOPHICAL CONTEXT OF TARSKI'S THEORY OF THE NATURE OF TRUTH-BEARERS

### 3.1. *Some candidates for truth-bearers*

A truth-bearer is that sort of entity about which one can say that it is true; or in other words, an entity of which the predicate 'is true' can be affirmed. This would be a linguistic definition of truth-bearer. Another formulation of the definition of the bearer for truth is: a truth-bearer is an entity which can have the attribute of being true or false. We can call this an ontological definition of truth-bearer. The two definitions are not equivalent. One entity can be a truth-bearer in the sense of the first definition, and not fulfil the second one. Brentano's act of judging (or a judging person) for instance fulfils the linguistic definition but a true judging act (or true judging person) is not true in the attributive sense — 'to be true' is, according to Brentano, not a feature of the judging act (judging person) like 'to be evident' or 'to be apodictic'.

In the Brentanian School and in the Lvov-Warsaw School we have several candidates for truth-bearers. The following — most probably incomplete — list is quite long and comes from the members of these two schools:

concrete psychological entities:

1. an act of judging (Brentano, Twardowski)
2. a belief (Brentano)
3. a supposition or assumption (Meinong) or a presentation of the act of judging (Twardowski)
4. a judging person (Brentano)

abstract psychological entities:

5. an act of judging as a psychical product (Twardowski, Czeżowski)
6. a content of an act of judging (Marty)
7. an act-species (Husserl)

psycho-physical entities:

8. an utterance as a psycho-physical product (Twardowski, Kotarbiński, Tarski)
9. an inscription as a psycho-physical product (Twardowski, Kotarbiński, Tarski)
10. an inscription as a general name (Kotarbiński, Tarski)
11. an inscription as an individual name (Tarski)

abstract psycho-physical entities:

12. an utterance-type as a psycho-physical product (Tarski)
13. an inscription-type as a psycho-physical product (Tarski)

physical entities:

14. an utterance as a physical object (Tarski)
15. an inscription as a physical object (Tarski)
16. a judging person (Kotarbiński)

abstract entities:

17. an utterance as a judgement in the logical sense (Łukasiewicz)
18. an inscription as a judgement in the logical sense (Łukasiewicz)

ideal entities:

19. a sentence in itself (*Satz an sich* — Bolzano)
20. an ideal meaning (Husserl)
21. an *Objectiv* (as an ideal meaning or as a state of affairs — Meinong)

So, following the Brentanian tradition and the Lvov-Warsaw School, we can choose the bearer of the truth from at least twenty one alternatives. In what follows I omit most of the arguments for the choice of the entities for truth-bearers given by the above-mentioned authors. Some of the arguments will be mentioned in the next paragraph.

According to early Brentano and the so-called gnosiological conception of truth there is only one entity that can serve as a truth-bearer: an act of judging.[20] If we are interested not in the act of judging itself but, for example, in one of the results of the judging act (like beliefs or statements) we can choose, for example, 2 or 6 from the list above. If we wish to avoid the charge of psychologism but still wish to orient our theory towards mental acts we should take the matter or content of the judgement or a psychological product of the judgement as bearer of truth (alternatives 5–7). If we want to have something physical, which has a good structure to describe it and which is able to be interpreted as the expression of a psychological event, then we can adopt for such entities an utterance or sentence-inscription as psycho-physical product of the judging act (alternatives 8–13). If we are not interested at all in the expressions of the psyche then we can take into account such entities as an utterance as a physical object or as a sentence-inscription (alternatives 14–15). One of the ways to avoid the problem of multiplication of truth-bearers is to choose as truth-bearer the type-utterance or type-sentence or sentence as name (alternatives 10–13). Meinong's *Objectiv* or *Sachverhalt* or a proposition, or a sense of an expression or judgement in the logical sense or *Sätze an sich* of Bolzano will serve for Platonists who want to find something unchangeable and objective to serve as truth-bearers. For reists there is a judging person who can serve as truth-bearer. Note that the list given above includes also information about the ontological status of the given truth-bearer: the entities are events (like an act of judging), processes (like an utterance), bodies (like an inscription) or ideal objects.

### 3.2. *How the choice was made in the Lvov-Warsaw School*

The first group of alternative bearers of truth (numbers 1–4) was rejected by Polish philosophers because of their psychological connections. All of the mentioned philosophers of the Lvov-Warsaw School adopted the position of antipsychologism. Jan Łukasiewicz had done so clearly in his (1907) paper 'Logic and Psychology'[21] and Twardowski in 1912 in his paper about acts and their products.[22] However, Twardowski never rejected the view according to which the judgement is the truth-bearer. Moreover, he thought that the theory of truth must include a good theory of judgement.[23] After his antipsychologistic turn Twardowski saw the correct bearer of truth in the product of the act of judging. Also, Tadeusz Czeżowski — the most Brentanian figure in Polish philosophy — held the view that the products of judging acts should be seen as the truth-bearers. He tried for example to interpret propositional logic in terms of judgement-theory: He interpreted, for instance, the value of arguments of truth-functions as elementary judgement products and the domain of the propositional variables as a set of judgement products.[24]

But even treating the psychic products of judging acts does not guarantee an antipsychologistic view. Only if they become consolidated in permanent psycho-physical products can we go out of the psyche. A product of a judging act can be fixed in an utterance. A short-lived non-permanent psycho-physical product like an utterance can be then fixed in an inscription which is a permanent psycho-physical product in the primary sense. So the products of acts of judgement can be expressed in inscriptions. In the primary sense it is the judgement (as an outcome of thinking) which is the truth-bearer, but the mental entity can be intentionally expressed in written language.[25]

One of the problems with this view is this: how are we to know what an inscription expresses? Stanisław Ossowski suggests[26] that instead of seeing inscriptions as a product of thought in which a given thought of Mister X becomes fixed intentionally, we should read an inscription with the semantical attitude. The inscription has meaning if we see it intentionally as the expression of someone's act of judgement. It does not matter whether or not the sign is intentionally produced as a meaningful sign which expresses a thought. Maria Ossowska goes one step further and suggests that we can take an inscriptional sign as a sign with the function of presentation only. The function of an expression of a language should be preserved for non-intended functions of signification of thought (Edmund Husserl's *Anzeichen*).[27] Thus for semantics the function of expressing is replaced by the presentational function of the written or uttered sign.

Jan Łukasiewicz saw a judgement directly as a sequence of words or other signs which says that an object has or does not have a given feature.[28] However, judgement is here understood as judgement in the logical sense; a judgement is not a psychical belief, nor is it an ideal meaning of a sequence of words. Each judgement refers to the objects of the world or to a state of affairs. This relation is the relation of the concordance or non-concordance. The relation between belief (and for Łukasiewicz this does also mean judging act) and state of affairs is intentional in Brentano's sense.[29]

The last group of truth-bearers (numbers 17–21) were rejected for ontological or metaphysical reasons. As Peter Simons points out[30] Kotarbiński, Leśniewski and Tarski were all nominalists in one of the senses of the word. And because of the nominalistic preferences of these philosophers of the Lvov-Warsaw School all of the ideal entities were rejected as bearers of truth. Nominalism was a result of the discussion about general objects which began in 1913 with Leśniewski's paper about the logical law of excluded middle.[31] In 1927 Leśniewski wrote:

> At the time I wrote that passage I believed that there are in existence in this world so called features and so called relations, as two special kinds of objects, and I felt no scruples about using the expressions 'feature' and 'relations'. It is a long time since I believed in the existence of objects which are features, or in the existence of objects which are relations and now nothing induces me to believe in the existence of such objects ... [32]

Some years later Kotarbiński added some modifications to Leśniewski's 1913 arguments and also presented new arguments for rejecting ideal objects. Both Leśniewski and Kotarbiński discussed this problem in the terms of the views which came from the Brentano School e.g. in relation to the idealisation of mathematical and logical objects, to the existence of relations and to intentional objects.[33]

One result of such views about general objects is Leśniewski's formulation of the concept of a language. An expression of a language is, he holds, composed of only a finite number of concrete parts such as inscriptions or utterances. Two sequences of inscriptions (utterances) are two different expressions. We can call this interpretation of language 'inscriptional syntax' or 'constructional nominalism'.[34] The bearers of truth were for him the finitely many existing inscriptions.

Another strong position was taken by Kotarbiński. Kotarbiński's reism was a result of Leśniewski's ontology.[35] For his reistic view a person can be a truly or falsely judging person[36] or an inscription (sentence) can be seen as true or false. The choice of inscriptions (sentences) for the bearers of truth on the part of Kotarbiński is made because of the definition of truth-bearers which was implicitly assumed in the Lvov-Warsaw School: The truth-bearer is an entity of which the predicate 'is true' can be attributively affirmed. To omit equivocation of the predicative use of the word 'true' we should say about the truly judging person that he is a right judging person.[37] The difference mentioned above between the modificational and attributive use of adjectives comes from Twardowski and belongs to his more interesting results.[38]

Another topic of importance for the choice of the truth-bearer is the problem of the objectivisation of truth. The first remarks referring to this problem in the Lvov-Warsaw School were made by Twardowski in his article 'On So-called Relative Truths' in 1900.[39] Twardowski here shows how to preserve the objectivity of truth without an assumption of ideal objects like ideal meanings. The problem of the objectivity of truth expressed in language was seen by Twardowski as the problem of decontextualisation or de-indexicalisation of sentences. The objectivity of truth was taken in the Lvov-Warsaw School as an axiom. The outcome of the discussion between Leśniewski and Kotarbiński about the eternity of truth[40] and of Maria Kokoszyńska's investigation referring to the objectivity of truth[41] was a relativisation of the truth-definition to a given language within which all expressions are seen as being univocally meaningful.[42]

The last topic which should be mentioned with reference to the problem of choice of truth-bearers is the issue of the theory of truth itself. The classical formulation of the truth-definition, the differentiation of language and metalanguage belong to the well-known elements of the truth-theory of the Lvov-Warsaw School. There is no space here to repeat these very well-known views. I would mention only that the conception of truth as a predicate in the metalanguage has its roots in Leśniewski's discussion of the definition of logic in his article (1913a) about the principle of the excluded middle.[43]

So, the middle group of candidates for truth-bearer (numbers 5–16) was taken seriously by Polish philosophers. The strongest position, mentioned above, was taken by Leśniewski. The most psychological view was taken by Twardowski and Czeżowski. For Tarski, given his philosophical background knowledge, a good candidate for the truth-bearer would be an inscription or an utterance in the sense of a psycho-physical product. The structure of such bearers is given by the syntactical definition of the sentence as a sentence-function without variables.[44] But for metalogical reasons truth-bearers should be types or names.[45] And because we need infinitely many types of inscriptions the latter should be taken as physical bodies: it is not possible for people to produce infinitely many psycho-physical products. But it is reasonable to say there are infinitely many physical bodies.[46] Thus the truth-bearer is a sentence of a given meaningful language. The sentence can be defined in the following way: it is a sentence which is determined by the syntax of the language, which is a finite inscription and which is a physical body, and which can be taken with the semantical attitude.

### 3.3. *What is Brentanian in these alternatives?*

The Brentanian influence in determining the choice of bearers of truth in the Lvov-Warsaw School can be see in following contexts:

a) There is a turn from the gnosiological conception of truth in the Brentano School to the classical formulation of the concept of truth in the Lvov-Warsaw School. The background of this turn belongs to the descriptive-psychological and logical investigation referring to judgement (Brentano, Husserl, Marty, Twardowski, Czeżowski, Leśniewski).

b) The analysis of the use of the adjectives 'true' and 'false' in ordinary language. This analysis is the starting point for both the epistemological and the classical conceptions of truth (Brentano, Meinong, Marty, Twardowski, Kotarbiński).

c) The turn from descriptive psychology to the antipsychological attitude in epistemology. The antipsychologistic position comes from Bolzano's *Wissenschaftslehre* and from Husserl's *Logical Investigations* and was affirmed by all Polish philosophers in Twardowski's school.

d) The philosophy of language. On the one hand the bearers of truth could be expressions for psychological events. This kind of understanding of language came down from Brentano and Marty to Twardowski and so to his students. On the other hand not the intentionality of language but rather its presentational function can be seen as primary (Kotarbiński, Leśniewski, Ossowska, Ajdukiewicz, Tarski).

e) Theory of meaning. For example: the turn from the interpretation of the sign as an outcome or product of thinking to the semantical attitude which relates to the receiver of the sign (Brentano and his pupils, Czeżowski, Ossowski, Ajdukiewicz). The interests in semantics in the Lvov-Warsaw School can be seen in context of their discussions of the problem of intentionality.

f) The nominalistic position can be seen as an outcome of the discussion of ideal objects and states of affairs (Bolzano, Twardowski, Husserl, Meinong, Leśniewski, Kotarbiński).

g) The objectivisation of truth. The question of whether the chosen truth-bearers serve the process of the objectivisation of truth (at least in the sense of making truth independent of person, time and circumstances) belongs here, for instance.

One can ask whether the philosophical views of Tarski are at all important to his logical and metalogical investigations. For methodological reasons, Tarski was not asking for an ontology of language or for a theory of language. For example: if the methods of logic were useful, the questions about the existence of universals were not mentioned — for Tarski the use of set-theoretical methods does not require that we accept such entities as types. But on the other hand the philosophical questions were very important in the first period of his work, even if in the late period of his activity he saw the philosophical comments only as private ones.

*Institute of Philosophy, Jagiellonian University*

## NOTES

* During the preparation of this paper the author was a Fellow of the Foundation for Polish Science.

[1] *Pojęcie prawdy w językach nauk dedukcyjnych*, Tarski (1933).

[2] Tarski (1933), p. 16; Tarski (1956) pp. 165–166.

[3] Tarski (1933), p. 25; Tarski (1956), p. 174.

[4] Tarski (1933), p. 25; Tarski (1956), p. 174.

[5] Tarski (1933), p. 29; Tarski (1956), p. 178.

[6] Tarski (1933), pp. 5, 16; Tarski (1956), pp. 156, 165–166.

[7] Tarski (1956).

[8] Kotarbiński (1929).

[9] See Kotarbiński (1913), Leśniewski (1913).

[10] Ajdukiewicz (1931), (1934).

[11] Ajdukiewicz (1988), (1993).

[12] More information about the relation between Meinong and Łukasiewicz is included in Simons (1992). Leśniewski's and Kotarbiński's discussion of the existence of general objects also has its roots in the theory of objects in the Meinongian sense. See below.

[13] Łukasiewicz (1910).

[14] "... franz. Philosoph (Toulouse), der sich auf erkenntnistheoretisch-logischem Gebiet in einem der Logik Husserls verwandten Sinn betätigt" Schmidt (1934).

[15] Kotarbiński (1967), pp. 4–5; also Kotarbiński (1933).

[16] Ajdukiewicz (1931).

[17] Husserl (1900/01), Marty (1908).

[18] Of course on the other hand one should take into account that there was a second very influential current of ideas which came from Frege and Russell, and later from the Vienna Circle. But this is not the subject of this paper. Also the mathematicians Wacław Sierpiński, Kazimierz Kuratowski, Zygmunt Janiszewski and Stefan Mazurkiewicz should be mentioned in the context of Tarski's scholarly background.

[19] Tarski (1992), p. 20 (letter to O. Neurath dated 25.04.1930).

[20] See e.g.: Brentano (1930), Kastil (1934).

[21] Łukasiewicz (1907). It needs to be mentioned that this paper was influenced by Husserl.

[22] Twardowski (1912).

[23] Twardowski (1975), pp. 246–250.

[24] See Czeżowski (1927).

[25] Twardowski (1912).

[26] Ossowski (1926).

[27] Ossowska (1928), (1931).

[28] Łukasiewicz (1910), p. 12.

[29] Łukasiewicz (1910), pp. 28–30.

[30] Simons (1993).

[31] Leśniewski (1913a). The discussion about the theory of objects began in the Brentanian tradition with Twardowski's Habilitationsschrift (Twardowski 1894).

[32] Leśniewski (1992), p. 198.

[33] See Kotarbiński (1921).

[34] See Woleński (1985), (1987), (1988), Simons (1993).

[35] See e.g.: Woleński (1990), Woleński (ed.) (1990).

[36] Here we can distinguish two kinds of judging person: the soul (Brentano), and the body (Kotarbiński). See e.g.: Smith (1994).

[37] Kotarbiński (1926), (1929), (1930).
[38] Twardowski (1894), (1927), Smith (1994).
[39] Twardowski (1900).
[40] Kotarbiński (1913) and Leśniewski (1913).
[41] Kokoszyńska (1936), (1936a).
[42] On the problem of objectivisation of truth see Rojszczak (1994), (199?).
[43] Leśniewski (1913a).
[44] Tarski (1933), pp. 12–13, 27–29, 63–65; Tarski (1956), pp. 162–164, 176–178, 212–214.
[45] Tarski (1933), p. 5; Tarski (1956), p. 156.
[46] Tarski (1933), p. 25; Tarski (1956), p. 174.

## REFERENCES


Ajdukiewicz, K.: 1931, 'O znaczeniu wyrażeń', in *Księga Pamiątkowa Polskiego Towarzystwa Filozoficznego we Lwowie* (A Memorial Book of the Polish Philosophical Society in Lvov), PTF, Lvov, pp. 31–77. Eng. trans. as 'On the Meaning of Expressions' in Ajdukiewicz (1978), pp. 1–34.

Ajdukiewicz, K.: 1934, 'Sprache und Sinn', *Erkenntnis* **4**, pp. 100–138.

Ajdukiewicz, K.: 1978, *The Scientific World-Perspective and other Essays*, Kluwer Academic Publishers, Dordrecht.

Ajdukiewicz, K.: 1988, 'Wykłady Kazimierza Ajdukiewicza z teorii poznania w roku akademickim 1930-31' (Kazimierz Ajdukiewicz's Lectures on Epistemology in the 1930-31 Academic Year), in *Edukacja Filozoficzna* **8**, pp. 447–474.

Ajdukiewicz, K.: 1993, 'Kategorie syntaktyczne i antynomie logiczne' (Syntactical categories and logical antinomies), in *Filozofia Nauki* **1**, pp. 163–182.

Brentano, F.: 1930, *Wahrheit und Evidenz*, Meiner, Hamburg.

Czeżowski, T.: 1927, *Klasyczna nauka o sądzie i wniosku w świetle logiki współczesnej* (The Classical Theory of Judgement and Conclusion from the Standpoint of Contemporary Logic), Księgarnia Zawadzkiego, Wilno.

Husserl E.: 1900/01, *Logische Untersuchungen*, Niemeyer, Halle.

Kastil, A.: 1934, 'Ontologischer und gnosiologischer Wahrheitsbegriff', in *Zur Philosophie der Gegenwart*, Brentano- Gesellschaft, Prag, pp. 23–34.

Kokoszyńska, M.: 1936, 'Über den absoluten Wahrheitsbegriff und einige andere semantische Begriffe', in *Erkenntnis* **6**, pp. 143–165.

Kokoszyńska, M.: 1936a, 'W sprawie względności i bezwzględności prawdy' (On Relativity and Absoluteness of Truth), in *Przegląd Filozoficzny* **39**, pp. 424–425.

Kotarbiński, T.: 1913, 'Zagadnienie istnienia przyszłości' (The Problem of the Existence of the Future), *Przegląd Filozoficzny* **16**, pp. 74–92.

Kotarbiński, T.: 1921, 'Sprawa istnienia przedmiotów idealnych' (The Problem of the Existence of Ideal Objects), in *Księga Pamiątkowa ku uczczeniu 25-letniej działalności nauczycielskiej na katedrze filozofii w Uniwersytecie Lwowskim Kazimierza Twardowskiego* (Essays in Honour of Prof. K. Twardowski on the Occasion of His 25th Year of Teaching at the Lvov University), Uniwersytet Lwowski, Lvov, pp. 149–170.

Kotarbiński, T.: 1926, *Elementy logiki formalnej, teorii poznania i metodologii — skrypt autoryzowany* (The Elements of Formal Logic, Epistemology and Methodology — Authorised Script), Wyd. Koła Filozoficznego i Koła Przyrodników, Warsaw.

Kotarbiński, T.: 1929, *Elementy teorii poznania, logiki formalnej i metodologii nauk* (Elements of Epistemology, Formal Logic and Methodology of Science), Ossolineum, Lvov. Eng. trans. as Kotarbiński (1966).

Kotarbiński, T.: 1930, 'Uwagi na temat reizmu' (Remarks on Reism), in *Ruch Filozoficzny* **1–10**, pp. 1–7.

Kotarbiński, T.: 1933, 'Grundlinien und Tendenzen der Philosophie in Polen', in *Slavische Rundschau* **4**, pp. 218–229.

Kotarbiński, T.: 1966, *Gnosiology: The Scientific Approach to the Theory of Knowledge*, Pergamon, Oxford.

Kotarbiński, T.: 1967, 'Introduction. Notes on the Development of Formal Logic in Poland in the Years 1900–39', in McCall S. (ed.), *Polish Logic 1920–1939*, Clarendon Press, Oxford, pp. 1–14.

Leśniewski, S.: 1913, 'Czy prawda jest tylko wieczna, czy też i wieczna i odwieczna?' (Is All Truth Only True Eternally or Is It Also True Without a Beginning), in *Nowe Tory* **18**, pp. 493–528. Eng. trans. in Leśniewski 1992, pp. 86–114.

Leśniewski, S.: 1913a, 'Krytyka logicznej zasady wyłączonego środka' (The Critique of the Logical Principle of the Excluded Middle), in *Przegląd Filozoficzny* **16**, pp. 315–352. Eng. trans. in Leśniewski 1992, pp. 47–85.

Leśniewski, S.: 1992, *Collected Works* edited by S.J. Surma, J.T. Srzednicki, D.I. Barnett, V.F. Rickey, Kluwer Academic Publishers, Dordrecht.

Łukasiewicz J.: 1907, 'Logika a psychologia' (Logic and Psychology), *Przegląd Filozoficzny* **10**, pp. 489–492.

Łukasiewicz J.: 1910, *O zasadzie sprzeczności u Arystotelesa*, PAU, Krakow. German version: 'Über den Satz von Widerspruch bei Aristoteles', in *Bulletin international de l'Academie des Sciences de Cracovie, Classe de philosophie.*

Łukasiewicz J.: 1970, *Selected Works*, North-Holland Publishing Company, Amsterdam, PWN, Warsaw.

Marty, A.: 1908, *Untersuchungen zur Grundlegung der allgemeinen Grammatik und Sprachphilosophie*, Niemeyer, Halle.

Ossowska, M.: 1928, 'O pojęciu wyrażania' (On the Notion of Expression), in *Przegląd Filozoficzny* **31**, pp. 145–147.

Ossowska, M.: 1931, 'Słowa i myśli' (Words and thoughts), *Przegląd Filozoficzny* **34**, pp. 203-258.

Ossowski, S.: 1926, 'Analiza pojęcia znaku', (Analysis of the Concept of Sign), *Przegląd Filozoficzny* **29**, pp. 29–56.

Pelc J. (ed.): 1979, *Semiotics in Poland 1894–1969*, Reidel, Dordrecht.

Rojszczak, A.: 1994, 'Über die Korrespondenz von Tarskis Definition der Wahrheit', in Meggle G., Wessels U. (ed.), *Analyomen 1. Proceedings of the 1st Conference 'Perspectives in analytical Philosophy*, de Gruyter, Berlin/New York, pp. 539–543.

Rojszczak A.: 199? (forthcoming), 'Od sądów do zdań. Nośnik prawdy a obiektywizacja wiedzy' (From the acts of judging to the sentences. The bearer of the truth and the objectivisation of knowledge), in *Filozofia Nauki*. An English version is available.

Schmidt, H.: 1934, *Philosophisches Wörterbuch*, Kroener, Leipzig.

Simons, P.: 1992, *Philosophy and Logic in Central Europe from Bolzano to Tarski*, Kluwer Academic Publishers, Dordrecht.

Simons, P.: 1993, 'Nominalism in Poland', in Cognilione F., Poli R., Woleński J. (eds.), *Polish Scientific Philosophy: The Lvov- Warsaw School*, Rodopi (Poznań Studies vol. 28), Boston, pp. 207–231.

Smith, B.: 1994, *Austrian Philosophy*, Open Court, La Salle.

Tarski, A.: 1933, *Pojęcie prawdy w językach nauk dedukcyjnych*, Tow. Naukowe Warszawskie, Warsaw. Eng. trans. as 'The Concept of Truth in Formalized Languages' in Tarski 1956, pp. 152–278.

Tarski, A.: 1956, *Logic, Semantics, Metamathematics*, Oxford University Press, Oxford.

Tarski, A.: 1992, 'Alfred Tarski: Drei Briefe an Otto Neurath', translated into English by Jan Tarski, in *Grazer Philosophischen Studien* **42**, pp. 1–32.

Twardowski, K.: 1894, *Zur Lehre vom Inhalt und Gegenstand der Vorstellungen. Eine psychologische Untersuchung*, Hölder, Vienna.

Twardowski, K.: 1900, 'O tak zwanych prawdach względnych', in *Księga Pamiątkowa Uniwersytetu Lwowskiego ku uczczeniu pięćsetnej rocznicy fundacji Jagiellońskiej Uniwersytetu Krakowskiego* (Commemoration Book of the Lvov University Celebrating the 500–th Anniversary of the Foundation of the Jagiellonian University in Krakow), Senat Akademicki Uniwersytetu lwowskiego, Lvov, pp. 25. Eng. trans. as 'On So-called Relative Truths' in Brandl, J., Woleński, J., *Kasimir Twardowski. Selected Writings*, Rodopi, Amsterdam (forthcoming).

Twardowski, K.: 1912, 'O czynnościach i wytworach' , in *Księga Pamiątkowa ku uczczeniu 250-tej rocznicy założenia Uniwersytetu Lwowskiego przez króla Jana Kazimierza* (Commemoration Book Celebrating the 250–th Anniversary of the Foundation of the Lvov University by the King Jan Kazimierz), Uniwersytet lwowski, Lvov, pp. 1–33. Eng. trans. as 'Actions and Products' in Pelc (ed.) (1979), pp. 13–27.

Twardowski, K.: 1927, 'Z logiki przymiotników', *Przegląd Filozoficzny* **30**, pp. 292–294. Eng. trans. as 'Issues in the Logic of Adjectives' in in Pelc (ed.) (1979), pp. 28–31.

Twardowski, K.: 1975, 'Teoria poznania. Wykład czterogodzinny lato 1924–1925' (Epistemology. Four-hours Lectures Summer 1934–1925), *Archiwum Historii Filozofii i Myśli Społecznej*, vol. 21, pp. 241–299.

Woleński, J.: 1985, *Filozoficzna szkoła lwowsko-warszawska*, PWN, Warsaw. Eng. trans. as Woleński (1989).

Woleński, J.: 1987, 'Tarski jako filozof' (Tarski as a philosopher), *Wiadomości Matematyczne* **27**, pp. 249–259. Eng. trans. as Woleński (1993).

Woleński J.: 1989, *Logic and Philosophy in the Lvov-Warsaw School*, Kluwer Academic Publishers, Dordrecht.

Woleński, J.: 1990a, *Kotarbiński*, Wiedza Powszechna, Warsaw.

Woleński J. (ed.): 1990b, *Kotarbiński: Logic, Semantics, and Ontology*, Kluwer Academic Publishers, Dordrecht.

Woleński, J.: 1993, 'Tarski as a Philosopher', in Cognilione F., Poli R., Woleński J. (eds.), Polish Scientific Philosophy: The Lvov- Warsaw School, Rodopi (Poznań Studies vol. 28), Amsterdam, pp. 319–338.

MAREK ROSIAK

# TWARDOWSKI AND HUSSERL ON WHOLES AND PARTS

It is well known what use was made of the part-whole notions in the metaphysics of Franz Brentano and his followers. They were preferred to the traditional metaphysical notions of genus-species, matter-form, substance-accident, etc. as intuitive, universal and, it seemed likely, clearer than the former ones.[1] The notions of part-whole, performing the main role in metaphysical analyses, subsequently were brought into focus themselves. The importance attached to them, common in Brentano's School, is evident in Twardowski's influential Habilitationsschrift *Zur Lehre vom Inhalt und Gegenstand der Vorstellungen.* The theoretical interest of these notions culminates in the third part of Edmund Husserl's *Logical Investigations, Zur Lehre von den Ganzen und Teilen.* Husserl, in his outline of a strict theory of parts and wholes, on the one hand proposes something better than the tentative statements of Twardowski, but on the other – very likely unconsciously — uncovers a certain essential difficulty contained in the whole attempt at 'mereologising' metaphysics.

This paper starts with the presentation of both the use made of part-whole notions in Twardowski's dissertation and the Husserlian critique of the very interpretation of Twardowski's notions. It then comes to certain problems contained in Husserl's own interpretation and via this in any 'combinatorial' approach to ontology. Finally it tries to strictly interpret Husserlian Part-Whole Theory and to set out towards its possible reconstruction.

## 1.

In his dissertation Twardowski applies the notion of part to analyse and compare both the content and the object of a presentation. Let us pay attention to the initial description of the notion of part, comparing it with the corresponding description of Husserl:

> The word 'part', 'constituent', has to be taken in its widest sense. Not only in what ordinary language or the mathematician calls a part is meant by the word, but generally everything that can be distinguished in or about the object of a presentation, irrespective of whether one can speak of a real analysis into the distinguishable parts or merely of an analysis in thought.[2]

> We interpret the word 'part' in the widest sense: we may call anything a 'part' that can be distinguished 'in' an object, or, objectively phrased, that is 'present' in it. Everything is a part that is an object's real possession, not only in the sense of being a real thing, but also in the sense of being something really in something, that truly helps to make it up: an object in itself, considered in abstraction from all contexts to which it is tied, is likewise a part.[3]

The 'derelativising' addition at the end of the Husserlian definition should be read as a departure from Twardowski's following explanation:



*K. Kijania-Placek and J. Woleński (eds.),*
*The Lvov–Warsaw School and Contemporary Philosophy,* 85–100.
© 1998 *Kluwer Academic Publishers.*

[The totality] of what one calls a part in the usual sense [ . . . ] is customarily called the material [Stoff] of which the whole consists[4] [ . . . ] These are in part [ . . . ] relations between the object and other objects.[5]

So, as we see, Twardowski counts not only the cover and pages but also lying on the table as parts of the book. Husserl does not go that far — for him, as for most of us, the relations of the book to its surroundings are not parts of the book itself. Why do we hesitate to acknowledge these relations as parts of the book? For one thing, should we have acknowledged them as parts of the table as well, we would in consequence find no boundary between the book and the table. This consequence seems to go too far beyond our common intuitions to convince us of such a definition of a whole. In Husserl, however, the relations between the book and the table are not homeless. They are regarded as parts of a whole containing in addition both the book and the table as separate parts.

The aforementioned general definition of a part is in Twardowski preceded by an essential remark; as will be shown, Husserl's criticism is directly aimed at certain of its consequences.

[ . . . ] whenever one speaks of the parts of a complex object, one has to consider, in addition to what one calls a part in the usual sense, also the relations which obtain among these parts and which are no less constituents of the complex whole.[6]

Twardowski counts as parts of a whole not only the relations just mentioned. Relations between parts of the whole and the whole itself are for him parts of the very whole too. He calls them primary formal constituents while the former relations get the name of secondary formal constituents.[7] Among the primary ones he distinguishes relations which make up a whole:

[ . . . ] the relations by means of which the parts are parts of this particular whole. [ . . . ] We call these relationships [ . . . ] 'primary formal constituents in the strict sense'.[8]

One can easily guess that after including relations among parts of a whole, nothing can prevent Twardowski from iterating this procedure and including still more and more relations which hold between objects which are already parts of that whole:

We call [ . . . ] 'relations of second degree' [ . . . ] all relations whose terms are relations.[9]

Moreover, relations themselves can be divided into parts-relations. Twardowski states an interesting condition of this division:

[ . . . ] more distant formal constituents result from an analysis of the closer formal constituents. [ . . . ] whenever analysis is possible, the complex relation does not appear as composed of relations which have the same terms as the complex relation; rather, the analysis of the complex relation involves an analysis of one or both of its terms.[10]

An immediately given illustration of such a divisible relation is the similarity between two partially identical objects: it can be divided into an appropriate number of identities between correspondent properties.

To make an order in this variety of relations contained in a whole, Twardowski distinguishes between the degree, rank and order of a relation.[11] The principles of classification can be briefly stated as follows:

The *Degree* of a relation depends on how many levels of other relations, serving themselves as arguments for subsequent relations, lie underneath.
The *Rank* of a relation depends on the order[12] of material constituents serving as its arguments.
The *Order* depends on the stage of another relation dissociation on which the given relation is reached.[13]

In addition to all that Twardowski introduces yet another distinction: the relation can be *primary* or *secondary*. It depends on arguments of the relation; whether they hold between proper parts of the whole or between a proper part and the whole itself. If we take as an example the aforementioned case of similarity (let us assume that the corresponding properties of both objects are material constituents of the first order), the identity relations between these properties (so these relations are neither primary nor secondary) will be formal constituents of the first degree but have rank of second order. Only now is Twardowski ready to say that:

This seems to exhaust what needs to be said in general about the kinds of primary formal constituents and the relations which are possible among them.[14]

And the reader, remembering that before making all these distinctions he confessed that

It is not our task here to investigate what kinds of parts and what kinds of wholes there are. [ . . . ] We are here only interested in what is common to all kinds of parts and to all forms of composition of parts, that is, in the type to which every synthesis belongs, and which is the basis for the various ways in which a whole can be composed.[15]

can ask now to what extent all the distinctions serve the purpose. And more so, as Twardowski in his analyses openly introduces an infinite regress — *horror philosophorum*. Husserl, in any case, seems to be dubious about this approach to the problem:

The view against which our objections are directed is based on the apparently plain truth that, wherever two contents form a real unity, there must be a peculiar part, a 'moment of unity', that binds them together. If then U is the moment of unity of A and B, there must be a new 'moment' of unity U1, for A and U (since these two are unified) and again a new U2, for B and U; and just so new 'moments' U11, and U21 for U and U1, and for U and U2 respectively, and so on *in infinitum*. If [ . . . ] one then takes the boundlessly varied conceptual differences that are possible a priori, and which ideal principles allow one to complicate indefinitely, and put all such differences into objects as their real moments, one gets the analyses which Twardowski has offered us in his 'psychological' researches, and which are as subtle as they are queer.[16]

These "queer analyses" lead Twardowski to the opinion that an infinite regress of parts not only is not suspicious, but on the contrary plays a decisive role in the formation of a unified whole. He comes to this conclusion in two steps: first he stresses a particular role played by the primary formal constituents in the strict sense in a whole. Because of them parts are parts of the given whole and the whole consists of just these parts. But at the same time Twardowski is aware of a very basic problem connected with the function of those constituents:

As important as this primary formal constituent is [ . . . ] as great are the difficulties which surround its concept. [ . . . ] We have always talked about relations between the whole and its parts. But the whole already contains the part. [ . . . ] The solution of this difficulty has to be the task of special investigations.[17]

The problem is clear: the relation which unifies parts into one whole, which makes the whole of different parts, has already as its relatum something which it just constitutes. This is clearly a *petitio principii*: the function of the property relation is already assumed as its own argument. Such 'looping' of a relation with its own arguments cannot be accepted in a standard set theory — but if we do not accept the Axiom of Foundation it becomes quite acceptable. From a philosophical point of view the problem is rather not in the loop itself but in the order of constitution: on the one hand the whole seems to be constituted by the property relation but on the other the whole itself makes the property relation possible, as it is one of its arguments. Not the relation is therefore suspect, but rather its function of constituting the whole. If we took the property relations (Twardowski calls them usually simply 'properties' — *Eigenschaften*[18]) to be not the reason but the consequence of the existence of the whole, the problem would disappear. Thus it would be natural to look for another factor constituting a whole. This is Husserl's choice, as we shall see later on. But Twardowski is not going to abandon the first approach and he comes to the next stage of his investigation, ready to admit that perhaps the problems mentioned will never be solved.[19]

But if the relations of 'having' which obtain between a whole and its parts are in turn parts of the whole — [ . . . ] formal constituents of the object — then these relations are had by the object no less then the material constituents. But now there arises an infinite complication in that, because these second primary formal constituents are likewise had by the whole.[20]

It is exactly this 'infinite complication' which Husserl called 'queer analyses.' He took it as an indication that the factor unifying parts into a whole cannot be found here. Twardowski however, to the contrary:

Perhaps it is just this infinite nesting of primary formal constituents which contains the key to the answer to the question concerning the nature of the relation which holds the parts together in a whole.[21]

Twardowski's opinions on the infinite complication of relational facts, albeit controversial, found an adherent in another Brentanist — Alexius Meinong. He did not hesitate to accept the regress in parts.[22] Recognition of this regress was, incidentally speaking, *locus communis* of the metaphysics at the end of the century. F.H. Bradley, e.g., took it as an argument against the reality of a pluralistic universe and for his absolute monism.

Husserl, criticising Twardowski's standpoint, rejects the regress. It does not really exist, he thinks, because

Nothing really exists — in the sense of being a possible object of sense-perception — beyond the aggregate of a whole's 'pieces', together with the sensuous forms of unity, which rests on these pieces conjointly. Unity is conferred on the 'moments' in the 'pieces', as also on the 'moments' of unity and the pieces, by the foundational relations in the sense of our definition.[23]

That a whole exists because the relation of foundation holds among its parts is stressed even more in the definition:

[...] one could then give the following noteworthy definition of the pregnant concept of Whole by way of the notion of Foundation:
By a Whole we understand a range of contents which are all covered by a single foundation without the help of further contents. The contents of such a range we call its parts. Talk of the singleness of the foundation implies that every content is foundationally connected, whether directly or indirectly, with every content.[24]

Foundations are things (in the widest sense) which unite a whole. They too are relations but of the formal sort as opposed to the material ones which are the above mentioned 'unifying moments'. In Husserl we find two definitions of foundation:

If a law of essence means that an A cannot as such exist except in a more comprehensive unity which associates it with an M, we say that an A as such requires foundation by an M or also that an A as such needs to be supplemented by an M.[25]
A content of the species A is founded upon a content of the species B, if an A can by its essence (i.e. legally, in virtue of its specific nature) not exist, unless a B also exists: this leaves open whether the coexistence of a C, a D etc. is needed or not.[26]

When it comes to the formal character of this relation, it is outlined as follows:

[...] the concepts or propositions which have content [...] we sharply distinguish from purely formal concepts and propositions, which lack all 'matter' or 'content'. To the latter belong the categories of formal logic and the formal ontological categories[27] [...] which are essentially related to these, as well as to all syntactical formations they engender. Concepts like Something, One, Object, Quality, Relation, Association, Plurality, Number, Order, Ordinal Number, Whole, Part, Magnitude etc., have a basically different character from concepts like House, Tree, Colour, Tone, Space, Sensation, Feeling etc., which for their part represent genuine content.[28]

One can say that purely formal concepts differ from material (having content) ones in lacking characteristics corresponding to the question 'What is it like?' And it does not mean that formal concepts lack any characteristics—they have their 'what', their *genera*. But they are the highest *genera* and their characteristics have no common element. There is nothing separable in their natures, nothing which could have qualified anything more general. This highest *genus* of a thing tells, Husserl thinks, not what is in a thing but only how this thing is composed, how its constituents or elements are organised into one whole. He calls this mode of organisation the form, and what is organised he calls the matter of a thing.[29] What Husserl regards as a form Twardowski calls formal constituents of a whole and as he regards them too as parts of the whole, an infinite regress of parts results. Husserl, for his part, regarding relations of foundation as the formal parts of a whole, makes a sharp distinction between them and material relations of nonrelational elements of the whole. He regards only material relations as real parts of a whole. Such relations (relation-instances, to be exact) cannot hold without corresponding arguments which are parts of the same whole. This fact means for Husserl that between this relation and its arguments still another relation holds — the formal relation of foundation. It is formal because it has no 'content' as, e.g., 'being greater', 'standing to the left', etc. All its nature, this 'something' which holds between its arguments, consists in 'being *conditio sine qua non* of existence'. This is no more than the way of coexistence of the condition (*fundamentum*) and what is conditioned (*fundatum*).[30] We cannot say that between this formal relation and its arguments, *fundamentum* and *fundatum*, still another relation holds. A relation, material or formal, can hold only between materially determined elements. In this way Husserl avoids Twardowski's regress.

To conclude this section the following can be said: Husserl undertakes Twardowski's task of finding the factor which unifies many parts into one integral whole. His answer differs from Twardowski's tentative solution and avoids the regress it involves. According to Husserl the factor unifying a whole is the relation of foundation holding between its parts — the relation which itself is not a part of the whole but only a connection of parts. Property relations between parts and the whole result from this connection, not vice versa. Perhaps the essential difference between material relations being parts of the whole and formal relations of foundation which are never parts can be put into relief as follows. Each material relation has two aspects: objectual — what it is like, and factual — that it holds. When it holds between parts of a whole, it is itself a part of this whole but as an object and not as a fact. A formal relation is deprived of this objectual aspect, so even when it holds between parts of a whole there is no extra object which could have been added to the whole as a new part.

## 2.

Now we come to another problem involved in the definition of a whole in Twardowski's and Husserl's writings. It has been remarked that Husserl's definition seems to be too wide for at least two reasons. First, the relation of foundation appears to bind certain objects which do not belong to the same whole. It is most evident in the case of any comparative relation, like e.g. 'being greater than'. This relation holds between an elephant in Africa and a flea in Krakow but this pair of creatures does not form any reasonable whole.[31] Can such an objection apply to Twardowski's definition of a whole? It looks as if more modest intentions protected him from the troubles which threatened the Husserlian attempt to give a general and precise definition. For Twardowski starts from the intuitive notion of part, as if from the initial step of an inductive definition and adds to these parts — in a quasi inductive step — relations between them:

[ . . . ] one has to consider, in addition to what one calls a part in the usual sense, also the relations which obtain among these parts and which are no less constituents of the complex whole.[32]

As a consequence of this second step he cannot avoid the regress but at least, in the beginning, he does not accept any objectionable complexes as wholes. Let us notice however that for Twardowski an infinitely complicated network of relations between parts is not a necessary but sufficient condition of a whole. Otherwise he would have been looking for another unifying factor. On the other hand, the regress in relations occurs if one accepts Twardowski's arguments, anywhere. Hence, any 'uncontroversial' whole can be 'inflated' unlimitedly. In conclusion we must admit that both Husserl's and Twardowski's definitions are too wide.

Another reason for regarding the Husserlian definition as too wide is the following: The relation of foundation in the sense of Husserl holds between individual objects whenever they belong to the appropriate species. We have many individual instances of a species colour or extension. Our definition rendered literally forces us to regard every instance of colour connected with every instance of extension and in consequence to take all of these individuals as parts of one comprehensive whole. Husserl, as if trying to avoid such an absurd consequence, says:

And even where a law tells us of the impossibility [ . . . ] of an association [ . . . ] our case still reduces to one of nonindependence. For an A can only exclude a B, if both exclusively require the same thing. A colour excludes another colour, but only if both aim to cover an identical piece of surface, and both cannot do so completely. To each essential, law-bound exclusion of a determinate characterisation, there corresponds a positive law-bound requirement of a corresponding characterisation and vice versa.[33]

The relation of exclusion has the aim of preventing the above mentioned individual instances of colour and extension from melting into one whole. However, it is not introduced in the definition of a whole. The first sentence of the last quotation seems to suggest that exclusion (impossibility of connection) is somewhat reducible to dependence (hence foundation). But obviously the relation of foundation alone is not enough to define the relation of exclusion.[34] It seems that Husserl borrowed this somewhat sketchy treatment of the notion of exclusion from Twardowski or from some source common to them. For in Twardowski we find the following:

The so-called 'incompatibility' of marks is based on this; for, two properties or the like can only be called incompatible insofar as they are conceived of as parts of one and the same object.[35]

The relation of exclusion taken as a primitive notion could be profitably used to form an additional condition in a definition of a whole. As such it could 'stop' inappropriate connections of parts. But even after such an improvement (and complication on the other hand), our problems do not disappear. Since the relation of exclusion cannot help us with the case of an object founded in some object of a particular kind, let us take two wholes which are just two coloured fragments of extension. The definition explains why they are wholes: colour requires extension as its supplementation (extension is the foundation of colour). But the same definition tells us that not only colour $c_1$ with extension $e_1$ and colour $c_2$ with extension $e_2$ form wholes. The same relation of foundation makes wholes of $c_1+e_2$ and $c_2+e_1$. Of course, we realise that it is just the whole $c_1+e_1$ which exists and not $c_1+e_2$. But there is nothing in the definition which reflects this difference. Will it help to recall that Husserl's Part-Whole theory, belonging to his ontology — the science of pure possibilities, does it not deal with questions of existence and nonexistence? Presumably not, as our problem is not to explain what an alleged 'existence property' of an object consists in, but rather to show a factor present in the whole consisting of $c_1$ and $e_1$ and lacking in the pair $c_1$, $e_2$ — the pair whose elements also exist but do not form one whole. This factor is simply the connection of elements and we see that it cannot be just the relation of foundation — both pairs are related by it. We can remark here that the present objection does not apply to Twardowski but not because it is better conceived — rather vice versa — because it is not developed enough to meet any detailed criticism.

From now on we will devote most of our attention to Husserl's theory. Even if it is open to doubt, it is such because it is formulated precisely enough. Its criticism can lead to definite conclusions. Let us concentrate now on the following problematic aspect of the foundation relation. What does a dependent part of a whole exactly require? This particular instance of colour does not require any particular fragment of extension. It could cover any of them. When we look closely once again at the two definitions of foundation quoted above, we can see that neither of them points to a particular object as a foundation of a given part.[36] Both definitions claim that for our part it is necessary to coexist with any object of a correlated species. But 'any object' of a given species is not a name of anything particular which could exist as a part of a real whole. Thus the foundation of the part under consideration is neither any particular individual nor the totality of them. What is it then? Let us look at the continuation of the definition of foundation:

> If accordingly $A_0$, $M_0$ are determinate instances of the pure kinds $A$ or $M$, actualised in a single whole, and standing in the relations mentioned, we say that $A_0$ is founded upon $M_0$, and that it is exclusively founded on $M_0$, if $A_0$'s need for supplementation is satisfied by $M_0$ alone. This terminology can of course be carried over to the Species, by a quite harmless equivocation.[37]

Husserl is saying here that among many instances of the species $M$ the foundation of $A_0$ is just this one which belongs to the same whole. But it is exactly the relation of foundation which constitutes the whole! Thus once again we find a vicious circle in an attempt to define a whole. This time it does not occur in the very definition of a whole, like in Twardowski, but in the sequence of two definitions — one of foundation and the other of a whole.

How could we avoid this error? Maybe we should try to establish another foundation for the dependent part we were talking about. If it can be neither a particular bit of extension nor the entire extension of the universe, let us try something of a totally different status — the extension as a universal: not this or that particular but extension as such. It seems reasonable to say that our part (a particular shade of colour) requires extension as such. But there is another trouble in such an approach. Aren't we forced now to agree that a *universal* appears to be a part of the same whole as our dependent individual part? In the second *Logical Investigation* Husserl explicitly rejects such an ontological mixture.[38] Roman Ingarden, who investigated the problem of an ontologically heterogeneous whole, did not reject the case of a real whole having some only possible parts but definitely excluded a real whole having ideal parts.[39] What is Twardowski's standpoint in this case? He clearly states that "genus and differentia appear as parts of one and the same uniform thing".[40] Actually Twardowski is only quoting Trendelenburg; nevertheless the lack of any critical comment on it leads us to suppose that Twardowski shares this opinion. Further on we find Twardowski's statements implying that he acknowledged certain general objects as parts of a real object:

> [ . . . ] certain characteristics belong to all objects of whatever kind [ . . . ] certain constituents supposedly belong to every object, and these constituents are always and necessarily conceived of, so that a presentation is not even possible, unless these constituents are presented by means of it.[41]

Still further Twardowski repeats, after Lotze, some examples of such constituents; they are: unity, identity and difference (with respect to other objects). All three he calls metaphysical parts of an object explaining that:

> By metaphysical part, one understands what can be distinguished in or about a whole by means of the ability to abstract, but cannot really be separated from the whole. [ . . . ] everything called a property in ordinary discourse — are metaphysical parts of the objects which have these properties.[42]

The parts mentioned above are not 'ordinary' *universals*, they are *transcendentals*. They are not the only kind of universal parts. In the last paragraph of his dissertation, devoted to so called general presentations, Twardowski claims that there exist particular objects of these presentations. This is the consequence of his famous statement that every presentation has its object — in this case the object is just the common part of corresponding individual presentations:

> [ . . . ] what is common as such to all [relevant] objects of individual presentations[43] [ . . . ] a group of constituents [ . . . ] presented as a whole [ . . . ] stands in the relation of equality to certain parts of objects of other individual presentations[44] [ . . . ] is in a certain way a metaphysical constituent of the individual objects which are subsumed under it.[45]

Twardowski is saying that such a general object corresponds to a Platonic idea and, although it is an object of a presentation, does not exist or rather

> [ . . . ] it exists in a sense that it can be found in the objects of the corresponding individual presentations in a form which is modified by the individual characteristics of these individual presentations.[46]

It is evident that Twardowski sees no problem in recognising general parts of individual objects. Why is it so then, that in Husserl's book, where we often find references to Twardowski and in controversial cases always a critical discussion, this time we find only a direct rejection? The reason probably is that the rejection of any existential connections between the real and the ideal sphere logically precedes problems handled in the dissertation on parts and wholes. Their characters differ too much to make any interface possible. Husserl writes:

> [ . . . ] there is a fundamental categorial split in our unified conception of being [ . . . ] we take account of this split when we distinguish between ideal being and real being;[47]

This is ultimately justified by an insight into the essence of an individual object — detailed analyses of which we find in Ingarden.[48] Thus we cannot doubt that the rejection of any possible junction of real and ideal in one object is definite for Husserl. However, one of his basic ontological concepts is hardly consistent without an assumption that the foundation of a dependent part is a general object — *universal*. Maybe Twardowski's last quoted remark that a general object when incorporated into a real whole needs to be modified. In case of, say, extension, modification consists in that a particular shade of colour requires extension in general but in a concrete whole is unified not with extension as such but with a particular bit of extension. So, it seems, extension (and other fundaments) actually assume two different forms in Husserl's theory: the general and the particular. An attempt to reduce one of them to the other results in absurdities and in fact ruins the theory. Apparently this duality lacks an explanation or even a stress laid on it. On the contrary, Husserl says permutably of the relation of foundation holding between individuals and between correlated genera that it is "a quite harmless equivocation".[49]

3.

Let us see now if this duality can be consistently incorporated into Husserl's theory. Surely we will have to give up the underlying paradigm of the resolvability of a whole into pieces. According to this paradigm, a whole can be actually or mentally decomposed into parts which keep their identity in this state of separation. It is most probably an echo of the Cartesian method of analysis based in its way on a model of, say, LEGGO bricks. However, it happens sometimes that, after separation, elements of a whole change and what we find in them does not correspond to what really is present in the integral whole.

Let us consider once again the example of a colour founded on a fragment of extension. If we imagine both separately what we get is not a transparent bit of extension but an indeterminate with respect to colouring or a transparency of particular shape. The shape is here the factor determining extension as such to a particular extended fragment. We should always remember that extension as such is not like an infinite empty space from which particular shapes could be as if cut off. It neither is, nor contains, a definite shape. It is just what is common to all possible shapes plus an empty infinite space — the extension. Its universal status consists in this and as *universal* it is the sole candidate for the foundation of colour. But the relation of foundation which obtains between the object and a *universal* does not hold any more inside, so to say, the whole. In the whole we find not the universal but a particularised extension. Thus Husserl should not have told us that parts forming a whole are bound together by the relation of foundation.

And what role does the shape play between the colour (a particular) and extension (a universal)? Its relation to extension differs from the relation between the colour and extension, for it does not seem possible to separate (even in mind) the shape from its extension. The shape is conceivable only as extended. On the other hand the colour does not require a shape — one can imagine an extended and coloured surface without any boundaries. So the shape in an extended object must be a part independent of the colour and vice versa while its relation to extension is apparently analogous to the relation, mentioned by Husserl, of, say, redness with colour in general:

> [...] if we are asked what must be added to Colour to produce its species Redness we can only answer 'Redness'.[50]

It is the problem of special parts of a whole which, being proper parts, nonidentical with the whole, leave no 'remainder' after subtracting from the whole. Brentano had already pointed to this peculiar case in his *Theory of Categories*.[51] Simons explicitly expressed and investigated the opposed supposition, the so called Weak Supplementation Principle, saying that for every proper part of a whole there is a complement part forming with it together the total whole.[52]

Already after this superficial overview of parts of some simple whole one can see that the paradigm of analysis collapses. What could be a substitute for it in a would-be Part-Whole Theory? First of all let us remark that the relation of foundation, after clearing up all doubts, could be of service. This relation, holding between a particular and a universal, does not constitute a sufficient condition of the existence of a whole as the universal can be instantiated (particularised) in many ways and only one instance actually joins the founded particular. However, we can adopt it as the necessary condition: if it holds between two given objects neither directly nor mediated by other objects the two in question do not form a whole.[53] Husserl describes a certain phenomenon which appears to fit better to the role of a unifying factor in a whole. He calls it *Verschmelzung*[54] and says that it consists in a particular interpenetration of parts. What is important for us is that it cannot be doubted that this relation holds exclusively between particulars and only among these which actually form one whole.

How can we describe its essence? In the case of a uniformly coloured shape it is not difficult. If a colour is blended with extension it is impossible to find in the phenomenon of the thing anything coloured but not extended or vice versa. The shape in turn is the shape of this particular fragment of extension which has it. We have remarked already that there is not a relation of foundation between a shape and extension: extension is present in a shape in such a way that one cannot find there a plurality of parts. Maybe it would be relevant here to mention that any complicated shape is something simple, without parts. Of course, it is possible to divide any given shape into many other shapes which, taken together, make the original shape again. But neither shapes of components nor components themselves are parts of the original shape. Otherwise we would have the 'big' shape wherever the 'small' shapes are. We do not have the former because there is something lacking among the plurality of the other ones. And what is lacking is just the original shape — the proper arrangement of all components. What results out of this consideration is once again the observation that the deconstruction of a thing does not always result in getting its actual parts. Two fragments of a broken pencil are not really its parts — they exist just because it does not exist any more.

Let us try to analyse the more complex phenomena of fusion. Take a thing like a table which has distinct parts. They differ in their shapes (which they retain in the whole) and maybe in colour or texture. So the fusion of all components here is out of question. Now take for comparison a totality of elements, let them be the parts of our table but joined otherwise so that they no longer form a table. At once a question arises: Is this another whole or only a pile of elements? The answer will depend on a point of view. If we limit ourselves to what is sensually given only, it is possible to treat the totality, if only its parts are somehow joined, as a whole — a spatial continuum. If, on the other hand, we also take into account the function of the given object, i.e. its formal or final cause, the table with, say, horizontally attached legs loses its original function and most probably does not assume any other. However, in this decomposed table the same spatial continuum is still present as in its purely sensual aspect. This non-monotonicity of the whole character points to an important fact: the same element does not always decide the unity of a whole. In a richer structure the element which could serve on the former stage may appear as non-sufficient as an integrator. What is more, its non-sufficiency can regard not only new elements added to the whole but the old ones as well. It is possible that different levels occur in a thing, e.g. what is given or what is meant, and on these levels different unifying factors are required. In the case of our first table it seems that the unifying role is played by its function as a piece of furniture. This 'layer' or aspect is neglected when only sensual data are taken into account. This is why another unifying factor acts at this level: spatial continuity. In this last respect both our objects do not differ. If a function of the object is, however, a stronger unifying factor than the continuity itself (as the presence of the second cannot make up for the lack of the first) we can expect that where the second factor is lacking the presence of the first will be enough to establish the whole. In fact, let us take as an example a set of certain objects, e.g. tools, usually regarded as a whole. The mark of this is the case of its breaking up, as if when certain parts are lost. Another mark is that its instrumental, not commercial, value exceeds the sum of the relevant values of the single components (in certain cases the latter can even amount to 0). Is it possible to account for this unifying factor which a function of an object is in terms of foundation or fusion? The function itself can clearly exist only where the correlated object exists. On the other hand it can be abstracted, thought of, without this substratum. It is as if we had the relation of foundation here, but all its old and some new problems in addition. Here again it cannot supply more than the necessary condition of a whole. Does it make sense to talk of fusion of the function of an object and its material components? Would it not make of this function a kind of part of an object? Such widening of the notion of part should be rather avoided.[55] And maybe this factor unifies plurality of parts without their blending (alone or with anything else)? Let us look at the relation of foundation here again. No doubt, the role played by a set of spanners can be played by tools belonging to other sets. But can it be played by the tools situated in different, distant places? The function we consider requires the presence of all tools at hand. But if we say that two sets placed together can play the function, the actual number of sets we have at our disposal is not 2 but 2 to the power of $n$, where $n$ is the number of tools in each of the identical sets. The following conclusion results: When a unifying factor is the function played by the thing (or set of things), the relation of foundation itself constitutes the whole. But if we recall that the relation of foundation obtains between parts of the whole we arrive at the troublesome conclusion that the function is a part of the whole. Maybe we should modify Husserl's theory and acknowledge objects of different categories as parts of one whole? In the given case the 'what for' of a thing would be founded in it as a material 'core'. But then the next question occurs: the function played by different but identical sets of tools is the same. It is then founded in every set but separately, not in all together, for sets do not form any 'superset of tools' and stay separate. And in the case of a whole deprived of a specific function (a simple stone, e.g.) — does its spatial continuity depend on fusion of any parts? It seems that here certain 'partial' (i.e. belonging to parts only) shapes and colours blend with corresponding parts of total extension of the whole.

How could we explain the diversity of ways of unifying things belonging to the two kinds considered here? Why in the case of a functional object is the relation of foundation obtaining between its function and material parts not only the necessary but sufficient condition of a whole? If a founded thing does not have a spatial character, it cannot multiply in different instances and keeps its uniqueness in all connections. In this way the trouble of matching appropriate *fundata* to their *fundamenta* is avoided. On the other hand it cannot happen that things from different categories, a function and a material object corresponding to it, blend together. In a thing having no function it is hard to find an immaterial factor which could play an analogous role. Its founded constituents are thus material, having many instantiations. That is why relations of foundation alone do not establish a whole here. It is produced only with the help from other factors like e.g. fusion. We must remember however, that such fusion will not occur unless the relation of foundation holds between appropriate *specii*.

We have considered above only some specially chosen examples of wholes. The conclusions of these investigations cannot then be accepted without restrictions or taken as a sufficient empirical basis to construct a general theory. They rather only uncover certain new problems and perspectives associated with attempts to establish a theory of Husserlian origin but able to overcome the problems which met the original theory. Husserl's theory is not the last but certainly it is a basic, indispensable stage of such a theory.

*Department of Logic, University of Łódź*

## NOTES

[1] E.g. Brentano tried to define the relation between substance and its attributes in terms of part-whole. Cf. Brentano, F., *Theory of Categories*, trans. by R.M. Chisholm, Nijhoff, The Hague 1981, Part II, Ch. II.C, p. 115f. and Translator's Introduction, p. 7f.

[2] Twardowski, K., *Zur Lehre vom Inhalt und Gegenstand der Vorstellungen*, W. Hölder, Vienna 1894, §9, p. 48f. (G 46f.) (G means the pagination of the English translation: Twardowski K., *On the Content and Object of Presentations*, trans. by R. Grossmann, Nijhoff, The Hague 1977, p. 46f.)

[3] Husserl E., *Logische Untersuchungen*, 2 Band, I Teil, Husserliana XIX/1, Nijhoff, The Hague 1984, §2, B1 228 (F 437) (B1 means the pagination of 1913 edition, given in the present volume, F—J.N. Findlay's English translation, 1st ed., vol. 2, Routledge & Kegan Paul, London 1970).

[4] Twardowski, op. cit., §9, p. 48 (G 46)

[5] Ibid., §13, p. 88 (G 84)

[6] Op.cit. §9, p. 48 (G 46)

[7] Op. cit. §10, p. 52 (G 50)

[8] L. cit.

[9] Op cit. §10, p. 53 (G 51)

[10] Op. cit. §10, p. 53f. (G 51f.)

[11] In the original: *Grad, Range, Ordnung*

[12] Analogously to the order of a relation it depends on subsequent partitions: the whole into parts, these parts into parts of parts, etc. See op. cit. §9, p. 49 (G 47)

[13] Op. cit. §10, p. 53 (G 51)

[14] Op. cit. §10, p. 55 (G 52)
[15] Op. cit. §9, p. 48 (G46)
[16] Husserl, op. cit., B1 280 (F 478f.)
[17] Twardowski, op. cit. §10, p. 56 (G 53f.)
[18] L. cit.
[19] L. cit.
[20] Op. cit. §10, p. 59 (G 56)
[21] L. cit.
[22] Meinong A., *Gessammelte Abhandlungen*, Barth, Leipzig 1914, vol. 2, p. 390f. An interesting critique of this standpoint can be found in Grossmann R., *Meinong*, Routledge, London 1974, p. 66f.
[23] Husserl, op. cit., §22, B1 280n. (F 479)
[24] Op. cit. §21, B1 275n. (F 475)
[25] Op. cit. §14, B1 261 (F 463)
[26] Op. cit. §21, B1 275 (F 475)
[27] Connections between formal logic and formal ontology are discussed in my Polish paper 'Husserlowska koncepcja . . . ' (Husserlian Conception . . . ) *Kwartalnik Filozoficzny* (forthcoming).
[28] Op. cit. §11, B1 252 (F 455)
[29] These things are not very clearly stated in *Logische Untersuchungen* but compare Ingarden's reconstruction in *The Controversy over the Existence of the World* §34f.
[30] There is a close resemblance between this formal relation of foundation and the Ingardenian *momentum existentiae* of self-sufficiency. See Ingarden, op. cit. §11–16, esp. §14
[31] An objection of this kind was raised by Prof. J. Srzednicki at the Conference of the History of Logic in Krakow in 1994.
[32] Twardowski, op. cit. §9, p. 48 (G 46)
[33] Husserl, op. cit., §10, B1 251 (F 455)
[34] One can draw an analogy between relations of exclusion and foundation on one hand and propositional functions of negated conjunction and implication on the other. See M. Rosiak, 'Formalizacja teorii części i całości opartej na pojęciu ufundowania' (A Formalization of a Part-Whole Theory Based on the Concept of Foundation), *Filozofia Nauki* 1, 1996, pp. 41–80. During our session in Lvov Kevin Mulligan made a suggestion that exclusion can be defined as follows: $A$ excludes $B$ iff $A$ and $B$ have the same foundation and it is impossible for them to coexist in the same whole. But if we adopt this suggestion, a vicious circle in definition arises, for we just need the relation of exclusion to define the whole.
[35] Twardowski, op. cit. §10, p. 62, in a footnote (G 59)
[36] Husserl, op. cit. §14, B1 261 (F 463)
[37] L. cit.
[38] Op. cit., *Untersuch. II*, §19, B1 155 (F 377). Kevin Mulligan expressed the opinion that acceptance of universal parts in individual objects would ruin the whole theory.
[39] Ingarden, op. cit. §11
[40] Twardowski, op. cit. §10, p. 61, in a footnote (G 57)
[41] Op. cit. §13, p. 87f. (G83)
[42] Op. cit. §10, p. 58, in a footnote (G 55)
[43] Op. cit. §15, p. 104 (G 99)
[44] Op. cit., p. 105 (G 100)
[45] Op. cit., p. 111 (G 105)
[46] Op. cit., p. 106 (G 101)
[47] Husserl, op. cit. *Untersuch. II*, §8, B1 125 (F 353)

[48] Ingarden, op. cit. Ch. VIII, X, XIII–XIV. See also Rygalski, A., 'Zmienna zawartości idei — koncepcja zmiennej w ontologii Romana Ingardena' (The Variable of a Content of an idea — a Conception of the Variable in Roman Ingarden's Ontology), *Kwartalnik Filozoficzny* **21**, 1993, pp. 23–39; also by the same author: 'Zmienna zawartości idei — próba interpretacji' (The Variable of a Content of an idea — an Attempt at an Interpretation), *Kwartalnik Filozoficzny* **21**, 1993, pp. 45–66.

[49] Op. cit., *Untersuch. III*, §14, B1 261 (F 463)

[50] Ibid. §10, B1 250 (F 454)

[51] Brentano F., Op. cit. p. 115

[52] Simons P., *Parts*, Clarendon Press, Oxford 1987, p. 27f.

[53] The first author who realized that foundation is not a sufficient but only a necessary condition for the existence of the whole was, to my knowledge, Peter Simons.

[54] Translated as 'blending' or 'fusion' by Findlay, see Husserl, op. cit., §§ 8, 9.

[55] But if we took this function as belonging to an ontological category of possibility we could find support from the side of Ingarden who accepted parts of this category in real wholes.

EVGENI G. VINOGRADOV

# THE RATIONALISTIC PARADIGM OF FRANZ BRENTANO AND KAZIMIERZ TWARDOWSKI

In the philosophical heritage of Kazimierz Twardowski the comparatively small work 'Franz Brentano and the History of Philosophy' is probably not very important. But the breath of a new day and new ideas make it available nowadays, a hundred years after its first publication in Vienna.[1] This work is a summary or a detailed review of Franz Brentano's paper 'The Four Phases of Philosophy and Its Contemporary Stage', in which the Brentanian historical-philosophical conception is put forward.[2]

Twardowski believes that we can use three methods to examine the history of philosophy — biographical, critical and constructive. The biographical method relies on acquaintance with personalities of these or those representatives of the philosophical sciences. The method of critical exposition means the comparison of some views and philosophical systems, and the estimation of their relative or absolute values. Finally, the constructive method is a search for laws governing the development of philosophy, and this method can be meant as a priori or a posteriori. The a priori method proceeds from a certain previously accepted view on philosophy, and tries to account for all its development according to an accepted a priori principle. The constructive method is a posteriori when the researcher attempts to find general laws of philosophical development in its real acts. The constructive method is the highest level of philosophical description and most perfect in its a posteriori mode, because the a priori mode always forces an artificial agreement between facts and a principle accepted in advance. "Suffice it to mention the name of Hegel" — Twardowski explains — "and the fallacy of a priori construction of historical facts would appear before everyone's very eyes". Twardowski gives the historical-philosophical conception of Franz Brentano as an example of the a posteriori constructive method. In this conception a division of European history into Ancient, Medieval and Modern philosophy is preserved. But within each phase he distinguishes four phases, repetition of which constitutes, according to Brentano, the general law of historical development in the field of philosophy.

The first phase is the phase of flourishing and is characterised by a truly theoretical interest in philosophical problems. That is Ancient philosophy in its development from Thales to Aristotle, Medieval philosophy from patristics to Scholastic philosophy and St. Thomas Aquinas, and Modern philosophy — Francis Bacon, Descartes, Locke, Leibniz. The next three phases of each epoch are the phases of decay.



*K. Kijania-Placek and J. Woleński (eds.),*
*The Lvov–Warsaw School and Contemporary Philosophy,* 101–104.
© 1998 *Kluwer Academic Publishers.*

In its second phase philosophy begins to serve practical goals and thereby loses in conscientiousness and accuracy of research, as with the ancient Stoics and Epicureans, medieval Scotists, and modern French and German rationalists.

The third phase appears as a reaction to the second phase: philosophy and, finally, the human mind in general loses confidence; they seem to be unable to acquire any knowledge. This phase of scepticism is represented by the Academics and Pyrrhonism in the Ancient period, the dominance of Nominalism (William of Ockham) in the Middle Ages, and the philosophy of David Hume in the Modern period.

In the fourth phase, as a reaction to the previous one, the thirst for knowledge, probably with feverish haste, creates new views, new philosophical systems. The human mind resorts to new putative sources of knowledge, to intuition and mystic forces, and luxuriates in the imaginary possession of a proclaimed truth going beyond any human understanding. There are the Neopythagoreans and Neoplatonists of Antiquity, the medieval mystics, Raymond Llull and Nicolaus Cusanus, and also — in the Modern period — the Scottish philosophy of common sense and, quite unexpectedly, Kant, Fichte, Schelling, and Hegel. 'The Great Germans', incidentally, caused the deepest crisis in philosophy.

As Twardowski concedes, one may have a great many objections to Brentano's views. It is easy to notice, for instance, that these four phases which so vividly clarify themselves in Ancient philosophy turn out to be not so clearly distinguished in Medieval and even less clearly in Modern philosophy. But this can be due to the fact that every epoch is under the influence of the previous one, and for this reason later periods can never be pure repetitions.

Another question arises: in this picture of historical development, where shall we place the Sophists, Spinoza, or Roger Bacon and others, who were not included in this scheme?

Twardowski stresses that no scheme aspires to be a complete presentation of a problem and it remains just a scheme, i.e., an outline of a general tendency of development. Naturally, he continues, the Brentanian historiography of philosophy is not the last word on this subject. But, when we speak of philosophy itself, not about its relation to other branches of human intellectual development, then the Brentanian account of the history of philosophy could be considered the best resolution of the difficulties which are piling up in this branch.

Immanuel Kant, of whom Twardowski said to be careful ("Ostrożniej z tem Kantem!"), once asked: Why do people wait for the end of the world at all?

One may ask a similar question: Why is the history of philosophy considered in categories of flourishing and decay? Even when criticising apriorism Twardowski could not avoid it.

There we find the mysterious structures of consciousness the presence of which was felt, probably, from the time of Plato. Generalizing the experience of Aristotle, Francis Bacon, Lichtenberg, Karl Popper, and Thomas Kuhn in this field we can call these structures *pre-cognitive sets*. They are the informative structures of consciousness which regulate and direct the thinking and cognitive activity of a human being. Among several sorts of them we are interested in the disciplinary-matrix group of sets, to use Kuhn's terms. They appear in the process of scientific cognition and are specific to each branch of research activity, although they have certain sources both in the socio-cultural environment and in the individual experience of every human being.

Owing to their very nature these sets are usually not realized by the individual as a priori or pre-requisites of his every cognitive act. The conviction that an origin as well as the ultimate end of philosophy should be scientific was not considered by Twardowski himself as a priori. But this conviction is absolutely necessary if we speak about 'flourishing' or 'decay', because we need to have something as a positive or a negative criterion. It is, in Twardowski's case, the apriorism of scientific knowledge, and scientific truth, almost making a fetish of knowledge of it.

Greek philosophers who, according to Twardowski, looked for knowledge just for the sake of knowledge, are ideals for him. The Stoics and Epicureans who first of all aimed to influence the practice of life "polluted the pure water" of scientific philosophy with ethical problems. Ancient philosophy in its fourth phase came to the deepest decay, "it ceased to be science", Twardowski writes, and we can believe that nothing could be more terrible for him. Finally, nevertheless, apriorism surfaces. Considering the situation after Leibniz, Twardowski writes: "Again they began to use philosophy with goals alien to it".

But the goals can be divergent. If the logician's ideal is scientific knowledge then the following phases would be the phases of decay. If the four Brentanian phases were considered by a philosopher devoted to ethics, then the phase of scientific cognition would be the formation of ethical knowledge. If, in the end, the mystic used this scheme, then all three previous phases would be the preparation for true knowledge — the mystical. And nobody knows which position would be more convincing.

> More than ever mankind stands helplessly at the end of the century and does not know where to turn. Therefore it is absolutely essential at least for philosophers not to lose their heads and to know where it is worth going. And they will learn it from the philosophy of the history of philosophy

— these are the last lines of Twardowski's paper.

Today, after a hundred years have passed, we can say that the choice of the path of development never was a single, rational act. Attempts to make such a choice in our recent past led to a serious crisis in philosophy. People can sometimes be helpless, but mankind, fortunately, can never be. And it is never waiting until philosophers or politics will find where it is worth to go. And this can be the guarantee of at least a partial liberation from the domination of our pre-cognitive sets and the guarantee of further 'decays' and 'flourishings'.

*Lviv State University, Ukraine*

## NOTES

[1] K. Twardowski, 'Franciszek Brentano a historya filozofji' (Franz Brentano and the History of Philosophy), *Przełom* **1**, No. 11, 1895, pp. 335–346. Our examination of Twardowski's work is based on B. Dombrowski's translation of K. Twardowski's *Rozprawy i artykuły filozoficzne* (Philosophical Dissertations and Articles), Lvov, 1927.

[2] F. Brentano, 'Die vier Phasen der Philosophie und ihr augenblicklicher Stand', Verlag der J.G. Gotta'schen Buchhandlung, Stuttgart, 1895.

URSZULA ŻEGLEŃ

# ŁUKASIEWICZ'S INTERPRETATION OF ARISTOTLE'S CONCEPT OF POSSIBILITY

I would like to devote my short paper to Jan Łukasiewicz, whose works made an important contribution to the achievements of the Lvov-Warsaw School. Łukasiewicz belonged to those eminent representatives of the School who joined their philosophical research with research in logic, proclaiming the program of rigorous philosophy.

In my paper I want to focus on just one topic which is especially important for philosophers as well as for logicians, namely, the problem of modality. I am going to deal only with one small part of the broad studies undertaken by Łukasiewicz and I restrict my paper to only some aspects of Łukasiewicz's interpretation of Aristotle's concept of 'possibility'.

Łukasiewicz develops his account in an analysis of Aristotle in whose work the role of 'possibility' is undertaken by two verbs: *ἐνδέχεται* and *δύναται*, each of which is used in many different senses. Łukasiewicz's interest in the problem of modality is strictly connected with the controversy between determinism and indeterminism, which he included in a list of fundamental philosophical issues to be resolved by the strict method of logical analysis.

Łukasiewicz presented his analysis of the concept of possibility at a meeting of staff members of the Jan Kazimierz University in Lvov on 5 June 1920, and then published it as a brief report in the same year in the fifth volume of *Ruch Filozoficzny*. The more detailed works on Aristotle were published much later. First of all we must include among them Łukasiewicz's famous paper 'Philosophical remarks on many-valued systems of propositional logic' (published originally in German in 1930), as well as his famous works from the 1950s on Aristotle's modal logic.

I shall begin my paper with a presentation of Łukasiewicz's remarks made in both the earlier works mentioned above. Among different concepts of possibility Łukasiewicz refers to Aristotle's concept of 'pure possibility' (*δύναται*), according to which *only what does not have to happen can happen*, and in the interpretation of that concept defends Aristotle's indeterministic position against the deterministic view of Chrysop and the Stoics. Following Aristotle, in 1920 he placed the notion of 'possibility' in modal sentences having the form 'S can be P', 'S can be non P' and so on, and in 1930 he linked modal items with propositional variables (without having any interest in the inner structure of a sentence), as for instance, M$p$, MN$p$ (or NM$p$, NMN$p$).[1] Further he said



*K. Kijania-Placek and J. Woleński (eds.), The Lvov–Warsaw School and Contemporary Philosophy*, 105–111.
© 1998 *Kluwer Academic Publishers.*

that a sentence of the form 'S can be P' is true if 'S can be non-P' is true too, and vice versa.

I will not discuss here the traditional problem of *de re* or *de dicto* modality, as I wish to focus on some points which are more important for Łukasiewicz. Nevertheless it can be seen from a formal point of view that Łukasiewicz could agree on the interpretation of *de re* modality into *de dicto*, and he omitted here a particular question as to whether modal words should be linked with a predicate or only with a connective. From his later writings it is known that he agreed with this traditional view according to which the basic modal words (such as 'necessity' and 'possibility') are linked with the quantified words (such as 'universal' and 'particular').[2]

However, if we write 'S can be P' as M$p$, and 'S can be non P' as MN$p$, then we see that Łukasiewicz's interpretation that M$p$ is true if MN$p$ is true too, and vice versa is given on the basis of his three-valued logic which he finds adequate for studying Aristotle. The illustration for Łukasiewicz's interpretation of Aristotle's notion of 'possibility' in the sense of two-sided possibility is Aristotle's famous example of a cloak (*De Interpr.* IX 19a 9–16):

It is possible (*δυνατόν*) [ . . . ] that this cloak should be cut, but it is also possible that it should not be cut.

Thus, if we denote Aristotle's notion of 'possibility' (*τὸ δυνατόν*) as $\mathrm{M}_1$, then we have its characterization given by the following formula:

(I) $\mathrm{M}_1 p \equiv \mathrm{M}p \wedge \mathrm{M}\neg p$

just in case $p$ takes the third value of 1/2.[3]

For Łukasiewicz such a concept of possibility (treated as two-sided possibility) is deeply rooted in ordinary ways of thinking and speaking. Łukasiewicz's interpretation is in conflict with the principle of bivalence, according to which every proposition (Aristotle's term was *ἀπόφανσις*) is either true or false. This point is very important for Łukasiewicz because according to him the rejection of that principle for sentences about future events allows for the defense of indeterminism. Formula (I) is even treated as the thesis of indeterminism.[4]

Why, however, does Łukasiewicz treat the above notion of possibility as pure possibility if it is a philosophically biased notion? Łukasiewicz's answer is that it is pure possibility because it is not implied by necessity. But on the other hand the connection with necessity is seen in that sense for which what is possible is not necessary (*ἀναγκαῖον*), i.e.

(II) $\mathrm{M}_2 p \equiv \mathrm{M}p \wedge \neg \mathrm{L}p$

If we take into account the basic sense of necessity (*ἀνάγκη*) given by Aristotle in his *Metaphysics* (Δ 1015a 35–1015b 3) where necessary (*ἀναγκαῖον*) is said "if something cannot be otherwise", then just as in standard modal logic we have $\mathrm{L}_1 p \equiv \neg \mathrm{M}\neg p$. Furthermore, here, by the definition of L and the law of double negation, we come back to (I).

However, we can also take into account the sense of Aristotle's possibility regarded as contingency (ἐνδέχεται), according to which what is possible (contingent) is what is not necessary, i.e.

(III) $M_3 p \equiv \neg L p$.

The term 'contingent' is adequate for Aristotle's notion ἐνδεχόμενον. This is how the notion is interpreted in Ross' English translation (whose interpretation Łukasiewicz accepted).

By 'contingent' is understood something which is not necessary, but, being assumed, results in nothing impossible (*De Interpr.* XIII, 22b 36). Thus, in addition to (III) we can list

(IV) $M_4 p \rightarrow \neg I p$

(I for impossible).[5] If we replace '$M_4 p$ by $\neg L p$ and define 'L' by '$\neg M \neg$', applying also the law of double negation, then we get

(V) $M \neg p \rightarrow M p$

which again can characterize two-sided possibility. It is seen that both of Aristotle's notions of 'possibility': δύναται and ἐνδέχεται, have been characterized in the same way.

A stronger sense than that in (V) is given by (III). Let us now try to pay attention briefly to it and the first formula. Let us return to the following formulae:

(I) $M_1 p \equiv M p \wedge M \neg p$
(III) $M_3 p \equiv \neg L p$

The first interpretation given by Łukasiewicz has also been given by Bocheński, Hintikka, Seel and Hartmann on the basis of *De Interpr.* and *Analitica Priora*; the second interpretation has been given by Bocheński, Czeżowski, Maier, Ross and Seidl (on the basis of *Analitica Priora*).

It is worth mentioning here Czeżowski's interpretation in his 1936 essay 'Arystotelesa teoria zdań modalnych' (Aristotle's Theory of Modal Sentences) and comparing it with Łukasiewicz's interpretation. Czeżowski characterizes Aristotle's concept of possibility (ἐνδέχεται) as metaphysical possibility and distinguishes it from the concept of metaphysical necessity. According to Czeżowski the foundation of the distinction between both concepts is a metaphysical distinction between matter and form. What *belongs to form is necessary* for substance, and *what belongs to matter, occurs*, but *is not necessary* for substance. It is only contingent for substance.

If we take Łukasiewicz's interpretation (II) and compare it with Czeżowski's interpretation, then we see that although linguistically we have the same characterization of the notion of 'possibility', philosophically each interpretation stresses different points. While Łukasiewicz joins his interpretation with indeterminism, Czeżowski treats Aristotle's concept of possibility as metaphysical one, and he considers its interpretation on the basis of Aristotle's fundamental composition of being which in its inner structure necessarily consisted of two elements ὕλη and μορφή.

To further compare Łukasiewicz's and Czeżowski's interpretation we can say that Łukasiewicz's interpretation deals with Aristotle's concept of possibility from the existential and external point of view, while Czeżowski's interpretation deals with it from the essential point of view. Although the essential aspect is fundamental for Aristotle's metaphysics, Łukasiewicz's interest is quite different and is rather closer to Aristotle's physics than his metaphysics (if metaphysics is taken in a narrow sense of the account of being).

In Łukasiewicz's analysis of Aristotle, modal sentences about events which occur in the real physical world determined in time deserve special attention. Thus, we can say that in Łukasiewicz's analysis possibility is taken as empirical possibility related to time. But then we should also pose the question of actuality, or the question of the relation between necessity and actuality or at least to inquiry about the relation between necessity, actuality and possibility.

In the case of necessity one of the answers given on the basis of Aristotle's texts is very simple: "whatever occurs necessarily when it occurs, and whatever does not occur, necessarily does not occur when it does not occur" (*De Interpr.* IX 19a, 24–25).[6] Łukasiewicz also expressed it by the traditional formula: *Facta infecta fieri non possunt.* In this sense, at the linguistic level, we have to deal with necessary sentences about events which have already happened. Aristotle's formulation can be written down in the form of the thesis known as one of the theses of determinism, $Pp \rightarrow \neg M \neg p$ which in the history of logic was, however, ascribed to the Megarians in their polemics with Aristotle.[7] Should we, thus, repeat with Hintikka "Aristotle in the snares of determinism"?[8]

While I shall not delve into an exposition of Łukasiewicz's considerations on determinism, we should recall his formulation of the thesis of determinism, the affirmation of which is, Łukasiewicz claims, supported by the affirmation of Aristotle's principle of the excluded middle on the one hand, and on the other, by the affirmation of the Stoic principle of causality.[9] Instead, I want to point out Aristotle's other modal concepts which allow us to find some connections between them and which allow us to defend Aristotle's indeterministic position.

In the same passage which has been mentioned above from *De Interpr.* Aristotle also said that not always what happened, happened of necessity, because to say that whatever occurs necessarily when it occurs is not the same as to say simply (ἁπλῶς) that it is of necessity (*De Interpr.* IX 19a 24–25). We have here two senses of Aristotle's notion 'necessary' given by two Greek terms: ἀναγκαῖον and ἁπλῶς.

Let us denote the class of events described by necessary sentences in the sense given above by the term ἁπλῶς (simple necessity) by $L_1$, and the class of events described by necessary sentences in the sense given above by the term ἀναγκαῖον by $L_2$. The class of actual events (already realized) let us denote by A. Then we can distinguish between the following cases:

$\neg L_1 \subset A$

$L_2 \subset A$

What would be the account of possibility, that is, the account of the events described by sentences qualified as possible ones in the senses given before, according to Łukasiewicz's interpretations? We are returning to the question which is important for the solution of the controversy about determinism, of whether there is a concept of Aristotelian possibility according to which what is possible needs to be actualized. At the linguistic level it is the question of the dependencies between possible sentences (problematic ones) and affirmative sentences (assertoric ones).

Łukasiewicz's analysis allowed us to see that the "future is open" for realization, and from the present moment of time we do not prejudge for something what is possible now, whether it will happen or not. Łukasiewicz shows that in the case of negative sentences or on the objective level inactualization of any event in the future would lead to omitting the concept of impossibility. The logical consequence of such solution would be equivalence between possible sentences and impossible sentences and that consequence is not required.

Łukasiewicz's interpretation also seems to show that what is possible in regard to the future is not yet actual: this is the sense of two-sided possibility. Let us mark the class of events described by modal sentences in that sense of two-sided possibility as $M_1$. Thus we have:

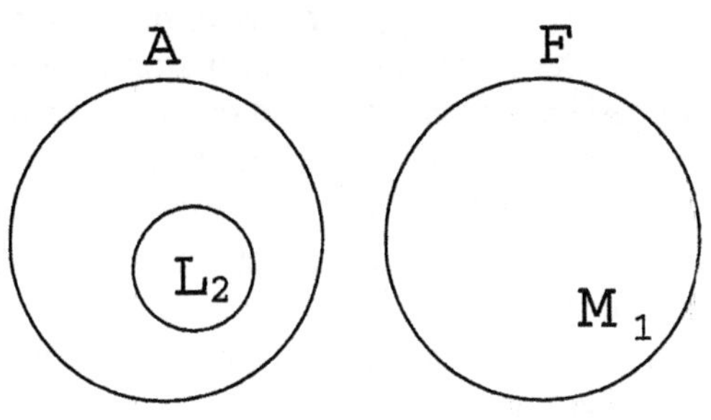


But we also have what is possible and not necessary in the sense of contingency, and from Aristotle's metaphysics we can ascribe this sense to events which have already happened, for example my standing here is not (metaphysically) necessary and this event can be qualified as contingent, although it is still actualized. Thus we have another class of events $M_2$ described by possible sentences where 'possible' has the sense of contingency:

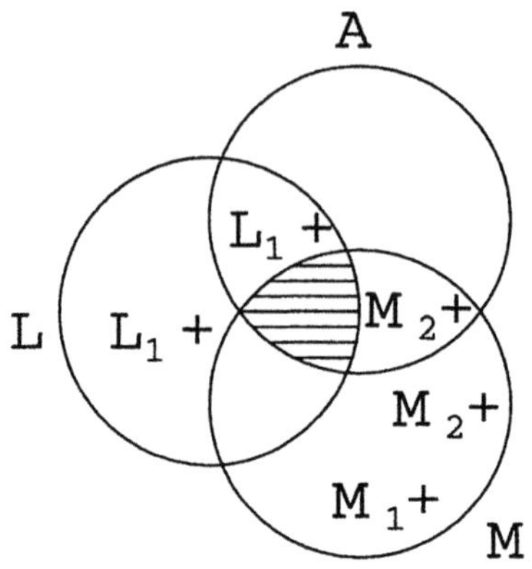


These remarks lead us very close to considerations which are made today on the ground of modal theories. However, I shall not develop this point. Instead I would like to conclude with the following remarks:

1. Łukasiewicz's interpretation allows us to discuss metalogical questions such as the applicability of the fundamental principles of classical logic in Aristotle's theory of modal sentences, such as the principle of bivalence, the principle of excluded middle and the principle of noncontradiction.

2. Łukasiewicz's interpretation allows us to ascribe different senses (although not all different) to Aristotle's notion of 'possibility' and compare this interpretation with other interpretations given by historians of logic and philosophers interested in Aristotle's philosophy.

3. Łukasiewicz defends indeterminism on the ground of his three-valued logic. This question can be developed formally in the discussion of the determinism-indeterminism controversy by looking for formal theories which defend Aristotle's position (here — the possibility of discussion of Aristotle's view with comparison with Megarians and Stoics), and open up the possibility of comparing Łukasiewicz's interpretation with Hintikka's view and others. The development of this issue can be also given on the ground of temporal logic.

4. Łukasiewicz's analysis also allows the development of studies on the ground of contemporary modal theories (such as possible world theories) where we can form the universe of worlds and study the relation between possibility and necessity (included the notion of 'actuality').

These general concluding remarks confirm the significant role of that part of Łukasiewicz's research in logic and philosophy of logic, which can be undertaken today in different perspectives.

*Department of Philosophy, Nicolaus Copernicus University, Toruń*

## NOTES

[1] Łukasiewicz (1930), p. 51.

[2] Łukasiewicz (1939), p. 221.

[3] The formula, the same as other formulae which will be written below,y should not be treated as formal definitions or theses. I only write them to show some of Łukasiewicz's intuitions. The precise definitions and theses are to be found in Łukasiewicz's works, for instance Łukasiewicz (1930).

[4] Kotarbiński (1985), p. 120.

[5] The strongest sense (IV) as $M_4 p \rightarrow \neg Ip$ is given by Becker and Ross for the notion of 'possibility' (ἐνδέχεται and δύναται) in *De Interpr.*.

[6] Aristotle illustrates that by the example of Callistenes.

[7] See also Bocheński (1951), Ch. V.; Rescher (1963), p. 173f.; Hintikka (1977), especially Ch. VI.

[8] Hintikka (1977), Ch. II, p. 31.

[9] J. Łukasiewicz (1919/20), (1922).

## REFERENCES


Bocheński, I.M.: 1951, *Ancient Logic*, North Holland Publishing Company, Amsterdam.

Czeżowski, T.: 1936, 'Arystotelesa teoria zdań modalnych' (Aristotle's Theory of Modal Sentences), *Przegląd Filozoficzny* **39**, pp. 231–241.

Hintikka, J.J., Remes, U., Knuuttila, S.: 1977, *On Modality and Determinism*, North Holland Publishing Company, Amsterdam.

Kotarbiński, T.: 1985, *Z dziejów logiki* (From the History of Logic), Ch. XXI: 'Wielowartościowy rachunek zdań' (Manyvalued Propositional Calculus), PWN, Warsaw.

Łukasiewicz, J.: 1919/1920, 'O pojęciu możliwości', 'O logice trójwartościowej', *Ruch Filozoficzny* **5**, 169b–171b. Eng. trans. as 'On the Notion of Possibility', 'On Three-valued Logic', in S. McCall, (ed.), *Polish Logic 1920–1939*, Clarendon Press, Oxford 1967, pp. 3–18.

Łukasiewicz, J.: 1922, 'O determinizmie', in *Z zagadnień logiki i filozofii* (On Logic and Philosophy), PWN, Warsaw, 1961, pp. 114–126. Eng. trans. in *Jan Łukasiewicz: Selected Works*, North-Holland Publishing Company, Amsterdam, PWN, Warsaw, 1970, pp. 110–128.

Łukasiewicz, J.: 1930, 'Philosophische Bemerkungen zu mehrwertigen Systemen des Aussagenkalküls', *Comptes Rendes des Sciences de la Société des Sciences et de Lettres de Varsovie* **23**, pp. 51–77.

Łukasiewicz, J.: 1939, 'O sylogistyce Arystotelesa' (Aristotle's Syllogistic), Sprawozdanie PAU **44**, pp. 220–226.

Rescher, N.: 1963, 'Aristotle's Theory of Modal Syllogisms and its Interpretation', in M. Bunge, (ed.), *The Critical Approach to Science and Philosophy*, The Free Press of Glencoe, London, pp. 152–177.

# PART II

# LEŚNIEWSKI

ARIANNA BETTI

# *DE VERITATE*: ANOTHER CHAPTER THE BOLZANO-LEŚNIEWSKI CONNECTION*

In '*De Veritate*: Austro-Polish contributions to the theory of truth from Brentano to Tarski' Jan Woleński and Peter M. Simons related an intriguing story of the "Austro-Polish obsession with truth".[1] Woleński and Simons mention the Bohemian philosopher Bernard Bolzano several times, with particular reference to absoluteness and sempiternity of truth in Twardowski and Leśniewski. My aim will be to cast more light on the extent of Bolzano's legacy in Leśniewski's early theory of truth, though leaving aside the foregoing aspects of absoluteness and sempiternity of truth for simple reasons of space. As a premise, from a purely historical perspective the following should be taken into account. First of all, while a general consideration of Bolzano's indirect influence upon the Lvov-Warsaw School (mediated by Twardowski) now seems to be established at least among some eminent scholars,[2] it seems on the contrary very difficult to answer positively about any direct influence of Bolzano upon Leśniewski, namely whether Leśniewski was influenced by Bolzano through reading the *Wissenschaftslehre*. Moreover, it must be stressed that since some elements of this connection may be ascribed to traditional logic, it seems more than natural that these two authors have them in common. Actually in his *Wissenschaftslehre*Bolzano gave us also a powerful fresco of traditional logic, re-elaborated in an extremely refined and original way, while Leśniewski was to describe his Ontology as "a certain kind of modernized 'traditional logic' ".[3]

In the following I wish to point out three issues. First, in the so-called *prelogistic* writings the early Leśniewski defines truth of sentences in such a way that truth conditions are the same — *mutatis mutandis* — as Bolzano's. Secondly, from this point of view the links between the early and the late Leśniewski, in this case between some parts of his early writings and some aspects of Ontology, are closer than they are commonly believed to be. Thirdly, in this perspective it can be shown that some of Bolzano's views come near to Leśniewski's Ontology. In discussing Bolzano's views I shall mostly follow Casari's reading of Bolzano's *Wissenschaftslehre*. The works of Leśniewski with which I am concerned are essentially 'An Attempt at a Proof of the Ontological Principle of Contradiction' (1912) and its Russian version published in *Logical Studies* (1913), the translation-revision by Leśniewski himself of the 'Attempt' and of his first article, 'A Contribution to the Analysis of Existential Propositions' (1911).[4]



*K. Kijania-Placek and J. Woleńsky (eds.),*
*The Lvov–Warsaw School and Contemporary Philosophy*, 115–137.
© 1998 *Kluwer Academic Publishers.*

## 1. OF BOLZANO'S LOGICAL UNIVERSE AND THEORY OF MEANING

For Bolzano, there are objects which have the property of existence and objects which do not.[5] There are *propositions-in-themselves* (*Sätze an sich*), or simply propositions: these are objects which are non-existing (akin to the Stoics' λεκτά)[6] and composed of parts. Parts of propositions which are not propositions are called *ideas* (*Vorstellungen an sich*). I shall call Bolzanian non-existing objects (ideas and propositions) 'lektological objects'. Lektological objects, although non-existing, subsist as a certain something.[7] They are the matter of psychic acts as well as the sense or the meaning in a restricted sense of linguistic expressions and they subsist independently of the latter. An idea may refer to objects. From this point of view it may be non-empty or empty: non-empty ideas, which may be singular (e.g. Socrates) or common (e.g. man), refer to at least one object, while empty ideas do not refer to any object. Examples of empty ideas are [circular polygon], [nothing], [golden mountain].[8] Objects may be subdivided into *qualities* (*Beschaffenheiten*), i.e. any object that can belong to at least one other object, and *pure objects*, objects which are not qualities. Every quality is an object while not every object is a quality.

Bolzano's Universe[9]

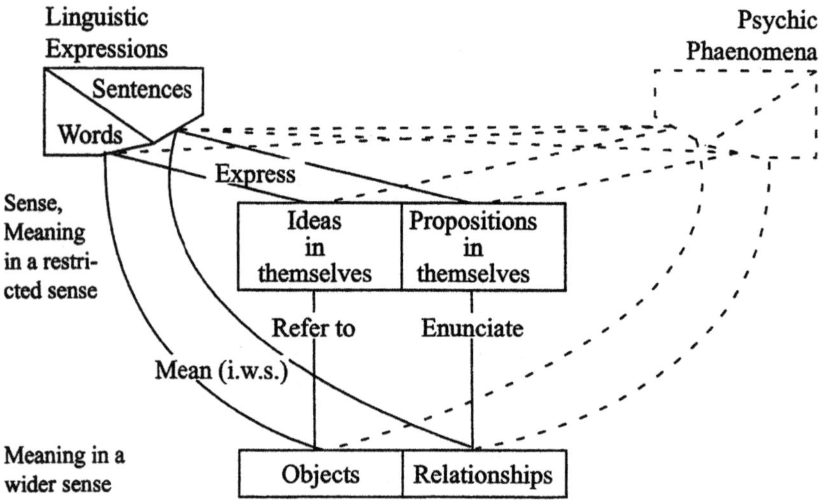


## 2. TRUTH CONDITIONS OF PROPOSITIONS IN BOLZANO

The fundamental structure of propositions is [*a* has (*b*)-ity] where [*a*] is an idea, [(*b*)-ity] is an idea (of quality), and moreover, [has] is an idea, so that a proposition is intended to be an exact compound of these three ideas. For instance [man has mortality] is a proposition, formed by the ideas [man], [has], [mortality]. Bolzano claims that every proposition has this form.[10]

Now, given ideas [*a*], [has], [(*b*)-ity], the proposition [*a* has (*b*)-ity] is true if and only if

B1 [*a*] is non-empty;
B2 [(*b*)-ity] is an idea of quality (and is non-empty);
B3 To any object which [*a*] refers to belongs at least one of the qualities which [(*b*)-ity] refers to.[11]

Thus, in Bolzano truth-bearers are propositions, while truth-makers are the relationships (*Verhältnisse*) that a proposition enunciates, where such a relationship is the belonging of a quality (which [(*b*)-ity] refers to) to an object (which [*a*] refers to).[12]

## 3. OF LEŚNIEWSKI'S EARLY SEMANTIC CONCEPTS[13]

For Leśniewski, linguistic expressions may or may not have *symbolic function*, that is the property to symbolize or not an object. They may be also *connotative*, i.e. they may connote some properties, or, equally, signify something, or not. Connotative expressions are those expressions that can be defined *per genus proximum et differentiam specificam*. Below are some examples given by Leśniewski himself:

| Expressions | Connotative | Non-connotative |
|---|---|---|
| with symbolic function | 'man'<br>'green'<br>'the possessing by every man of the property of mortality' | 'object' |
| without symbolic function | 'round square'<br>'centaur'<br>'the possessing by every man of the property of immortality' | 'for'<br>'abracadabra' |

Expressions may or may not have the property of *symbolic disposition*: an expression which has this property only *seems* to symbolize something, no matter whether it does or not. Sentences (*zdania*) are all those expressions that have the disposition of symbolizing relations of inherence, but only true sentences actually symbolize relations of inherence.

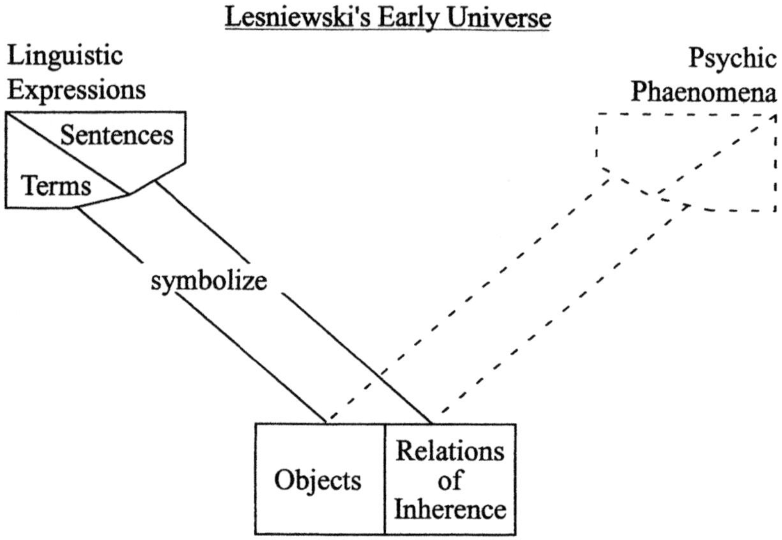


For instance, the sentences

(1) ‘Every man is mortal’
(2) ‘A hippocentaur possesses the property of horseness’

are linguistic expressions with symbolic disposition, that is both have the disposition of symbolizing a relation of inherence, i.e. the possessing by every object denoted (symbolized) by the subject of the properties connoted by the predicate, but (1) has symbolic function, while (2) does not.

## 4. TRUTH CONDITIONS OF SENTENCES IN LEŚNIEWSKI'S EARLY WORKS

The canonical form of a sentence is ‘*a* is *b*’,where *a*, *b* are names and ‘is’ is the copula.[14] Now, the sentence ‘*a* is *b*’ is true if and only if

L1 *a* has symbolic function;
L2 *b* is connotative;
L3 the object(s) denoted by *a* possess(es) the properties connoted by *b*.[15]

Thus, truth-bearers in Leśniewski are sentences, while truth-makers are relations of inherence i.e. the possessing by the object(s) denoted by *a* of the properties connoted by *b*.

## 5. A COMPARISON

### 5.1. *Truth-makers*

Both philosophers share the same view regarding truth-makers, upholding the classical theory of truth as correspondence with reality. Truth-makers, relationships in Bolzano and relations of inherence in Leśniewski, are not only for both existing objects (with the exception for Bolzano of symbolic predications, that is predications about lektological entities),[16] but, most importantly, the same objects. For example, imagine there is a singular idea [*a*] and a singular name *a*. Morscher wrote that in Bolzano the truth-maker of [*a* has (*b*)-ity] is the object which [*a*] refers to which has the properties which [(*b*)-ity] refers to, which here I have above called the belonging of (at least one of) the properties (which [(*b*)-ity] refers to) to the object (which [*a*] refers to).[17] It is easy to observe that in Leśniewski

(3) 'the possessing by the object denoted by *a* of the properties connoted by *b*'

*corresponds* to

(4) 'the object *a* which has the properties connoted by *b*',

which means, in Leśniewski's terminology, that (3) and (4) denote the same object and the same properties.[18]

### 5.2. *Truth-bearers*

The main difference between the Bohemian and the Polish philosopher is substantially a question of the objects that play the role of truth-bearers (and their parts), marking the distance between the world of the logical realist (or logical objectivist) Bolzano and that of the early Leśniewski, indicating the radical nominalist he was to become. Truth-bearers (and their parts) in Bolzano are lektological objects, propositions composed of ideas, while in Leśniewski truth-bearers (and their parts) are linguistic entities, sentences composed of terms. Bolzano allows us to speak of sentences only as *secondary* as far as the former are *primary* bearers of truth.[19] In the latter case, '*a* has (*b*)-ity' means in a wider sense (a) relationship(s).

### 5.3. *Objects*

First we may ask which objects are non-existing objects. For Bolzano existence is a quality like any other quality: 'to have existence' or 'reality' means 'to be present in the causal chain as cause or effect'. Thus, Bolzano's universe is populated with existing and non-existing objects: lektological objects themselves are examples of non-existing objects. In Leśniewski's universe there are no non-existing objects, nor are there contradictory and general ones, the latter being a special kind of the former. Nor are there fictitious ones. There are only existing objects.[20] Moreover, the terms 'object', 'something', 'existing' or 'being' are all synonyms in Leśniewski.[21] Of course this does not hold in Bolzano: 'object' is synonymous with 'thing' and 'something' but not with 'existing', 'being' or 'real'.[22]

Bolzano, in any case, does not claim the round square to be an example of a non-existing object: the round square is not even an object, or rather, if one looks at the matter from a semantic point of view, the idea of the round square, as of any other contradictory or fictitious object, is empty. This point was a crucial one, the watershed between Bolzano and some Brentanians, such as Meinong or Twardowski, Leśniewski's supervisor. Leśniewski was nearer to Bolzano's position: for him the name 'round square', like 'hippocentaur', does not symbolize anything.

In Bolzano the idea-subject of a true proposition — a lektological and therefore non-existing object — may be *symbolic*, that is can refer to another lektological object (non-existing, too). This does not hold in Leśniewski for the very simple reason that, since there are no non-existing objects, the subject of a true sentence (an existing object) can refer only to existing objects.

Second, as regards properties and relations: as already noted, for Bolzano, qualities are objects. Qualities are divided into internal and external: Bolzano calls internal qualities properties (*Eigenschaften*) and external ones relations (*Verhältnisse*), where relations are qualities of a whole.[23] Leśniewski speaks of properties (*cechy*) and relations (*stosunki*), and properties and relations were at that time considered also by him to be special (existing) objects. One can refer to the following passage from 'On the Foundations of Mathematics':

At the time I wrote that passage [of the work 'Critique of the Logical Principle of Excluded Middle', A.B.] I believed that there are in existence in this world so called properties and so called relations, as two special kinds of objects, and I felt no scruples about using the expressions 'properties' and 'relations'. It is a long time since I believed in the existence of objects which are properties, or in the existence of objects which are relations and now nothing induces me to believe in the existence of such objects [ . . . ].[24]

Besides, even if we do not believe Leśniewski, consistently with his earlier views one is driven to assume at least the existence of relations, i.e. we have to admit that these are a special kind of object and moreover existing objects, as suggested by the following passages taken from the 'Attempt':

§9: [ . . . ] the expression 'every man is mortal' [ . . . ] symbolizes the possessing by every man of the property of mortality [ . . . ]
§12: [ . . . ] the proposition 'A hippocentaur possesses the property of horseness' [ . . . ] does not possess a symbolic function because *no object is the possessing, by a hippocentaur, of the property of horseness;* thus *no object is the only one that could be symbolized by the expression 'a hippocentaur possesses the property of horseness'* [ . . . ].[25]

One must here remark that for Leśniewski 'No object is $A$' expresses adequately '$A$ does not exist', so that in the example above he stated in this way that the relation of inherence which is the possessing by a hippocentaur of the property of horseness and which is the sole object that "could be symbolized by the expression 'a hippocentaur possesses the property of horseness' " does not exist.[26] But there is also another important passage to be considered in which Leśniewski writes that 'to connote' is 'to symbolize properties':

Properties $p_1, p_2, p_3, \ldots, p_n$ are *objects* which the expression *w symbolizes* and by which it is defined, and they are said to be connoted by $w$, so that 'man' connotes (given its definition) the properties of mammality, two-leggedness, two-handedness.[27]

I have to stress here again that in his extremely precise terminology, Leśniewski uses the verb 'to symbolize' very carefully and only when he wants to describe the relation between words and objects that really exist in the world. Thus, relations as well as properties are existing objects: first, this can be easily derived from the clear definitions given by Leśniewski and from his strictly consistent use of technical terms like 'symbolic relation' and 'symbolic disposition'; secondly, it follows from the fact that Leśniewski does not allow reference to non-existing entities *à la* Bolzano.[28] Since sentences and terms are existing objects, in case of reference to other sentences and terms the *denotata* of the former are nothing but existing objects as well.[29] The last passage quoted seems to be particularly valuable for its source, as it did not appear in the original Polish version (1912), but was included only in *Logical Studies* (1913). This later translation contains a few major changes, therefore it is worthwhile noting that Leśniewski was convinced of the necessity of the addition. So Bolzano's ontology is much richer than early Leśniewski's, because in the latter's universe the entire lektological sphere is absent. However, both include relations and properties among objects.

### 5.4. *Form of truth-bearers*

Both Bolzano and Leśniewski developed some rules for reducing sentence occurrences to a canonical form.[30] As a general rule in Leśniewski, any sentence can be transformed into a sentence of the form '$a$ is $b$'.[31] If one applies the rule to the Bolzanian scheme

(5) '$a$ has ($b$)-ity',

(5) becomes

(6) '$a$ is having-($b$)-ity' or '$a$ is something-which-has-($b$)-ity'

From Leśniewski's definition of synonymous sentences it stems that (4) is synonymous with

(7) '$a$ is $b$'.

The same holds for negative sentences: any negative sentence can be transformed into a sentence of the form

(8) '*a* is non-*b*'.

In Bolzano (Bolzano's Conjecture) any proposition, even the most complicated, has the form

(9) [*a* has (*b*)-ity]

or

(10) [*a* has lack-of-(*b*)-ity].

Forcing the situation to some extent one can say that for Bolzano any sentence can be reduced *salva significatione* to the one that (to use Casari's words) "optimizes the correspondence between the mereological structure of the expression and the mereological structure of the sense":[32] the sentence

(11) '*a* is *b*'

has the same sense as

(12) '*a* is something which has *b*-ity'

which again means the same as

(13) '*a* has *b*-ity'

which clearly does the job, as it mirrors the proposition [*a* has (*b*)-ity]. With the proviso that the proposition [*a* has (*b*)-ity] is true, the idea [(*b*)-ity] is an idea of quality, an *abstractum*. For Bolzano any *abstractum* [(*b*)-ity] has a corresponding *concretum*: the idea [something which has (*b*)-ity]; for instance the *concretum* of the *abstractum* [mortality] is [something which has mortality].[33] *Abstracta* can be simple, while *concreta* are always complex ideas, i.e. they are always composed[34] (in this case they are composed of the immediate idea-components [something][which][has][mortality]), and the word 'mortal' is only the linguistic expression of [something which has mortality]. *Abstracta* and *concreta* differ in their extension, i.e. the set of the objects which they refer to. In fact *abstracta* extensions have only qualities as members, while this is not true of *concreta*: for instance [virtue] has as referents veracity, charity etc., while [virtuous person] has as referents Socrates, Aristides, etc.[35]

Leśniewski claims that any sentence could be reduced to the canonical form '*a* is *b*'; yet, since truth conditions impose that *b* must be connotative, in order to see if it is so, *b* should be itself defined canonically as 'something which has $(c_1, \ldots, c_n)$-ity and $(d_1, \ldots, d_n)$-ity', where $(c_1, \ldots, c_n)$-ity and $(d_1, \ldots, d_n)$-ity are the properties of the *genus proximum* and the *differentia specifica* of *b* in respect to the *summum genus* Being. Leśniewski thinks that this step allows one to say that *b* connotes the properties $(c_1, \ldots, c_n)$-ity and $(d_1, \ldots, d_n)$-ity. Despite the fact that the canonical form of sentences in Leśniewski's early works appears different from Bolzano's canonical form of propositions, this difference in Leśniewski is completely removed by the fact that he looks at the names $a$ and $b$ in '$a$ is $b$' in an 'asymmetric' way, I would say: $a$ must be denotative but need not be connotative, while $b$ must be connotative, but need not be denotative. Hence, even though Leśniewski seems to treat '$a$ is $b$' sentences in a formally different way from '$a$ has ($b$)-ity' ones, for him they are, in actual fact, synonyms. What is more, L3 states clearly that '$a$ is $b$' is true if and only if $a$ has ($b$)-ity.

However, while the Leśniewskian canonical form of truth-bearers is synonymous with Bolzano's, Bolzanian truth-bearers are not so — so to speak — flexible. Bolzano's Conjecture, even in the weakened form found in the 'Bolzanian' claim "it is possible that not every proposition is basic; in any case, every proposition is equivalent to a basic one"[36] seems to bar the way. I shall come back to this problem later.

### 5.5. *Reference to properties*

As already stated, for both Bolzano and Leśniewski the predicate of a proposition/sentence must be intensionally non-empty as it must refer to at least one property. Bolzano's requirement that the idea-predicate must be non-empty means that there are in his universe ideas of qualities which are empty (e.g. [square circularity]), just as for Leśniewski there are non-connotative names. Yet reference to properties made by Bolzano's ideas of quality and by Leśniewski's connotative names is not identical. For Leśniewski, as previously said, a name is connotative if and only if it is definable *per genus et differentiam*. It should be noted in this respect that definitions in Leśniewski deal with the content of names, and that the content of a name is its intensional meaning, the properties it connotes. From the definition of connotative name it follows that '$A$ is a round square' satisfies L2. In fact, in accordance with his own premises, Leśniewski considers the name 'round square', given its definition as 'something that has the properties $p_1, p_2, \ldots, p_n$, roundness and squareness', connoting among others the property of roundness *and* the property of squareness (and *not* the property of round squareness, one imagines). On the contrary, for Bolzano the proposition [A has round squareness] does not satisfy B2 because [round squareness] is an empty idea of quality. Moreover, Leśniewski claims 'existing' — and its synonyms — to be non-connotative (a peculiar Leśniewskian position), so that 'existing' does not satisfy L2.[37] Thus, while for the early Leśniewski any existential sentence of the form

(14) 'a is existing'

cannot ever be true, for Bolzano the proposition

(15) [a has existence],

i.e. the translation of (14) into Bolzanese, may be true.

Even if all this should be the case, in some sense Leśniewski's position has much in common with another of Bolzano's issues, although all the elements I am going to consider now are also traditional elements of a common logical heritage. Indeed, the concept of content in Leśniewski, as treated right from his first paper on existential statements, is more or less the same as Twardowski's content of a presentation in *On the Content and the Object of Presentations*. In Bolzano the content of an idea is the sum of the parts of which it is composed. Now, Twardowski's content can be considered extremely close to Bolzano's, and Twardowski's components[38] of the content of an idea very close to Bolzano's components of an idea. The issue is, however, much more complicated than it may appear, since the common logical heritage intertwines itself with Bolzano's complicated mereology. The relationship between Twardowski's content and Bolzano's matter of psychic acts/content would require an entire paper to discuss and would go beyond the aim of the present. Therefore, I must terminate by saying that, when speaking of names, Twardowski says that the content of a name is its intensional meaning, i.e. the properties which it connotes.[39] Although Leśniewski uses the concept of content less and less after his first paper, the concept of definition, a keystone of the theory of truth in his early writings, is directly linked with the content of a name as its intensional meaning. Leśniewski's claim that there is no ontological commitment in defining 'round square' as 'something' or even '*existing* possessing the properties $p_1, p_2, \ldots, p_n$, roundness and squareness'[40] was evidently a result of Twardowski's position.

### 5.6. *Truth conditions of sentences/propositions*

As one can easily see, L1–L3 and B1–B3 are substantially identical. A sentence/proposition is true if and only if the subject is non-empty, the predicate refers to at least one property, and to the objects which the subject refers to belong at least one of the properties (B3)/'the properties' (L3) which the predicate refers to. The difference between L3 and B3 is easily explained: the issue is connected with Bolzano's highly refined theory of *composition* and *extension* of ideas, which in Leśniewski is not present to the same extent (for this reason I use 'the properties' in L3 more ambiguously). As Casari points out, B3 is the correct formulation, *contra* Berg's "to any object which *a* refers to belongs the quality which *b* [here: [(*b*)-ity)], A. B.] refers to." Bolzano claims that, despite the fact that in the proposition [Caius has intelligence] the predicate-idea is not singular, for the truth of this proposition it is sufficient for Caius to have at least *one kind* of intelligence.[41] Berg's definition fits well only if [(*b*)-ity] is a singular idea of quality.

### 5.7. *Non-emptiness and truth*

Both in Bolzano and Leśniewski there is a close connection between true propositions/sentences and non-empty ideas/names, as there is for false propositions/sentences and non-empty ideas/names. In Bolzano a proposition is non-empty if and only if its subject is non-empty.[42] Consider the following scheme, taken from Berg:[43]

1. If [*a* has (*b*)-ity] or [*a* has lack-of-(*b*)-ity] are true then [a] is non-empty.
2. If [a] is non-empty then [*a* has (*b*)-ity] and [*a* has lack-of-(*b*)-ity] are non-empty.
3. If [a] is empty then [*a* has (*b*)-ity] and [*a* has lack-of-(*b*)-ity] are empty.

This concept of empty proposition has more or less a counterpart in Leśniewski (although he does not use the term), if we agree to term 'empty sentences' those which have an empty subject and a non-connotative predicate. And with the proviso of putting names in place of ideas and adding a request for connotativity of predicates in 2., the corresponding issues in Leśniewski function exactly as in 1. 2. 3. There are some really interesting features linked with empty propositions/sentences. I am aware that these logical matters are backed by age-old traditions, from Aristotle onwards, but I shall mention them all the same. There is a particular 'historical' reason to do so, at least in the case of Leśniewski: these positions pervade all his work and, being also the basis for his views on absoluteness and sempiternity of truth, they came into play during the debate on determinism and the excluded middle, which took place around 1913 and also involved Łukasiewicz and Kotarbiński. Besides, the particular interest in cultivating and improving the classical heritage seems to bring Leśniewski nearer to Bolzano than any other philosopher of the Austro-Polish tradition. And it should not to be forgotten that Bolzano's and Leśniewski's greatness lay also in the goals they were able to achieve by making the best of these traditional tools. As a general assumption, Leśniewski and Bolzano share the Principle of Bivalence[44] and the Principle of Contradiction. From this point of view, a notable consequence of the theory of empty propositions is that in general for Bolzano, *mutatis mutandis*, a true proposition (supposedly singular) *p* (of the form [*a* has (*b*)-ity]) is not contradictory with [*a* has lack-of-(*b*)-ity], but with

(16) [the proposition *p* has falsity].

Consequently, in Bolzano one has:

(17) The propositions [*p* has truth] and [*p* has falsity] cannot both be true,

where [*p*] is a symbolic idea, and has a proposition as referent. In Leśniewski one can reconstruct, given a sentence *p* of the form '*a* is *b*':

(18) The sentences '*p* is true' and '*p* is false' cannot both be true,

where *p* is the name of a proposition. Consider the following:

(19) 'Pegasus is dead'
(20) 'Pegasus is not dead'.

In Bolzano (19) (20) should be reduced to sentences that mirror the propositions

(21) [Pegasus has death]
(22) [Pegasus has lack-of-death].

(21) and (22) are both empty, hence false. But

(23) [the proposition (21) has falsity],

the contradictory of (21), is true. So, (21) and (23) are not both false.

Empty sentences are the basis of Leśniewski's important early paper 'Is the Class of Classes not Subordinated to Themselves Subordinated to Itself?' (1914). One point to consider here is that Leśniewski never rejected this paper,[45] which was against Russell's paradox and is part of the pre-history of Mereology. The question from which Russell's paradox springs was called by Leśniewski a *pseudoproblemat* (to use Leśniewski's former terminology),[46] because the subject of the sentence

(24) 'The class of classes not subordinated to themselves is subordinated to itself',

i.e. the nominal expression 'the class of classes not subordinated to themselves' is empty.[47] Unlike Bolzano, Leśniewski did not invoke an alternative definition of 'contradictory sentences': his answer to the problems linked with empty sentences and contradiction was to put a restriction on *tertium non datur (tnd)*, i.e. the convention

K: For any non-empty sentence *p*, if *p* is not true, then its singular contradictory is true.[48]

*Tertium non datur* clearly does not apply to empty sentences, because if *a* is empty, 'Every *a* is *b*' and 'A (certain) *a* is non-*b*' are both false. Therefore, in 'The Critique of the Logical Principle of Excluded Middle' (1913) Leśniewski, although accepting what he calls the Ontological *tertium non datur*, declares that he is against the Logical *tnd*. He claims that the latter is not true just because the principle is false in the case of empty sentences. On the contrary, his acceptance of the Ontological version of the principle is consistent with all that has been said above:

(25) (Ontological *tnd*) Every object must either possess or not possess the property (*b*)-ity.
(26) (Logical *tnd*) At least one of two contradictory sentences must be true.[49]

To sum up, the difference here lies in the definition of contradictory sentences: Leśniewski's traditional understanding of them drives him to refuse the Logical *tnd*, while Bolzano can save it thanks to the definition of contradictory proposition he accepts; Leśniewski, however, would have certainly refused such a position, since (21) and (23) have different subjects.[50]

## 6. TRUTH CONDITIONS OF SENTENCES IN ONTOLOGY AND LEŚNIEWSKI'S EARLY WRITINGS

As is known, the 1920 sole axiom of Ontology is not a definition of $\varepsilon$, but states truth conditions of sentences of the form $A \varepsilon b$:

[AO] $\forall Ab(A\varepsilon b \leftrightarrow \exists x(x\varepsilon A) \wedge \forall xy(x\varepsilon A \wedge y\varepsilon A \rightarrow x\varepsilon y) \wedge \forall x(x\varepsilon A \rightarrow x\varepsilon b))$

According to Kotarbiński's reading, quoted also by Leśniewski,[51] one can read:
For any $A$ and $b$, $A \varepsilon b$ is true if and only if

1. for some $x$, $x$ is $A$
2. for any $x$ and $y$, if $x$ is $A$ and $y$ is $A$, then $x$ is $y$
3. for any $x$, if $x$ is $A$, then $x$ is $b$

Consider now the following definitions:

| | |
|---|---|
| *ex* $(A) \leftrightarrow \exists x(x \varepsilon A)$ | (The objects denoted by) $A$ exist |
| *sol* $(A) \leftrightarrow \forall xy(x \varepsilon A \wedge y \varepsilon A \rightarrow x \varepsilon y)$ | (The object denoted by) $A$ is one and only one |
| $Aab \leftrightarrow \forall x(x \varepsilon A \rightarrow x \varepsilon b)$ | Any (object denoted by) $A$ is (denoted by) $b$. |

Hence [AO] can be rewritten as

$$\forall Ab(A \varepsilon b \leftrightarrow ex(A) \wedge sol(A) \wedge Aab).$$

With respect to truth conditions of '$a$ is $b$' in his early works, Leśniewski introduces two new elements into the truth conditions of $A \varepsilon b$ in the formal system of Ontology: he restricts $a$ to singular subjects — *sol* (A) — and gives up intensional concepts, i.e. connotativity conceived as definability of terms, which L2 dealt with. The truth conditions of Ontology follow from conquering a completely extensional perspective, by extending such perspective to predicates: while $ex$ $(A)$ corresponds to L1 right from Leśniewski's very first article, where he claims that the sole adequate linguistic formulation of '$A$ exists' is 'A (certain) object $x$ is $A$', *Aab* corresponds to L3 providing an extensional understanding of properties of '$a$ is $b$'. It was Kotarbiński who stressed how near the truth conditions of '$a$ is $b$' in Leśniewski's early writings were to the singular sentence '$A \varepsilon b$' of Ontology:

From the old, pre-logistic texts, he retained, as results that followed the reading of the initial parts of Mill's *Logic*, the definition of truth conditions of the simple subject-predicate sentence: this sentence is true if and only if the object denoted by the subject possesses the properties connoted by the predicate. The revolutionary turn consisted in abandoning properties and relations of connoting [ . . . ] [therefore] we obtain the following formula: a subject-predicate sentence is true if and only if the object denoted by the subject is the object denoted by the predicate. It is easy to note that adding a restriction to singular subjects and inscribing everything with the help of symbolic logic, we obtain the statement which Leśniewski accepted as the sole axiom of ontology:

$a \,\varepsilon\, b \leftrightarrow \forall x(x \,\varepsilon\, a \rightarrow x \,\varepsilon\, b) \wedge \exists x(x \,\varepsilon\, a) \wedge \forall x, y(x \,\varepsilon\, a \wedge y \,\varepsilon\, a \rightarrow x \,\varepsilon\, y)$.[52]

But Kotarbiński misses the point. In Mill's view the sentence 'A centaur is a fiction of poets' is *true*,[53] while in Leśniewski it is *false*, for L1. So, if we start from Mill, in addition to dropping connotation and adding a restriction to singular subjects, we should also require a condition like L1, i.e. the existence of the object denoted by $a$.

Among the many important logical materials of Ontology that appeared even in Leśniewski's earlier writings, I shall quote at least the general rule cited previously, according to which the sentence 'Socrates runs' becomes 'Socrates is running', just as 'Socrates has mortality' becomes 'Socrates is having mortality': Leśniewski will redeem it in Ontology as the definition

D1: $\forall A\phi\,(A \,\varepsilon\, A \wedge \phi(A) \leftrightarrow A \,\varepsilon\, trm < \phi >)$

## 7. BOLZANO AND LEŚNIEWSKI'S ONTOLOGY

Of the features of Bolzano's logic which can be placed near Ontology I should mention at least the concept of extension of ideas, incidentally quoted previously, together with Bolzano's remarks about truth and well-formedness of propositions.

A well-known feature of Leśniewski's Ontology is that the *truth* of $A \,\varepsilon\, b$, a matter of semantics, requires that $A$ should be replaced by a singular non-empty name and $b$ by a singular or common non-empty name; nevertheless, as regards its *well-formedness*, a matter of syntax, $A \,\varepsilon\, b$ admits singular, common or non-empty names as substitutes for $A$ and $b$.[54] Now, with the proviso of giving due weight to the fact that this was also a traditional issue, it should be remarked that Bolzanian ideas, as regards their *denotata*, may be singular, common, or empty, just as Leśniewskian names may be. On the basis of this distinction, Bolzano speaks of the extension of ideas as the set of the objects to which an idea refers. So, one can borrow the famous graphic presentation of Lejewski's Ontological Table for Leśniewski's Ontology, and apply it to Bolzanian ideas:

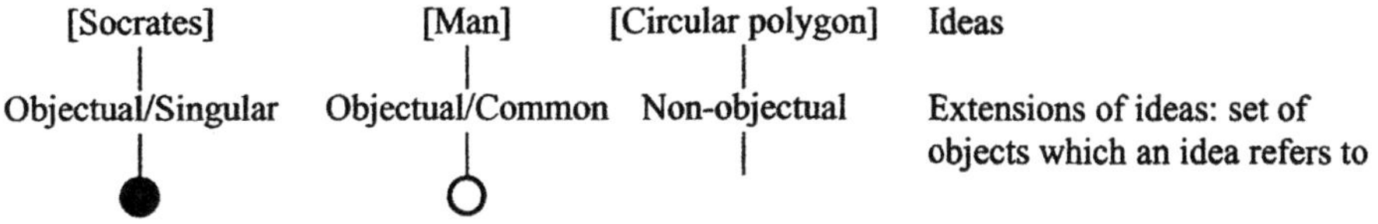


We must be careful here, because Lejewski's table might suggest that a name *denotes* its extension, while in Leśniewski a name does not *denote* its extension, but rather *has* it.[55] The same happens with ideas in Bolzano: circles might be read as graphic presentations of both the extension of an idea and what Bolzano called the *scope* of the extension of an idea. If the extension of an idea is the collection of individual referents a non-empty idea refers to distributively — i.e. it refers to each of them — its scope is the cardinality of that collection (note that an empty idea has no extension). Thus one could read the presence of a circle as the presence of the extension of a given idea, the black colour as indicating a scope = 1 and the white colour a scope > 1 of that extension. Of course the same applies to Leśniewski's names.

With what has been said about truth and well-formedness in Leśniewski in mind, consider now the following passage from Bolzano:

> Moreover, if the proposition in question is to be true, then it should be possible to prove that the two parts *A* and *b* must be ideas, indeed idea with referents, and that *b*, in particular, must represent an attribute. But if it is not required that the proposition '*A* has *b*' is a truth (and at present we merely speak about attributes which hold for all propositions, even false ones) then I do not see why *A* and *b* must always be ideas, or even referring ideas, and why *b* must be an attributive idea. Can we not call every compound of the form '*A* has *b*' a proposition, no matter whether the signs 'A' and 'b' designate mere ideas, or ideas of a certain sort, or even entire propositions?[ . . . ] *we may restrict the concept of a proposition to those cases where* A *and* b *are mere ideas, but we should in no case demand that these ideas should both have referents, and that the latter should be an attributive idea.*[56]

If, again, one replaces here 'ideas' by 'names', the similarity to Leśniewski becomes astonishing (even in the orthographic use of the letters *A*, *b*, although in Bolzano the former are used for *concreta* such as [man] and the latter for *abstracta* such as [mortality]). Now, coming back to Leśniewski's axiom, one should remember the frequency with which the issue of the 'intuitive' reading of $\varepsilon$ in $A \varepsilon b$ has been and is still being discussed. This is not a *façon de parler*: readings of this kind have been suggested over the years by the fact that many times Leśniewski himself was requested to explain the sense of $\varepsilon$ in $A \varepsilon b$:

> It often happened that someone with whom I spoke did not realize the sense in which I was using the sign '$\varepsilon$' in my ontology, and experienced interpretational difficulties in attempting to relate semantically the given sign to one or other signs handed down by the tradition of 'mathematical logic' and 'the theory of sets'. In the theoretical troubles in which, due to this state of affairs, I found myself, I shared the misfortune of all who are forced by circumstances to convey in 'their own words' the sense of the various primitive terms of deductive theories which they were constructing.[57]

Among the many readings that have been given for [AO], one can consider Jan Woleński's

By saying *Socrates est homo* he meant that Socrates is one of the human beings or *that the name 'Socrates' and the name 'man' are both applicable to a certain object.*[58]

And compare now the following passage taken from Bolzano:

Now if [ ... ] [a proposition, A.B.] is true (as for example the proposition: God has omniscience), then of course we can say that the object represented by $X$ (by the concept: God) is represented at the same time — not to be sure by the *abstractum* $b$ itself — but by its *concretum*, namely by the idea of 'something that has the property $b$'. But if we have a false proposition before us (for example the proposition: Man has omniscience), then it is not true at all that every object contained under $X$ (namely a man) is at the same time an object which the idea, 'something which has $b$ (something which has omniscience)', also includes.[59]

Here '$x$ represents the object $y$' means that $y$ is the referent of $x$. The last quotation, about the canonical form of propositions [$a$ has ($b$)-ity] and the *concretum* [something that has ($b$)-ity], immediately reminds one of the problem of interchangeability of truth-bearers in Bolzano and Leśniewski. Although Leśniewski's Ontology is based on singular sentences of the form $A \varepsilon b$, by means of the definitions D1 and the following D2 (where right and left sides of each formula are said to be equivalent),

D2: $\forall Ab\,(\{b\}(A) \leftrightarrow A \varepsilon b)$

one can obtain respectively non-abstract nominal expressions (n/(s/n)) from predicates (s/n) and predicates ((s/n)/n) from names (n).[60] While there is not a real difference in this respect between the early Leśniewski — trained in the schools of Marty and Husserl, in which the concept of *syncategorematicity* had an important place — and Leśniewski's later consideration of these transformations, Bolzano's claim that any proposition *has* the form [$a$ has ($b$)-ity] works, together with his proofs, on different levels.[61] Bolzano did not deal satisfactorily with the concept of syncategorematicity, which is a basic idea in Frege, for example.[62] The copula [has] may be considered an unsaturated part not 'closed' since it connects two other ideas, [$a$] and [($b$)-ity], but Bolzano, unlike Frege or the Stoics, did not include in the lektological sphere verb ideas like 'runs', 'eats', 'reads', etc. so that in his universe [Franz has runability] may be present, but [Franz runs] is not, and these forms are therefore not equivalent. For this reason one is forced to have doubts about the possibility of considering Bolzano's form of truth-bearers as completely interchangeable with the Leśniewskian one. It is true that Bolzano's operation of *concretization* illustrated briefly above guarantees that in his universe, unlike verb ideas, there are at least as many concrete ideas [something which has ($b$)-ity] as there are abstract ones [($b$)-ity] to which they correspond, and that Bolzano often seems to treat 'Leśniewskian' propositions like standard propositions; nevertheless, I think that Bolzano's Conjecture and his view of the notion of quality as a primitive concept[63] does not allow us to consider the Bolzanian and the Leśniewskian form of truth-bearers as being equal. I judge Bolzano's passages involving *concreta* to be explicit in this matter.[64]

Coming back to $A \varepsilon b$, one can, of course, also read it in other ways, for example

A sentence of the form '*A* est *b*' is true if and only if 'A' is a non-empty singular term and the object denoted by 'A' is one of the one or more denoted by 'b' (i.e. is one of the *b*s),[65]

and even in this case find a nice correspondence in Bolzano, although he does not consider exclusively singular sentences:

It is similarly obvious, given this assumption [B2, A.B.] that the *concretum* which belongs to the predicate idea of a true proposition must always be a referring idea which stands to the subject in the relation of inclusion (§95), hence which must be either equivalent to or higher than it. For if the proposition '*A* has *b*' is to be true, then the idea *B* or 'something which has *b*', must include every object that falls under the idea *A*.[66]

## 8. A HISTORICAL POSTSCRIPT

After having partially explored the elements of this truth connection, a few words should be spent on the question of Bolzano's 'purely historical' influence upon Leśniewski. As I said at the beginning of this paper, a direct influence seems to be impossible to ascertain, or at least it is impossible for me to answer such a question. In any case, one should consider at least the following facts. Bolzano's ideas were discussed in the Lvov-Warsaw School. Leśniewski took his doctorate under Twardowski in Lvov, and, ever since secondary school, Twardowski — Brentano's pupil — had been under the influence of Bolzano's *Herzensjunge* Robert Zimmermann, who was later his supervisor.[67] Bolzano was one of the interlocutors of Twardowski's masterpiece, *On the Content and the Object of Presentations*. Apart from being himself a logical objectivist, supporting Bolzano's tetrahedron, he very often quotes Bolzano's name, especially in treating "the so-called objectless presentations," i.e. Bolzanian empty ideas. Indeed, he investigates nearly all the concepts of part I 'On ideas in themselves' of the *Wissenschaftslehre*. Moreover, for a short period Leśniewski was fascinated by Marty, another pupil of Brentano's. Marty quotes Bolzano on truths-in-themselves, even if he shows an incorrect comprehension of Bolzano's views on the subject.[68] It is a hypothesis of mine that Leśniewski's giving up Marty's views (for instance *judgement-contents* as truth-makers) can be dated back to 1911.[69] In 1911 (the 'Attempt' was written in June 1911) Leśniewski elaborated for the first time his theory of truth in which truth-makers are relations of inherence rather than special entities like Marty's judgement-contents.[70] And it was in 1911 that Twardowski's *Actions and Products* had appeared, where the theory of meaning was genuinely Bolzanian in certain respects.[71] A pure coincidence? Moreover, in this Twardowski spoke of Bolzano in such enthusiastic terms, that in consequence Leśniewski was likely to have developed some interest in Bolzano's views. Of course this does not mean that Leśniewski had really read Bolzano. But that Leśniewski did not quote Bolzano's name until 1927 does not mean that he was totally unaware of his views.[72] Initially he had not read even Russell directly, although he had started to take Russell's antinomy into consideration at least from 1911. In the 1914 article which was entirely dedicated to Russell's antinomy, Leśniewski said clearly that the source of his information about Russell was Łukasiewicz (and also, I claim, Chwistek).[73] The 'Attempt' has two fundamental issues: the elaboration of semantic concepts on one hand and a subtle inquiry into the general theory of objects on the other. We should not forget that the paper concerned contradictory objects and was a reply to Łukasiewicz, who was at that time profoundly under the influence of Meinong.[74] Contradictory objects are the protagonists of the already mentioned chapter of Twardowski against Bolzano and his objectless presentations. The famous Leśniewskian critique against Twardowski's general objects was included only a little later in 'The Critique of the Logical Principle of the Excluded Middle' (1913): it appeared for the first time in *Logical Studies* (1913), where it occupies its most natural place. The issue is treated starting with the definition of metaphysics as a general theory of objects and as a *prote philosophia* in Aristotle's sense. Since for him general objects are contradictory ones, by proving the Ontological Principle of Contradiction Leśniewski proves that they do not exist, either. Bolzano's positions in terms of pure ontology, although not completely acceptable to the (still moderate nominalist) early Leśniewski, will probably have seemed to him more acceptable than Twardowski's, for instance, since the latter had been much freer than Bolzano as regards pure ontology, to say nothing of Meinong and the early Łukasiewicz.[75] The stream later followed by Meinong was originated in fact by Twardowski's original synthesis of Bolzanian *humus* and Brentano's basic idea of intentionality. Meinong's doctrine of the independence of being from being-so has its roots here, close to Twardowski's conception of object.[76]

*Institute of Philosophy, Florence*

## NOTES

* I am grateful to Ettore Casari, Kevin Mulligan and Peter Simons for helpful comments on an earlier version of this paper.

[1] Woleński-Simons (1988), p. 391.

[2] Cf. Simons (1992), Cap. 2, *Bolzano, Tarski and the Limits of Logic*, particularly p. 14 and n. 15.

[3] Leśniewski (1927/31), *Intr.*, p. 166 (Eng. trans. in Leśniewski (1991), p. 176).

[4] *Logical Studies* are not included in Leśniewski's *Collected Works* and not available in any West European language.

[5] I shall consider briefly only those elements which are essential for the discussion. See Casari's papers in the bibliography for a more exhaustive picture of Bolzano's views.

[6] A λεκτόν is "what subsists according to a rational presentation (φαντασία λογική)", Sext. Emp., *Adv. Math.*, VIII, 70.

[7] Cf. Bolzano WL §48 (Bolzano (1972), p. 62).

[8] Following Morscher (1972), (1990) I designate lektological objects with the help of square brackets, to distinguish them from linguistic expressions. With *a*, *b* I designate objects, with *(b)-ity* a quality.

[9] From Casari (1990/91), with changes and omissions: linguistic and lektological spheres are not properly related this way. The 'phantasmatic' part of the sketch can be neglected for the present purpose.

[10] The issue is connected with the so-called 'Bolzano's Conjecture', see Casari (1992), p. 75.

[11] Cf. Casari (1992), pp. 73-75. Bolzano WL §28, §131, §196 (for B2, cf. for instance Bolzano WL §80).

[12] Cf. most of all Casari's (1992) dense Appendix about the 'still open' semantic value of propositions, in which Casari uses a later terminology, introducing the ontological notions of *fact* and *situation*.

[13] Cf. Leśniewski (1912), pp. 212 ff. (Eng. trans. p. 31 ff.).

[14] Cf. for instance Leśniewski (1911a), p. 330 (Eng. trans. p. 3) and (1912), p. 203 (Eng. trans. p. 21).

[15] Cf. Leśniewski (1912), pp. 212-222 (Eng. trans. 31-42), especially *Uwaga II* to §18. (*Remark 2*, pp. 40-42).

[16] "If we say that a truth states something as it really is this phrase must also be taken in a figurative sense. This is so, because not all truths state something that really is (i.e. has existence). This holds especially for those truths that deal with objects that do not themselves have reality. Thus, the proposition 'A truth is not something that exists' does certainly not state something that exists, and yet it is a truth." Bolzano (1972), p. 34 (Bolzano WL §27). 'Truth' stands for 'truth-in-itself', a true proposition in itself. See also p. 120.

[17] Cf. Morscher (1972), p. 77.

[18] Cf. Leśniewski (1912), pp. 207-209 (Eng. trans. pp. 26-7).

[19] So it should be remarked that the truth of a sentence in any case depends on the truth of the proposition it expresses: Bolzano always grants a privilege to the relationships between lektological sphere and the other spheres, not to one side or to the other of his tetrahedron (see p. 116 above).

[20] Cf. Leśniewski (1994b), p. 139 ff.

[21] Cf. Leśniewski (1911a), p. 332 (Eng. trans. p. 4) and (1912), pp. 208-209 (Eng. trans. pp. 27-28).

[22] Cf. Bolzano WL §142 (Bolzano (1972), p. 189) and Bolzano WL §48 (Bolzano (1972), p. 62). Cf. also Morscher (1972), p. 71.

[23] Cf. Bolzano WL §80 (Bolzano (1972), p. 120).

[24] Leśniewski (1927/31), Cap. II, p. 183 n. 1 (Eng. trans. p. 198 n. 6., reproduced here with slight changes).

[25] Leśniewski (1912), §9 and §12, pp. 214-215 (Eng. trans. p. 34), my emphasis.

[26] Leśniewski (1912), §12, p. 215 (Eng. trans. p. 34).

[27] Leśniewski (1994b), *Remark II*, p. 143, my emphasis.

[28] See p. 120 above.

[29] This is cleared especially in Leśniewski (1913b).

[30] Cf. Bolzano WL §127 (Bolzano (1972), p. 173-179).

[31] I must warn the reader that Leśniewski only supposes this reduction to be a general rule: he actually elaborates only three special cases of such a reduction, cf. Leśniewski (1911a), p. 331 (Eng. trans. p. 3) and (1912), pp. 204-205 *Uwaga III* to §2 and pp. 207-209 *Uwaga I* to §3 (Eng. trans. pp. 22-24 and pp. 25-28).
[32] Casari (1992), p. 72.
[33] Cf. Bolzano WL §60, (Bolzano (1973), p. 92).
[34] Cf. Bolzano WL §58 (Bolzano (1972), p. 73).
[35] Cf. Bolzano WL §60 (Bolzano (1972), p. 76).
[36] Casari (1992), p. 77.
[37] Cf. Leśniewski (1994a), p. 132 n. 107.
[38] Cf. Twardowski (1894) p. 32, §6 (Eng. trans. p. 29).
[39] Cf. Twardowski (1894) p. 11, §3 (Eng. trans. p. 9).
[40] Leśniewski (1911a), p. 340 (Eng. trans. pp. 13-14), my emphasis.
[41] Cf. Bolzano WL §131 (Bolzano (1972), p. 180).
[42] Cf. Bolzano WL §146 (Bolzano (1972), p. 193).
[43] Cf. Berg (1962), p. 53.
[44] Cf. Bolzano WL §147 (Bolzano (1972), p. 193).
[45] Cf. Leśniewski (1927/31), Ch. II, p. 182 and nn. 1-5, p. 183 (Eng. trans. pp. 197-198).
[46] Cf. Leśniewski (1911b).
[47] Cf. Leśniewski (1914), p. 74 (Eng. trans. p. 127).
[48] Cf. Leśniewski (1912), p. 216, §16 (Eng. trans. p. 36).
[49] Cf. Leśniewski (1913b), for instance p. 340, §6 (Eng. trans. p. 73).
[50] Cf. Leśniewski (1913a), p. 525 (Eng. trans. p. 112).
[51] Cf. Leśniewski (1927/31), Cap. IX, p. 164 (Eng. trans. p. 376).
[52] Kotarbiński (1966), p. 160.
[53] See Mill (1843), p. 79 and 146; besides, see in particular his polemic with Brentano on the meaning of 'is', for example in Brentano (1874), part two, Ch. VII, pp. 266-306 (Eng. trans. pp. 62-70).
[54] Cf. Woleński (1986), p. 174.
[55] Cf. Küng-Canty (1970), p. 178.
[56] Bolzano (1972), p. 178 (Bolzano WL §127, 7), my emphasis. The following lines are no less interesting. As one can see, Bolzano terms *b* what I have chosen to indicate as [(*b*)-ity] in order to clarify his position in comparison to Leśniewski's.
[57] Cf. Leśniewski (1927/31), p. 163 (Eng. trans. p. 375) with slight changes in the translation.
[58] Woleński (1989), p. 154, my emphasis.
[59] Bolzano WL §66 (Bolzano (1973), p. 103).
[60] Cf. Henry (1962), p. 195.
[61] Cf. Bolzano WL §127, where he says, for instance, that all propositions have the [*A* has (*b*)-ity] form "although this is often obscured by their linguistic form", Bolzano (1972), p. 173.
[62] See Casari (1989), p. 337.
[63] See Casari (1996).
[64] Cf. Bolzano WL §196 (Bolzano (1972), p. 268) and Bolzano WL Note II to §148 (Bolzano (1972), p. 200), and pp. 122–123 above.
[65] Woleński-Simons (1988), p. 407.
[66] Bolzano WL §196 (Bolzano (1972), p. 268).
[67] Cf. Smith (1988), pp. 313-314: Zimmermann's *Philosophische Propädeutik* was indeed a *résumé* of Bolzano's logic.
[68] Cf. Morscher (1990), p. 189.
[69] Cf. Betti (1996a).

[70] Cf. Morscher (1990), p. 192.

[71] Cf. Twardowski (1911).

[72] Leśniewski (1927/31), Ch. III, p. 196 n. 1.

[73] Cf. Chwistek (1912).

[74] Cf. Betti (1996b).

[75] Cf. Casari (1986).

[76] Cf. Smith (1988), p. 315.

## REFERENCES


Berg, J.: 1962, *Bolzano's Logic*, Almqvist & Wiksell, Stockholm.

Betti, A.: 1996a, 'Il rasoio di Leśniewski' (Leśniewski's Razor), to appear in *Rivista di Filosofia*.

Betti, A.: 1996b, 'Łukasiewicz and Leśniewski on Contradiction', to appear in the volume of proceedings of the conference *Łukasiewicz in Dublin* (Dublin, 7–9 July 1996).

Bolzano, B.: 1972, *Theory of Science*, University of California Press, Berkeley/Los Angeles.

Bolzano, B.: 1973, *Theory of Science*, D. Reidel Publishing Company, Dordrecht (partial English translation based on one of Bolzano's own copies of *Wissenschaftslehre* with the author's corrections).

Bolzano, B.: 1837, WL *Wissenschaftslehre* in *Bernard Bolzano-Gesaumtausgabe*, I/II, 1–9, Frommann, Stuttgart/Bad Cannstadt, 1985–1994.

Brentano, F.: 1874, *Psychologie vom Empirischen Standpunkt*, Duncker & Humblot, Leipzig; page numbers refer to the 1924 edition; partial English translation in R. Chisholm (ed.), *Realism and the background of Phenomenology*, Atascadero, 1960.

Casari, E.: 1985, 'L'universo logico bolzaniano' (The Bolzanian logical universe), *Rivista di Filosofia* **LXXVI, 3**, pp. 339–366.

Casari, E.: 1986, 'I sistemi logici di Leśniewski' (The Logical Systems of Leśniewski), unpublished manuscript.

Casari, E.: 1989, 'Una fonte dimenticata? La teoria bolzaniana del significato' (A forgotten source? The Bolzanian theory of meaning), *Rivista di Filosofia* **LXXX, 3**, pp. 319–349.

Casari, E.: 1990/91, *L'oggettivismo logico da Bolzano a Husserl* (Logical Objectivism from Bolzano to Husserl), notes of the seminar 1990–91, Department of Philosophy, University of Florence (unpublished manuscript).

Casari, E.: 1992, 'An Interpretation of some ontological and semantic notions in Bolzano's Logic', in *Bolzano's Wissenschaftslehre 1837–1987*, International Workshop (Florence, 16–19 September 1987), Olschki, Florence, pp. 55–105.

Casari, E.: 1996, 'Some remarks on Bolzano's notion of a quality', to appear in *Proceedings of Bolzano's Congress*, Salzburg 1991, Kluwer Academic Publishers, Dordrecht.

Chwistek, L.: 1912, 'Zasada sprzeczności w świetle nowszych badań Bertranda Russella' (The Principle of Contradiction in the Light of Bertrand Russell's More Recent Inquiries), *Rozprawy Akademii Umiejętności, Wydział Historyczno-Filozoficzny* **2**, series 30, Krakow, pp. 270–334.

Henry, D.P.: 1962, 'An Anselmian Regress', *Notre Dame Journal of Formal Logic* **III, 3**, pp. 193–198.

Kotarbiński, T.: 1966, 'Garstka wspomnień o Stanisławie Leśniewskim' (Memories of Stanisław Leśniewski), *Ruch Filozoficzny* **24, 3–4**, pp. 155–163.

Küng, G., Canty, J.T.: 1970, 'Substitutional Quantifiers and Leśniewskian Quantifiers', *Theoria* **36**, pp. 165–182.

Leśniewski, S.: 1911a, 'Przyczynek do analizy zdań egzystencjalnych', *Przegląd Filozoficzny* **14**, pp. 329-345; Eng. trans. as 'A Contribution to the Analysis of Existential Propositions' in Leśniewski (1991), vol. I, pp. 1–19.

Leśniewski, S.: 1911b, 'Problemat istnienia w oświetleniu norm gramatycznych' (The Problem of Existence from the Point of View of Grammatical Norms), conference delivered at the Polish Psychological Society on the 12th of May 1911, *Ruch Filozoficzny* **1/6**, p. 133.

Leśniewski, S.: 1912, 'Próba dowodu ontologicznej zasady sprzeczności', *Przegląd Filozoficzny* **15**, pp. 202–226; Eng. trans. as 'An Attempt at a Proof of the Ontological Principle of Contradiction', in Leśniewski (1991), vol. I, pp. 20–46.

Leśniewski, S.: 1913a, 'Czy prawda jest tylko wieczna czy też i wieczna i odwieczna?', *Nowe Tory* **VIII, 10**, pp. 493–528. Eng. trans. as 'Is all Truth only True Eternally or is it also True without a Beginning' in Leśniewski (1991), vol. I, pp. 86–114.

Leśniewski, S.: 1913b, 'Krytyka logicznej zasady wyłączonego środka' *Przegląd Filozoficzny* **16**, pp. 315–352; Eng. trans. as 'The Critique of the Logical Principle of the Excluded Middle' in Leśniewski (1991), vol. I, pp. 47–85.

Leśniewski, S.: 1914, 'Czy klasa klas niepodporządkowanych sobie jest podporządkowana sobie?', *Przegląd Filozoficzny* **17**, pp. 63–75; Eng. trans. 'Is the Class of Classes not Subordinated to Themselves Subordinated to Itself?', in Leśniewski (1991), vol. I, pp. 115–128.

Leśniewski, S.: 1927/31, 'O Podstawach Matematyki', Ch. I–III, *Przegląd Filozoficzny* **30**, 1927, pp. 164–206, Ch. IV, *Przegląd Filozoficzny* **31**, 1928, pp. 261–291, Ch. V–IX, *Przegląd Filozoficzny* **32**, 1929, pp. 60–101, Ch. VI–IX, *Przegląd Filozoficzny* **33**, 1930, pp. 77–105, Ch. X–XI, *Przegląd Filozoficzny* **34**, 1931, pp. 142–170; Eng. trans. as 'On the Foundations of Mathematics' in Leśniewski (1991), vol. I, pp. 181–382.

Leśniewski, S.: 1991, *Collected Works*, 2 vol., ed. by S. Surma, J. Srzednicki, D.I. Barnett, V.F. Rickey, Kluwer Academic Publishers, Dordrecht/PWN, Warsaw.

Leśniewski, S.: 1994a, 'Przyczynek do analizy zdań egzystencjalnych' (A Contribution to the Analysis of Existential Propositions), *Filozofia Nauki* **2**, no. 1(5), pp. 117–134, Polish translation of the second part of 'Logičeskie Rassuždenija' (*Logical Studies*, in Russian), tip. A. Smolinskogo, Zagorodnyj pr. 74, s: IV, 88, St. Petersburg, 1913.

Leśniewski, S.: 1994b, 'Próba dowodu ontologicznej zasady sprzeczności' (An Attempt at a Proof of the Ontological Principle of Contradiction), *Filozofia Nauki* **2**, no. 2(6), pp. 117–147, Polish translation of the first part of 'Logičeskie Rassuždenija' (*Logical Studies*, in Russian), tip. A. Smolinskogo, Zagorodnyj pr. 74, s: IV, 88, St. Petersburg, 1913.

Mill, J. S.: 1843, *A System of Logic ratiocinative and inductive — Being a Connected View of the Principles of Evidence and the Methods of Scientific Investigation*, in *Collected Works of John Stuart Mill* VIII, ed. by J.M. Robson, University of Toronto Press, Routledge & Kegan Paul, London, 1973.

Morscher, E.: 1972, 'Von Bolzano zu Meinong: Zur Geschichte des logischen Realismus', in R. Haller (ed.), *Jenseits von Sein und Nichtsein — Beiträge zur Meinong-Forschung*, Akademische Druck — u. Verlagsanstalt, Graz, pp. 69–102.

Morscher, E.: 1990, 'Judgement-contents', in K. Mulligan (ed.), *Mind, Meaning and Metaphysics — The Philosophy and Theory of Language of Anton Marty*, Kluwer Academic Publishers, Dordrecht, pp. 181–196.

Simons, P.: 1992, *Philosophy and Logic in Central Europe from Bolzano to Tarski*, Kluwer Academic Publishers, Dordrecht.

Smith, B.: 1988, 'Kasimir Twardowski: an Essay on the Borderlines of Ontology, Psychology and Logic' in K. Szaniawski (ed.), *The Vienna Circle and the Lvov-Warsaw School*, Kluwer Academic Publishers, Dordrecht, pp. 313–375.

Twardowski, K.: 1894, *Zur Lehre vom Inhalt und Gegenstand der Vorstellungen — Eine psychologiche Untersuchung*, Vienna; page numbers refer to the unchanged reprint of the first edition with an introduction by R. Haller, Philosophia Verlag, Munich/Vienna, 1982. Eng. trans. as *On the Content and Object of Presentations — A Psychological Investigation*, ed. by R. Grossmann, Nijhoff, The Hague, 1977.

Twardowski, K.: 1911, 'O czynnościach i wytworach. Kilka uwag z pogranicza psychologii, gramatyki i logiki', Gubrynowicz i Syn, Krakow, 1911, (preprint of *Księga pamiątkowa ku uczczeniu 250-tej rocznicy założenia Uniwersytetu Lwowskiego przez króla Jana Kazimierza* (Commemoration Book Celebrating the 250–th Anniversary of the Foundation of the Lvov University by the King Jan Kazimierz), vol. II, Lvov, 1912) pp. 1–33. Eng. trans. of fragments in J. Pelc (ed.), *Semiotics in Poland 1894–1969*, D. Reidel Publishing Company, Dordrecht/PWN, Warsaw 1979, pp. 13–27.

Woleński, J.: 1986, 'Reism and Leśniewski's Ontology', *History and Philosophy of Logic* 7, pp. 167–176.

Woleński, J.: 1989, *Logic and Philosophy in the Lvov-Warsaw School*, Kluwer Academic Publishers, Dordrecht, 1989.

Woleński, J., Simons, P.: 1989, '*De Veritate*: Austro-Polish Contributions to the Theory of Truth from Brentano to Tarski', in K. Szaniawski (ed.), *The Vienna Circle and the Lvov-Warsaw School*, Kluwer Academic Publishers, Dordrecht, pp. 391–443.

ROBERTO POLI AND MASSIMO LIBARDI

# LEŚNIEWSKI'S CONCEPTION OF LOGIC

## 1. INTRODUCTION

The current availability in English of almost all of Leśniewski's works allows even those scholars unfortunate enough not to read Polish to gain a clearer picture of his ideas.[1] Even a brief reading of Leśniewski's works reveals that his main references are J.S. Mill, Austrian philosophy (Brentano, Marty, Husserl), J. Łukasiewicz and L. Petrażycki.

The first three (Mill, Austrian philosophy and Łukasiewicz) are well-known; what is new is the Russian-Polish thinker Leon Petrażycki.[2] His 1905 book, *Introduction to the study of law and morality: the bases of emotional psychology*, was quoted by Leśniewski in some central passages. Unfortunately the 1955 translation of Petrażycki's book into English does not include the logical part of his work.[3]

In the course of this paper, we propose to present a reconstruction of Leśniewski's position, gathering together and analyzing his general theories and — as far as possible — his explicit references to other authors.

Before proceeding with the presentation of our proposal, we should draw attention to the fact that Leśniewski mainly described his general position in his early works. It is well-known that Leśniewski formally 'repudiated' the early stage of his research.[4] However we should not be misled, as he actually declared that "the change to a symbolic way of writing, which constituted a far-reaching revolution in my scientific life in the field of symbolic technique, was not accompanied by any far-reaching parallel events in the domain of my 'logical' views" (FM, p. 366). The change, then, was limited to the decision to move from a presentation of his ideas in natural language to a presentation in symbolic language, without any real change in the underlying ideas.

For a good reconstruction of Leśniewski's logical thought[5] it is important to understand that he based his proposals on a different logical paradigm, that of traditional logic.[6] With this statement we are saying something significant and difficult. Each paradigm in fact brings with it a whole baggage of thoughts and mental habits. Paradigms work because they save us the effort of having to rethink each time the key concepts which are accepted as a common heritage. Changing a paradigm means rethinking the initial moves, the basic concepts on which it is constructed and which have become common sense, such as,



*K. Kijania-Placek and J. Woleński (eds.),*
*The Lvov–Warsaw School and Contemporary Philosophy*, 139–152.
© 1998 *Kluwer Academic Publishers.*

for example, the concepts of formalization, proposition, variable, etc. In effect, much of the 'difficulty' and 'incomprehension' regarding Leśniewski's 'technical' choices are in fact linked to the desire to understand them by using current logical orthodoxy, without fully exploring the different presuppositions of his logic.

## 2. THE SYSTEM OF THE FOUNDATIONS OF MATHEMATICS

In different passages Leśniewski declared that he was working towards a system of the foundations of mathematics, even if it is not wholly clear what he meant by 'foundations' and 'mathematics'. Regarding his idea of 'mathematics', the most explicit passage is perhaps the following: "mathematical sciences [ . . . ] serve to capture the various realities of the world in the most exact laws possible" (FM, p. 176). In other words, mathematics is a theory of reality. This is an interpretation that the formalistic frenzy of the last sixty years has considered to be of little importance.

If mathematics — as Leśniewski thought and as Brentano and his pupils firmly believed — is the formal theory of reality or of aspects of reality, then it makes no sense to speak of structures based on a so-called meaningless syntax. On this point Leśniewski is penetratingly ironic. In several passages he reiterates that "an unintuitive mathematics contains no effective remedy for any malady of the intuition" (FF, p. 413). Or: "Having no predilection for various 'mathematical games' that consist in writing out according to one or another conventional rule various more or less picturesque formulae which need not be meaningful, or even — as some of the 'mathematical gamers' might prefer — which should necessarily be meaningless, I would not have taken the trouble to systematize and to often check quite scrupulously the directives of my system, had I not imputed to its theses a certain specific and completely determined sense, in virtue of which its axioms, definitions, and final directives [ . . . ] have for me an irresistible intuitive validity" (FF, p. 487).

Leśniewski saw the problem of the antinomies as lying at the root of the divide between formal theories built on an underlying intuition and formal theories as pure syntactic games. "[The antinomies] encouraged the disappearance of the feeling for the distinction between the mathematical sciences, conceived as deductive theories, which serve to capture various realities of the world in the most exact laws possible, and such non-contradictory deductive systems, which indeed ensure the possibility of obtaining, on their basis, an abundance of ever new theorems, but which simultaneously distinguish themselves by the lack of any connection with reality of any intuitive, scientific value" (FM, pp. 177–178).

To sum up, Leśniewski proposed a system of the foundations of a theory of reality possessing, as a system, considerable intuitive force. This system is made up of the three theories that Leśniewski developed during his research work — protothetic, ontology and mereology — as the author himself reminds us in several passages.[7]

## 3. MODERNIZED TRADITIONAL LOGIC

It is well-known that Leśniewski defined his proposal as a system of modernized traditional logic (IR, p. 650). But for what reasons did Leśniewski feel that he had to turn to traditional forms of logic? Put another way, why did he never seriously consider some modern form of logic, even though he had studied it in depth? We are faced with the case of a scholar who had a good knowledge of the modern mathematical form of logic and who nonetheless decided to use the old form of two names plus copula. Obviously there must have been aspects of the modern form of logic which were radically unacceptable to him. The problem then is to discover what were these problematic aspects of the modern method of studying logic. Unfortunately, in this respect, Leśniewski is not very explicit. Nevertheless, some indications can be found, at least indirectly.

We know for example that Leśniewski devoted several years to a meticulous study of *Principia Mathematica* (FM, p. 175 ff. , p. 195).[8] We also know that he was extremely critical of both the *Principia*'s approach and the decidedly more rigorous approach of Hilbert and his school. Let us briefly consider Leśniewski's main criticisms of them.

First of all, Leśniewski found "shocking defects" in the formal instrumentation of the *Principia* and stated that the fundamental questions were "presented disastrously" (FM, p. 179), so much so that he was "unable to understand the terminology of Whitehead and Russell" (FM, p. 224). Now, it is widely acknowledged that the presentation of the expressive and deductive apparatus of the *Principia* is considerably lacking in precision. This issue is well-known and not worth dwelling on here. Indeed the limitations of the *Principia Mathematica* were highlighted by many scholars and within a decade it had become a widely shared view. What is special about Leśniewski's criticism is perhaps its extreme, even excessive, radical nature (FM, p. 181).

Leśniewski's second criticism was levelled at the formalistic approach which was emerging at the time. In other words he was attacking the distinction between the syntactic aspect, reduced to the calculistic and manipulatory dimension, and the moment of the interpretation. To quote Lejewski, "Leśniewski meant his logic to be a true, though very general, description of reality, a kind of *prote filosofia* [ . . . ]. Contrary to the fashion prevailing among mathematicians, Leśniewski insisted that only *true* propositions should be allowed as axioms of a deductive theory and that only those rules of transformation should be admitted which embodied intuitively *valid* rules of inference. He did not reject formalization as a method of presenting deductive theories. Far from it. But he maintained that a theory which was intuitively sound did not lose its meaning through being subject to a process of formalization. He consequently described himself as a confirmed intuitionist who at the same time was a radical formalist, and a better characterization of his standpoint could hardly be given."[9] For Leśniewski, then, formalization was an instrument for making precise and clear assertions which were already true and significant. We will come back to this point later on.

From this perspective the distinction between formalization and interpretation obviously makes no sense and there is therefore no point in considering formal systems as manipulations of arbitrary formulae. Here one of the fundamental characteristics of contemporary logic falls down — the distinction between syntax and semantics, and therefore the distinction between the level of correctness and the level of validity.

To sum up, logic is not an empty game of symbols to which an interpretation is added a posteriori, but an outlining of the rules for the use of terms which have been meaningful from the very start. This does not clash with exact formalization: "I see no contradiction [ ... ] in saying that I advocate a rather radical 'formalism' in the construction of my system even though I am an obdurate 'intuitionist' " (FF, p. 487).

But how do things stand with the problem of the antinomies? Before essaying a possible answer, it is worth noting that, among the different replies to the problem of the antinomies that were developed in the first developments of modern mathematical logic, those who chose to *accentuate* the distance from the level of intuition, from, let's say, the level of common sense, were the ones to prevail. In other words, the winning tendency was the exact opposite of the one chosen by Leśniewski. From this point of view the "obdurate 'intuitionist' " found no fellow travellers.

But to come back to the main point. A possible reply to the question of why the beginnings of modern logic fell such an easy prey to the antinomies may lie in its initial move of starting with terms (individual variables and constants) which had a denoting role only, without any connotative value and, from there, to constructing terms (predicative variables and constants) which were to play a vicarious connotative role. Leśniewski did not accept such a position, for various reasons. Some of his reasons take us back to the fact that a theory of this kind must adopt two different types of terms (one for individuals and one for properties) and that such a theory ends up by treating propositional and predicative calculi at distinct layers (thus distinguishing a further type of variable, the propositional one) and it is therefore difficult, if not impossible, to develop a *general* theory of linguistic signs.

From these preliminary notes, it becomes clearer that Leśniewski's diversity was not so much or not only of the technical kind, but rather of a paradigmatic nature. Before considering more closely what this means, we must eliminate a pseudoproblem.

It is usually thought that the difference between traditional logic (in its various forms) and modern formal logic depends on the fact that the latter — unlike the former — is a *mathematical* logic. In fact, it is perfectly possible to give a mathematical structure to (at least some of the versions of) traditional logic, as has already been done in several significant cases. Our vision of a mathematical logic results from certain choices which in general — and quite irrespective of our current reference to Leśniewski — deserve to be explored more thoroughly. Consider, for example, the concept of function, whose application to linguistic objects represents Frege's great intuition. We all know that a function is a many-one or one-one connection, where the fundamental point is the uniqueness of the value. In effect, however, it would be perfectly acceptable to weaken the requisite of the unicity of the value without leaving the sphere of mathematical rigour. Instead of functions, we could for instance use mappings, that is a many-many or one-many kind of connection. Who is to say that in this case we are no longer in the sphere of influence of logic? And we can consider other possibilities — which moreover in certain cases have already been tested — such as that of admitting operations which are not defined for all the objects of the domain. The point, therefore, once again, is not one of mathematical shaping.

The problem can be seen more clearly if we carry out a check on the mathematization of traditional logic effected from the point of view of modern formal logic. It is not hard to see that in many cases there is something that is not working properly. The issue has been raised many times but, it would seem, without coming to a conclusive answer. We believe that the difficulties depend on the fact that different forms of logic may use different semiotical frameworks.

Two of the fundamental components of every semiosis are the choice of the type of admissible variables and the expression of the distinction between reference (denotation) and meaning (connotation). Let us look at the latter.

If we go back to Mill's classic doctrine — the starting point for Leśniewski's reflections — there are four possible cases: terms which have denotation but no connotation; terms which have connotation but no denotation; terms which have connotation and denotation; terms which have neither connotation nor denotation. Let us stop for a moment to consider this situation. If we look at the 'standard' extensional viewpoint, we see at once the drastic nature of its starting point; in fact it admits only denoting expressions which have no connotation (names of individuals) and, from there, a different class of terms (the predicates) which should show a connotation (intended rather daringly as the class of bearers of the property) is constructed. However, the situation is very different if we start with natural language. For this reason Leśniewski moved on to the development of a more general framework, which we will now look at.

## 4. FORMS OF EXPRESSIONS

The starting point is the development of a general theory of expressions. From Mill and traditional logic Leśniewski took the idea by which all expressions can be divided into connoting and non-connoting expressions and into denoting and non-denoting expressions. According to Leśniewski, the expression 'connoting expression' denotes expressions that can be defined, whereas the expression 'non-connoting expression' denotes expressions that cannot be defined. The examples that follow come from Mill's *Logic* and are pretty clear: the expressions 'man', 'green', 'square circle' and 'centaur' are examples of connoting expressions; the expressions 'to a man', 'well', 'at', 'abracadabra', 'object', 'every man is mortal', etc. are examples of non-connoting expressions (PC, p. 31). In other words, all nominal expressions which have a superordinated genus are definable. The other kinds of expressions are not definable: non-nominal expressions (every man is mortal), nominal expressions without a superordinated genus (object), adverbs and prepositions (well, at), exclamations and interjections (abracadabra) and incomplete expressions (to a man).

Moving on to the other classification, we can say that "an expression which denotes something can be exemplified by the following: 'man', 'green', 'object', 'the possessing by every man of the property of mortality', 'every man is mortal', etc. The expressions which do not denote anything can be exemplified by the following ones: 'abracadabra', 'square circle', 'centaur', 'the possessing by every man of the property of immortality', 'every man is immortal', etc." (PC, p. 31).

To understand how this second classification works, it is essential to look at the world itself. Thus we can check whether there are things which are men, things which are green, things which are objects and whether the things which are men are also mortal things. Continuing in the same vein, we can check whether it is impossible to find things which are abracadabras or things which are square circles or things which are centaurs, just as we can check whether the things which are men are not immortal things.

At this point there are two fundamental observations to be made. The least we can say is that the distinction between nominal expressions and propositional expressions is not given *ab origine* but is constructed within the system.

It should also be acknowledged that checking how things stand in the world is not the task of logic but of science. We will come back to this aspect further on, when we address the problem of Leśniewski's theory of science and the relationship between scientific propositions and logical propositions. For now we will leave this aspect of the theory to one side. The last observation is that the two distinctions of 'denoting/non-denoting terms' and 'connoting/non-connoting terms' are orthogonal.

It follows "that there exist expressions which connote something [ . . . ] but denote [ . . . ] nothing; such expressions are, e.g. 'square circle', 'centaur'. On the other hand, there exist expressions which connote nothing and yet denote something, e.g. 'object', 'every man is mortal'. Besides, there also exist expressions which neither denote nor connote anything; example: 'abracadabra' " (PC, p. 32).

The expressions which denote everything and which connote nothing cannot, because of their very nature, be defined (EP, pp. 4–5; PC, p. 27).

To sum up, we can state that for Leśniewski some terms are denoting and connoting (man), some are denoting and non-connoting (object), some are non-denoting and connoting (centaur) and some are neither denoting nor connoting (abracadabra).

The distinction between expressions which denote and expressions which do not denote is the same distinction that we have when we distinguish those expressions which are symbols from those which are not. In other words Leśniewski thought that the distinction between expressions which denote something and expressions which denote nothing can also be formulated by resorting to the distinction between symbolizing something and symbolizing nothing, or expressions which are symbols and those which are not. To say that an expression denotes something or that an expression symbolizes something or that an expression is a symbol are different ways of formulating the same content.

For this reason we can say that "the expression 'square circle' does not possess a symbolic function because no object is a square circle; in other words there is no such object as could be symbolized by the expression 'square circle'; thus the expression 'square circle' symbolizes no object, in other words symbolizes nothing" (PC, p. 31–32).

If logic, at least for some of its components, is the discipline of rendering more exact natural language which operates from our linguistic instinct, it is necessary to introduce at least one further distinction between symbolic *disposition* and symbolic *function*. If we fail to be clear on this point, we run the risk of "treating expressions which are not symbols as expressions possessing a symbolic function" (PC, p. 33).

According to Leśniewski, an expression's property of being used as *if* it were a symbol, or as *if* it had a symbolic function, is called 'symbolic disposition'. Thus "not all of the expressions possessing a symbolic disposition possess also a symbolic function". The difference between a symbolic disposition and a symbolic function does not alter the opposition between connotation and non-connotation: just as the expressions possessing a symbolic function are or are not connoting, so the "expressions that possess a symbolic disposition can be connoting or non-connoting".

At this point it is possible to use the distinctions that have just been introduced to define new classes of expressions. Consider for example the class of expressions which are propositions. By 'proposition' Leśniewski intended those expressions "possessing the *disposition* to symbolize relations of inherence" (PC, p. 34, my italics), whereas the relation of inherence is "the relation between an object and a property which consists in this object possessing that property" (EB, p. 89).[10]

Once the initial framework has been outlined, the next step is to check whether a certain content is or is not adequately represented in a proposition. To be able to see this, we have to establish the criteria of adequacy (or of inadequacy). These criteria belong to the general theory of propositions (EP, p. 16). They concern "certain general conventional-normative schemas to embody the dependence of symbolic functions of propositions on the symbolic functions of their elements, and on the mutual relationship between these elements" (EP, p. 17). One of the first of Leśniewski's formulations was: "To express this dependence I have adopted the following normative schema: every proposition is to represent the possessing, by the object represented by the subject, of the properties connoted by the predicate" (EP, p. 17).

In light of what we have just seen, many things change. Among the aspects that deserve at least a brief discussion there are the theory of proper names, the problem of synonymy and the difference between analytic and synthetic propositions, which we will now look at.

## 5. PROPER NAMES, SYNONYMOUS EXPRESSIONS AND THE ANALYTIC/SYNTHETIC DISTINCTION

Let us start with Mill, quoting Leśniewski: Mill "says that not all names have connotations; to those which have no connotations belong, according to Mill, proper names such as, e.g. Paul, Caesar [ . . . ]. Yet even the names which I have mentioned, and which according to Mill have no connotation, in my opinion have connotation; proper names connote the property of possessing a name which sounds like the given proper name [ . . . ]. In fact the word 'Socrates', while denoting different objects, connotes always one property, that is the property of bearing the name 'Socrates' " (EP, pp. 5–6). Unlike Mill, therefore, Leśniewski asserts that proper names have a connotation: to be precise they connote the denoted object's property of having the name that they have. This position is extremely interesting, even if it seems to have gone completely unnoticed for at least eighty years. Placed alongside an opportune definition of synonymy, it enables us to eliminate a number of problems. To fully understand this, let us move on to look at how Leśniewski defined synonymy.

A preliminary condition of the possibility of checking whether two expressions are synonymous is that they be reduced to a canonical form which, according to Leśniewski, is that of "categorical propositions with positive copulas and predicates in the nominative". Once this reduction has been made it is possible to proceed. Leśniewski says: "I employ the expression 'synonymous propositions' to denote such propositions as ... possess adequate subjects and synonymous predicates. I employ the expressions 'adequate subjects' to denote such subjects as do not denote different (if only numerically) objects and do not connote different properties [ ... ]. I employ the expressions 'synonymous predicates' to denote such predicates as connote identical properties" (EP, p. 21). Let us look at an effective application of these instructions. In this respect Leśniewski states the following: "Łukasiewicz claims that the propositions 'Aristotle was the founder of logic' and 'The Stagirite was the founder of logic' are synonymous; I think this claim is false. These two propositions are not synonymous because they do not have adequate subjects; their subjects — 'Aristotle' and 'the Stagirite' — are not adequate because they connote different properties: the word 'Aristotle', e.g. connotes the property of possessing the name 'Aristotle' while the word 'Stagirite' does not connote that property". It is quite obvious that by adopting a similar perspective, some if not most of the difficulties related to certain aspects of the problem of opaque contexts are disposed of.

Moving on to the distinction between the analytic and the synthetic, we will begin with some much-needed definitions. A proposition in canonical form is analytic if it contains no predicates which connote properties that are not connoted by the subject. A proposition in canonical form is synthetic if it contains predicates which also connote properties that are not connoted by the subject (EP, p. 3). Examples of analytic propositions are the following: 'Paul bears the name 'Paul' '; 'Napoleon bears the name 'Napoleon' ' (EP, pp. 7–8). On the other hand, a proposition such as 'The victor of Jena bears the name 'Napoleon' ' is synthetic.

It is usually thought that analytic propositions are true a priori. In reality, given the definition adopted by Leśniewski, the distinction between analytic and synthetic interferes in a new original way with the distinction between a priori and a posteriori.

Leśniewski stated that "only the following two types of propositions can be true a priori: (1) propositions whose subject is the word 'object', and (2) propositions which [ ... ] are based on propositions whose subject is the word 'object' " (PC, p. 42).

Moreover, "true propositions whose subject is the word 'object' are always synthetic since their subject does not connote anything" and therefore the predicate always connotes properties which are not connoted by the subject. In brief, "only (the) following two types of propositions can be true a priori:

(1) synthetic propositions and (2) propositions which are based only on synthetic propositions. Hence it follows that all analytic propositions true a priori are based on the validity of synthetic propositions" (PC, p. 42).

Even if "in modern logic there is a widespread conviction that all analytic propositions are true [ ... ] the discussion (above) indicates that some analytic propositions might be true, while others might be false" (EM, p. 61). Put another way, Leśniewski's proposal radically upsets a picture which seemed to have reached a certain stability at the cost of much effort and obliges us to rethink once again the whole issue.

## 6. THE CONCEPT OF LOGIC

We wished to reconstruct Leśniewski's position in some detail in order to present some of his general theories with greater effect. In the light of the analyses that we have summarized, it is clear that at their basis lies a particular idea of what should be understood as logic. Leśniewski's answer was as follows: "I use the word 'logic' to denote the discipline which investigates the question which propositions are true and which are false" (EM, p. 53). It follows that "a logical proposition is not every proposition but only such a proposition about a proposition or propositions which asserts that the latter proposition (or propositions) is true or false. To every true proposition of the type '*A* is *b*' corresponds a true logical proposition of the type 'the proposition '*A* is *b*' is true' " (EM, p. 54). Thus a logical proposition is a proposition such as: 'the thesis that-*p* is true'. Let us say that it is logical because it concerns the proposition *p*.

In other words, propositions of the kind '*A* is *b*' are not always logical propositions. From Leśniewski's point of view, a proposition such as 'all bodies are heavy' is a proposition of physics, not of logic; while the proposition 'the thesis that all bodies are heavy is true' is a logical proposition.

Secondly, Leśniewski explicitly stated: "I call all propositions which possess a symbolic function, true propositions; I call all propositions which do not possess a symbolic function, false propositions" (PC, p. 35). It follows that the correspondence between a 'true proposition' and a 'proposition possessing a symbolic function' is the point of contact between the theory of logic and the theory of signs.

We can summarize Leśniewski's position in the following thesis: "(1) a true proposition always possesses a subject which denotes something; (2) a true proposition always possesses a predicate that connotes something" (EM, p. 57). In other words: (1) every proposition whose subject denotes nothing is a false proposition; (2) every proposition whose predicate connotes nothing is a true proposition.

From what has been said about the idea of logic as developed by Leśniewski, we can make several observations. Firstly we get an original interpretation of semantics, according to which the semantic relationship is not seen as a relationship between a term and its denotation in some universe of discourse. We should rather say that the relationship is between the terms which occur in logical propositions and the terms which occur in the corresponding propositions with which the logical propositions are concerned.

A reflection of this position is shown in the theory of definition. Leśniewski notes that mistakes are commonly made based on 'typical' symbolic inadequacies (EP, p. 18). The most common mistakes are those which in mediaeval terminology lead back to an erroneous use of the *suppositio materialis*. These are mistakes in which "instead of formulating a proposition about the expression we are going to define, we formulate a proposition about the object of which the expression in question can only be a symbol" (EP, p. 18). The definitions must proceed in such a way that their "subjects are symbols of the symbols of objects, that is symbols of words and not symbols of the objects themselves or their so-called concepts" (EP, p. 19).

## 7. THE THEORY OF SCIENCE AND METAPHYSICS

According to Leśniewski, a science is "a systematized aggregate of propositions which possess a symbolic function" (PC, p. 35). We know that possessing a symbolic function means that the corresponding propositions are true. A science then is a systematized aggregate of true propositions. "The aim of any scientific criticism of a proposition or of a system of propositions consists in answering the question whether these propositions possess a symbolic function, i.e. whether they are true. The purpose of a consistent scientific criticism is the exclusion of all propositions which do not possess a symbolic function" (PC, p. 35).

From what has been said up until now, there is nothing surprising in the theory that to understand the concept of metaphysics adopted by Leśniewski and to distinguish it from the concept of logic, the first step is to understand which propositions are metaphysical propositions and which are logical propositions. An early declaration goes as follows: "I employed the word 'metaphysics' to represent the system of true propositions concerning all objects in general" (EM, p. 48). To which we can add: "propositions [ . . . ] which refer to all objects in general are metaphysical propositions (I also call them 'ontological'). The propositions (which refer only to a certain group of objects) are logical propositions" (EM, p. 48). Given that "propositions concerning all objects in general cannot be formulated as conditions" (PC, p. 30), it follows that "if we apply the word 'metaphysics' to denote the system of true propositions concerning all objects in general, we shall thus obtain the following thesis: metaphysics can be constructed only as a system of categorical propositions" (PC, p. 30).[11]

## 8. CONCLUSION: AN INTERPRETATION OF LEŚNIEWSKI'S PROPOSAL

We can end our description of the general traits of Leśniewski's proposal with these observations. If we look at what Leśniewski accomplished in the course of his career as a scholar, it seems that the fundamental reason for his entire research was to progress towards a formalization and a 'modernization' of Aristotle's *Metaphysics*. Indeed, his early works are perfectly in keeping with the Aristotelian renaissance of the period. One only needs to think of the papers on the ontological principle of contradiction (PC) and on the principle of the excluded middle (EM), as well as those on existential propositions (EP) and on truth (EB). But the most telling fact comes from a possible and, in the end quite reasonable, interpretation of the structure of his system of the foundations of mathematics.

In this sense, it is not so difficult to see mereology as a system which proposes a formalization of the theory of substance (matter). If we continue in this fashion, we realize that the so-called implausibility of mereology immediately disappears. The results of mereology as a theory of substance correspond perfectly to the expectations of our intuition. The move towards ontology is even less problematic. For example, Leśniewski states: "I used the name 'ontology' to characterize the theory I was developing, without offence to my 'linguistic instincts' because I was formulating in that theory a certain kind of 'general principles of existence' " (FM, p. 374), from which one naturally draws the conclusion that ontology is a formalization of the theory of being. Finally, with protothetics, we have — and this is something we have already seen — a formalization of the theory of *verum*.

To these internal observations we can add some notes concerning its intellectual context. In those years, first at Lvov and later at Warsaw, Aristotle was studied in depth. In addition to the universally known works by Łukasiewicz, we should also mention Twardowski, the founder of the Lvov-Warsaw School and one of the most important pupils of Brentano. Then Kotarbiński is certainly no less important and we should at least mention his analysis of Aristotle's categories which he developed in *Elementy teorii poznania, logiki formalnej i metodologii nauk*, an analysis carried out along lines very similar to Leśniewski's approach.[12]

In brief, we would like to conclude by noting that, however surprising it may seem, the hypothesis that Leśniewski's fundamental intention was perhaps to seek to formalize the *Metaphysics* is perhaps the deepest and clearest explanation of his unshakeable faith in the superiority of the traditional theory of logic.

*University of Trento and Centro Studi per la Filosofia Mitteleuropea*

## NOTES

[1] S.J. Surma, J.T. Srzednicki, D.I. Barnett, V.F. Rickey (eds.): *Stanisław Leśniewski. Collected works*, 2 vol., Dordrecht, Kluwer Academic Publishers, 1992. The publication has been an arduous editorial undertaking that has occupied Jan Srzednicki for more than eighteen years. To him we owe our gratitude for the tenacity and the strength of will that enabled him to surmount the many and difficult obstacles that stood in his way. Unfortunately, the English translation is not always crystal clear and certain papers contain some irritating mistakes. For some general assessments, see the critical notice by M. Libardi, *Axiomathes*, 1993, pp. 105–129, and the review by P. Simons, 'Discovering Leśniewski', *History and Philosophy of Logic*, 1994, pp. 227–235. In our quotes, we will use the following initials: EP: 'A contribution to the analysis of existential propositions', pp. 1–19; PC: 'An attempt at a proof of the ontological principle of contradiction', pp. 20–46; EM: 'The critique of the logical principle of the excluded middle', pp. 47–85; EB: 'Is all truth true eternally or is it also true without a beginning?', pp. 86–114; FM: 'On the foundations of mathematics', pp. 174–382; FF: 'Fundamentals of a new system of the foundations of mathematics', pp. 410–605; FO: 'On the foundations of ontology', pp. 606–628; IR: 'Introductory remarks to the continuation of my article: 'Grundzüge eines neuen Systems der Grundlagen der Mathematik' ', pp. 649–710.

[2] For some information about Leon Petrażycki and for a wider analysis of Leśniewski in general, cf. R. Poli and M. Libardi, 'Logic, theory of science and metaphysics according to Stanisław Leśniewski' (submitted).

[3] L. Petrażycki, *Law and Morality*, Cambridge (Mass.), Harvard University Press, 1955. The Leon Petrażycki Society published a German translation in Paris in 1933.

[4] "Living intellectually beyond the sphere of the valuable achievements of the exponents of 'Mathematical logic', and yielding to many destructive habits resulting from the one-sided, 'philosophical'-grammatical culture, I struggled in the works mentioned with a number of problems which were beyond my powers at that time, discovering already discovered Americas on the way [ ... ] I formally 'repudiate' them [ ... ] affirming the bankruptcy of the 'philosophical'-grammatical work of the initial period of my work" (FM, pp. 197–198).

[5] It is well-known that Leśniewski defined ontology, one of his three foundational systems, as "a certain kind of modernized 'traditional logic' " (FM, p. 176). In this respect it is worth remembering that in the 1937/38 academic year Leśniewski taught a course called "Traditional 'formal logic' and traditional 'set theory' on the ground of ontology"; cf. J.T.J. Srzednicki and Z. Stachniak (eds.): *S. Leśniewski's lecture notes in logic*, Dordrecht, Kluwer Academic Publishers, 1988, p. 180. Given these premises, we can better understand his interest in the principles of non-contradiction (PC) and excluded middle (EM). Twardowski, the founder of the school, was also interested in traditional logic. As a lecturer at the University of Lvov, for many years he taught a course on *Attempts to reform traditional logic*, in which he outlined the theories of Bolzano, Brentano, Boole and Schröder; cf. I. Dąmbska: 'François Brentano et la pensée philosophique en Pologne: Casimir Twardowski et son école', *Grazer philosophische Studien*, 1978, pp. 117–129, quotation from p. 123. As confirmation of the fact that "the borderline between modern and traditional logics can hardly be drawn in the case of the Lvov-Warsaw Philosophical School" (V.L. Vasyukov, 'A Leśniewskian guide to Husserl's and Meinong's jungle', *Axiomathes*, 1993, pp. 59–74, quotation from p. 59), we may consider the scholars who in the early nineteen thirties formed the so-called 'Krakow Circle' to address some of the traditional logical and philosophical problems of Thomism and Scholasticism with the new methods of formal logic. The more important of them are Innocenty M. Bocheński (1902–1994), Jan Salamucha (1903–1944), Jan Drewnowski (1896–1978) and Bolesław Sobociński (the latter was one of the main characters linking the two schools). For the Krakow Circle, cf. the monograph section of *Axiomathes* 1993, pp. 169–226.

[6] The concept of traditional logic has not yet been defined in conceptually acceptable terms. For a minimalist perspective, consider the *Dictionary of Philosophy* by D. Runes, according to which traditional logic can be defined as those parts and that method of the treatment of logic which have come down to us essentially unchanged from the classical and mediaeval ages. In this sense, traditional logic highlights the distribution of propositions into subject and predicate and the related classification into the four forms A, E, I, O. For some further observations, cf. R. Poli, 'Twardowski's theory of modification against the background of traditional logic', *Axiomathes*, 1993, pp. 41-57 and M. Libardi and R. Poli, 'La polemica sulle nuove direzioni della logica' (unpublished), R. Poli and M. Libardi, 'Il concetto di logica tradizionale. Elementi per una riattualizzazione' (submitted).

[7] (FM, p. 176) In some other passages the system seems to be limited to only the first two of the above theories. Cf. for example (FO, p. 608). The logical basis of the system consists of protothetics (roughly a second order propositional calculus) and ontology (a calculus of predicates founded on the traditional structure of two names plus copula). Mereology, or the calculus of parts, is an extension of ontology based on the introduction of the 'part-of' functor or of similar functors.

[8] Cf. also C. Lejewski, 'Ricordando Stanisław Leśniewski', *Quaderni del Centro Studi per la Filosofia Mitteleuropea*, 1989.

[9] C. Lejewski, 'On Leśniewski's Ontology', in J.T.J. Srzednicki, V.F. Rickey and J. Czelakowski (eds.): *Leśniewski's systems. Ontology and Mereology*, Dordrecht, Kluwer Academic Publishers, 1984, pp. 123–148, quotation from pp. 123–24.

[10] For further details, cf. R. Poli and M. Libardi, 'Logic, theory of science and metaphysics according to Stanisław Leśniewski' (see note 2).

[11] The thesis according to which metaphysics refers to all objects in general is to be understood in a very precise sense: "the propositions about 'general' objects which are allegedly opposed to 'individual' objects cannot be metaphysical propositions because they do not refer to 'individual objects' and, thus, are not propositions concerning all objects" (EM, p. 50).

[12] With reference to expressions of the '*A* is *b*' kind, Kotarbiński's analysis of Aristotle's theory of categories distinguishes purely nominal categories of the *b* kind (the first six) from categories of the 'is-*b*' kind (the last four). Cf. T. Kotarbiński, *Gnosiology. The scientific approach to the theory of knowledge*, Oxford, Pergamon Press, 1966, pp. 49–50. The original edition was published in 1929.

VLADIMIR L. VASYUKOV

# NON-ELEMENTARY EXEGESIS OF TWARDOWSKI'S THEORY OF PRESENTATION

In spite of the historical proximity of S. Leśniewski to K. Twardowski, an attempt to look at Twardowski's heritage through Leśniewski's eyes leads to striking results. Firstly, it results in a wider framework than Leśniewski's Elementary Ontology and secondly, it involves a transition from Formal Ontology to Formal Phenomenology. In this paper an extension of Leśniewski's Non-Elementary Ontology is presented which is suitable for investigating Twardowski's Theory of Presentation.

## 1. GENERAL REMARKS ON NON-ELEMENTARY FORMAL PHENOMENOLOGY

Vasyukov (1993) presented some phenomenologically oriented formal languages based on Leśniewski's Ontology which allow us to obtain the 'subjective' vision of Husserl's and Meinong's jungles. Vasyukov (1993b) developed this Leśniewskian approach, demonstrating further applications of those languages by extending them to incorporate the Brentano-Husserl theory of time. The semantic considerations of those papers were of a phenomenological character. And it seems that in this connection one needs to speak of Phenomenology more than of Ontology, even though we wish to follow Leśniewski's terminology. But this transition seems to have been obscured and we wish to comment on it more profoundly.

The term 'Formal Phenomenology' has at least two distinct meanings. From the point of view of analytic philosophy Formal Ontology is that part of ontology which admits of investigation by formal logics. In Roberto Poli's opinion the principal exponent of this position is Nino Cocchiarella (Poli, 1992, p. 51). Accepting, in particular, that every particular science considers its specific 'modes of being', he claims that Formal Ontology studies the different formalizations concerning the systematic classification of all 'modes' or categories of being in general. Usually, every 'mode of being' is fit to some particular Formal Ontology and represents some type of variables for which the syntactic rule of their application reflects an ontological rule of the given 'mode of being' (Cocchiarella, 1974, pp. 29–30). It is clear that from such a point of view Formal Ontology studies logical characteristics of predications and different universals.



*K. Kijania-Placek and J. Woleński (eds.),*
*The Lvov–Warsaw School and Contemporary Philosophy,* 153–167.
© 1998 *Kluwer Academic Publishers.*

We find another definition of Formal Ontology in Husserl's works, in particular in *Logische Untersuchungen*, *Ideen*, *Formale und transcendentale Logik* and *Erfarhung und Urteil*. In this case Formal Ontology is regarded mainly as a theory of the parts and the whole together with their interrelationship.

At first sight Leśniewski's Ontology satisfies the first definition of Formal Ontology, while Leśniewski's Mereology falls precisely under Husserl's definition. This division seems to be accepted by default in investigations into ontological problems.

The main defect of such a tenet looms large when we pay more attention to the problem of a presentation and its object. The latter does not mean, loosely speaking, that the question is of objects in the mind and objects in reality, as it may seem at first glance. The point is that even when Husserl in his *Logical Investigations* rejects Twardowski's idea of the correspondence between the parts of an object and the parts of the content of a presentation, he nevertheless by no means casts doubt upon the notion of an object of presentation as such. Moreover, he speaks of intentional objects, noemata, and does not treat the simple parts of an object of presentation as the whole. We recall also that Brentano had already spoken of an object of presentation.

And from the point of view of the traditional problem of other minds a material object, while being the object of many minds' presentations, will possess, so to speak, a new dimension of its own entity. Perhaps, being simply of an epistemological nature, this dimension could be treated in the framework of Husserl's considerations of parts and wholes — but not necessarily yet.

The situation has an affinity with the treatment of modality. In this case we could proceed by identifying an intentional object with a 'necessary' object and a noema with a 'possible' object. The intuitive meaning of these concepts should be as follows.

When our intentional activity of consciousness is directed onto some object then only part of its meanings may be grasped in an act of presentation and, hence, we get only a possible image of the genuine object. And, in all acts of presentations of one and the same object its basic image always looms large in consciousness (as kind of background). Thus, its appearance is always necessary and always independent of our will.

We can easily adapt the possible worlds semantics of modal logics to this modal interpretation. It is enough to speak of a state of mind instead of a possible world in order to convey the sense of the concept of a 'possible' object as an object of presentation whose apprehension is connected with the intuitive conviction (or knowledge) of its apprehension in some situation (state of mind). Likewise, a 'necessary' object can be seen as an object of presentation whose apprehension is connected with the intuitive conviction of its apprehension in all situations (states of mind).

Thus, if we accept Cocchiarella's definition of Formal Ontology, then according to our consideration above 'the phenomenological mode of being' could be fitted to our particular Formal Ontology (setting aside the question of whether Phenomenology could be a particular science). There remains only the problem of the type of variables and the syntactic rules of their application suitable for our kind of Formal Ontology. And here those peculiarities of Leśniewski's Ontology mentioned above allow us to solve this problem.

An obvious question is that of the name of this kind of Formal Ontology. It seems very natural to speak in this situation of Formal Phenomenology: this name from the outset requires us to take into account the impact of phenomenological ideas on modern investigations. Below we shall use this name specifically as the name for those systems which are extensions of Leśniewski's Ontology.

We shall also, however, present systems that lie outside the scope of elementary Formal Phenomenology. Since even Słupecki (1984) does not contain a systematic presentation of non-elementary Ontology we use Stachniak's version of non-elementary Leśniewskian Ontology in our further exposition, also employing the terminology and notation of Stachniak (1981).

That the system of Twardowskian Formal Phenomenology is non-elementary is a consequence of the attempt to show that his teachings on presentation, which at first sight seem palpably mereological in nature, could be exposited only in Ontological terms, without any part-whole considerations. Thus, in particular, Husserl's criticism of Twardowski's ideas mentioned above, in some sense becomes stale.

Our considerations will now focus on objects of presentation having a composite inner structure and referring to other (not always simple) objects. Since the name-formative functors meet precisely the requirement of the correspondence to composite objects in Ontology, our phenomenological considerations lead to a shift to the level of non-elementary Ontology. This results in the non-elementarity of phenomenological operators introduced to interpret oblique presentations.

It worth stressing that at precisely this level of consideration, the formal proofs of some hypothesis and conjectures notwithstanding, the seeming redundancy of the operators allows us avoid the unwanted multiplication of entities.

## 2. THE LATE TWARDOWSKI ON PRESENTATIONS

Our considerations here will be based on Twardowski's unfinished and unpublished work *A theory of judgments*. This fragment (34 pages) was found by B. Dombrowski in the library of Lvov University (see Twardowski, 1991). Since Twardowski's book *Zur Lehre vom Inhalt und Gegenstand der Vorstel-*

*lungen* was published in 1894, one might expect that in some sense *A theory of judgments* includes a more complete and elaborated point of view, especially as concerns the theory of presentations. Alas, the work is too brief and, moreover, refers to Twardowski (1894) as more complete work. Nevertheless, below we shall propose a brief outline of Twardowski's theory of presentations according to (1991).

Besides judgments, Twardowski distinguishes three kinds of mental phenomena: presentations, emotions and desires. As to presentations we cannot judge without knowing the presentation of the thing under judgment. A presentation is a necessary condition for the existence of a judgment and therefore we should have a preliminary agreement on this matter in order to speak about judgments.

When we think about some thing, taking into account only our awareness of it, the presentation of this thing is within us. It is irrelevant whether the given thing exists or not, and one can conceive a thing of which we can say with certainty that it does not exist and, moreover, could not exist. It has been said that this kind of presentation is unaccomplished (*Unvollsichbare Vorstellungen*). But this is not correct, since the non-existence of some thing may not limit its presentation: whenever we read fairy tales, we always have our mind filled with the presentations of non-existent persons and things.

The mistake consists in speaking of the presentation in general instead of speaking of the presentation in the precise sense of the word, i.e. 'intuition' (*ogląd*). The latter presentations are those whose objects appear to be or could be the reasons for observation whether or not there was something perceptible. Usually we have intuitively graspable presentations for example, both of color, sound etc. and our own mental phenomena, our feelings, judgments etc.

When we imagine a figure with a thousand equal angles this presentation may be not intuitively graspable for even if this figure is painted, one cannot apprehend it at first glance, in one act of observation, in order to ascertain the number of angles. And the presentation of a single triangle may be intuitively graspable. However, this presentation appears to be not intuitively graspable if we imagine the triangle in general, when we have the general presentation of a triangle. This was the source of the erroneous opinion that there are no general presentations and that only general expressions are able to designate many similar ones, from a certain viewpoint. So, there are no general presentations which are intuitively graspable.

Intuitively graspable presentations will be called direct presentations and 'non-surveying' presentations will be called oblique. Concepts are of the latter kind and the problem of their nature as presentations can be resolved only after the investigation of judgments.

According to Twardowski the main difference between direct and oblique presentations is as follows: we manage to conceive some objects directly without any assistance from the presentations of other objects. But when we conceive an object obliquely we always need the assistance of another object's presentation by which, through the presentation of some relations established between the former and the latter object, we come to the oblique presentation. An auxiliary presentation ought to be intuitively graspable or, otherwise, the result relies again only on the assistance of the interrelation presentation. This law was expressed by Aristotle as follows: διὸ οὐδέποτε νοεῖ ἄνευ φαντάσματος ἡ ψυχή... (*De Anima*, III, 7, 431a, 16–17) where he meant by φάντασμα presentation and by νόημα — oblique presentation.

Three parts can be distinguished when speaking of presentations. However, one must take into account that they do not exist separately, but only together with the others, constituting the whole presentation. Twardowski called them metaphysical and wrote that in this way he named those parts which can be separated in the mind but not in reality. For example, we are able to distinguish the color and form of a sheet of paper when it lies before us, but we cannot separate them, because neither paper, nor color nor form, taken separately, exist. The situation is rather different when we cut the sheet into four parts: now we may not only distinguish these parts but also separate them, and each will exist independently (they are physical parts now). Even though the color and form may not exist in a self-contained manner we can nevertheless distinguish them. We can say something about one of them regardless of the others or the whole: we can claim that one color is brighter or darker than another, that the form is symmetrical or not, and so on.

Being also metaphysical, parts of presentations have the peculiarity that one of them sometimes becomes a physical part too. A presentation has the following parts: an act, content and object (*conceptus formalis, conceptus objectivus* and *ens* in medieval terminology).

Twardowski's argument that Wundt and his school are wrong when they teach that there is a difference between an act and content of presentation is as follows: When I think about a cow I imagine it. When I further think about a steam machine, Venus etc., I imagine a steam machine, Venus etc. It is obvious that in this case we speak only about presentations themselves. An easy observation confirms that the state of our mind remains the same if we stay passive for a long time, allowing the presentations of horse, cow etc. to pass through our mind. Always only the mental image, so to speak, changes, presenting at one time the horse and at another the cow. We call the common property of all these states of mind (when we imagine the horse, cow etc.) an act of presentation and that which allows us to differentiate the presentation of a horse from one of a cow etc. a content of presentation.

But to say that the horse, cow etc. are the content of this presentation does not require that the content of presentation be creatures and things existing beyond the limits of our mind. The real horse, cow etc. do not seem to be the content of presentations but rather their counterparts in our minds. Therefore there is always without exception a real content in every presentation, even if the real counterpart of presentation does not always exist. A content of presentation is what we usually call the mental image of some thing; in the same way as an act of presentation, a content of presentation is something that in all completeness exists only in our mind.

Apart from an act and content every presentation has an object. An object is what we mean when we imagine something. When I imagine a horse then the object of presentation is one horse or another, or a horse in general — but always something of which I am convinced does not depend on my mind. An object of presentation may exist or not. When I imagine the pen with which I write I imagine the existing object but when I imagine a 1000 meter high tree then the object does not exist. The latter is the best illustration of the difference between the content and object of presentation because it is evident that the object may not exist while the content always exists insofar as we have imagined something.

In Twardowski's opinion, consideration of the interplay between language and thought could lead to a better understanding of the difference between the three metaphysical parts of a presentation. The signs of a language which deal with presentations are nouns (or any part of language which may be substituted for them). In medieval terminology these expressions are called categorematic and contemporary grammarians call them names (*nomina*). Every name fulfills a threefold task. For example, when someone says 'the sun' then he first manifests that he is thinking about something representing the sun. Hence, the noun informs the reader or listener that in the mind of the speaker or writer an act of presentation has happened or is still going on.

Thus an utterance of 'the sun' impels the listener to an action of the mind and simultaneously brings the content with which this action ought to be filled (for if somebody will hear the words unuttered then he will not be imagining something but what this expression 'means', he will be imagining the sun). And this is exactly the second task of a name: it fills the mind of the listener with some content which is the same as the meaning of the expression. Therefore the content of presentation is called the meaning of the expression which here is the sign of what is going on in the mind.

However, when someone utters 'the sun' then he at the same time directs the listener's attention to this fiery body — the center of our planetary system. The expression 'sun' not only means something, *viz.* the content of the presentation 'the sun', but at the same time denotes the object. Thus every exposed mental act (*exponit actum*) signifies the meaning of this act, i.e. its content (*significat*) and denotes the object (*nominat*).

## 3. TWARDOWSKIAN FORMAL PHENOMENOLOGY: QUANTIFYING NAMES

Summarizing our brief outline of Twardowski's theory of presentations, we can illustrate the most interesting (for us) aspects with the following figures:

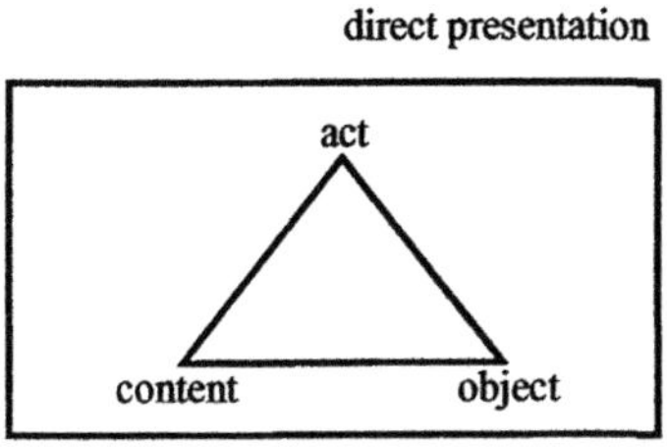


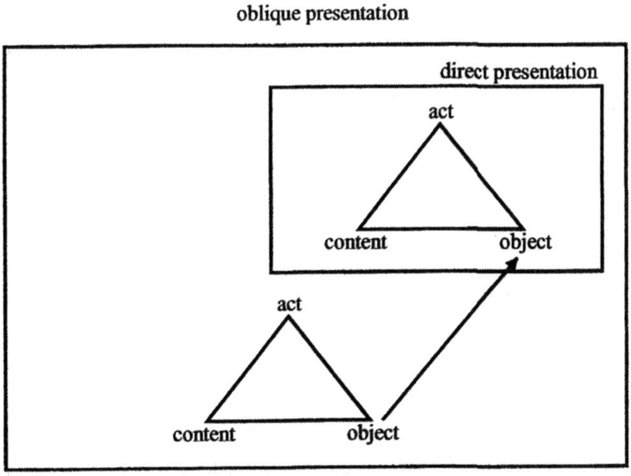


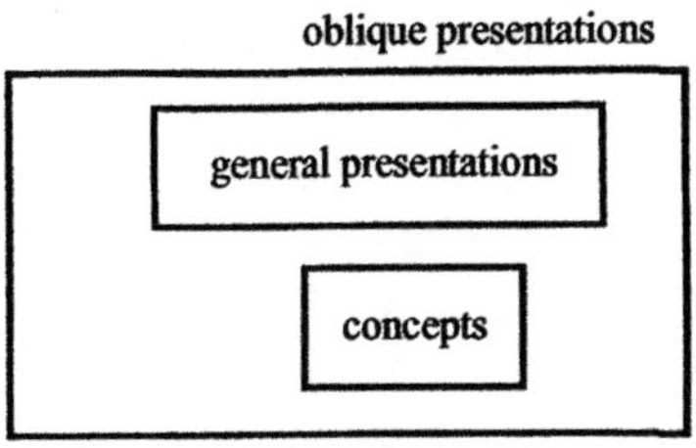


The problem of the parts of presentation, which Twardowski called metaphysical, is usually the main subject of investigation: analysis mainly concerns the aspects of interplay of the parts and wholes in Twardowski's triad act-content-object of presentation. For instance, Poli, in his *Formal Ontology*, considers Twardowski's theory in the section 'Parts and wholes' and then he naturally passes on to Leśniewski's Mereology. Most authors do likewise.

We shall proceed otherwise: our approach will be based on Leśniewski's Ontology (and not Mereology) and correspondingly we shall analyze only the wholes. According to Poli (1992, p. 440) there are four kinds of object in Twardowski's theory: 1. an object in general, which corresponds to Kant's transcendental object; 2. an ontological (real) object; 3. an intentional object (object of presentation); 4. a general object, which corresponds to the term. We shall concentrate on general objects, since these are very important for our further inquiries.

In Twardowski (1894) we find a detailed theory of a general object. As his main tool Twardowski uses the notion of a characteristic note (*Merkmal*). Loosely speaking, a characteristic note is the metaphysical part of an object, which in the case of a general object also appears to be the part of some subordinated object of presentation, i.e. the part of the general object which is in an equivalence relation to some part of objects of other singular presentations. Here we first ought to bear in mind Figure 1 and then, secondly, Figure 2: general presentations are a kind of oblique presentation and, according to the picture of links between objects of oblique and direct presentations, our equivalence relation coincides with the one from Figure 2. Thus, a general object in some way plays the role of the metaphysical component of subordinated singular objects.

It is worth noting that the lack of a direct presentation of a general object involves its being rendered as a secondary individual, a quasi-object. In formal terms, they are represented as corresponding to abstract singular terms. Hence, they are, in some sense, the result of abstraction. Let us look at a general object from this point of view.

According to Poli (1992, p. 111) an abstraction is an operation that takes place within a particular universe of discourse. To abstract, we select those individuals of the universe which possess one or more particular features in common (or which do not possess such features). In this sense, abstraction is an equivalence relation between the individuals of the universe of discourse.

In the case of a general object the given equivalence relation holds only between the metaphysical parts of singular objects and the general object. Thus, in this sense, abstraction allows us to classify objects of presentation through the subordination of singular objects to general ones.

But need we speak about the metaphysical parts at all? Maybe from the viewpoint of Ontology it is enough to fix the subordination between objects according to Leśniewski's conception of the singular proposition $x \varepsilon S$ as being true only if its subject '$x$' is a singular name while '$S$' is a general one. The analogy becomes more transparent if we take into account the relation the theory of the division of objects into logical types bears to the theory of semantic categories of Ontology. Słupecki writes: "The analogy between them stands out clearly if we assume that to every constant symbol of logic, excepting quantifiers and the constants of the calculus of propositions, corresponds an object the name of which is this symbol, and that two objects are of the same logical type if and only if their names belong to the same semantic category" (Słupecki, 1984, p. 71).

It seems very natural in this context to speak of the objects of presentation instead of logical objects, and hence of the names of the given objects respectively. The main trouble now seems to concern the direction of subordination: does Leśniewski's general name correspond to the general object of presentation? According to our considerations above, a general object corresponds to an abstract singular term and, as a consequence, in the singular proposition $x\varepsilon S$ the general object corresponds rather to '$x$' than to '$S$', which blurs the whole picture. Another problem is that we need to extend the conception of oblique presentation to avoid ambiguity: from Figure 2, drawn according to Twardowski, an oblique presentation can point to another oblique one as well, i.e., we may transform our diagram into a recursive one. In effect, this recursiveness results literally from Twardowski's example: we get an oblique presentation of the integer 1000 by starting from the intuitively graspable presentation of the integer 5 and then proceeding by getting an oblique not graspable intuitively presentation of the integer 10 in respect to 5, an oblique presentation of the integer 100 in respect to 10 and so on (Twardowski, 1991, p. 86).

As a consequence, a more promising candidate is the proposition

$$\forall y(y\varepsilon X \to y\varepsilon S)$$

i.e. $X \subset S$—every '$X$ is $S$' (Słupecki, 1984, p. 80). The advantage of doing so becomes more evident when we deal with objects of presentation corresponding to name-formative functors.

In this case an object of an oblique presentation corresponding to the name $\psi(x_1, \ldots, x_n)$ is linked by some relation with the objects corresponding to $\psi(a_1/x_1, \ldots, a_n/x_n), \psi(b_1/x_1, \ldots, b_n/x_n)$ etc. By Twardowski's method, as construed above, one may constitute a general object of presentation for the $\psi(a_1/x_1, \ldots, a_n/x_n), \psi(b_1/x_1, \ldots, b_n/x_n), \ldots$ which is described by a proposition like $\forall y(y\varepsilon X \to y\varepsilon\psi(a_1/x_1, \ldots, a_n/x_n))$ where $X$ is the name of given general object. Hence, we have that every '$X$ is $\psi(a_1/x_1, \ldots, a_n/x_n)$' and the import of this proposition seems very like that of the previously considered notion of abstraction. The only thing we need to do now is to propose some logical 'mechanism' for the choice of $X$. A good candidate seems to be the operation of name quantification, whose algebraic counterpart is the operation of getting the infimum of the subset of objects which plays the role of the range of values of the functor $\psi$, i.e. the subset constituted by elements like $\psi(a_1/x_1, \ldots, a_n/x_n)$.

Accepting this peculiarity we propose to extend the language of Leśniewski's Ontology by the addition of a more general (quasi) quantifier Π which acts only on names, i.e. expressions governed by this quantifier are to be name-formative functors. Thus, for example, the expression

$$\Pi\psi(x_0, x_1, \ldots, x_n)$$

will be a name, where the letters $x_0, x_1, \ldots, x_n$ are nominal variables and $\psi$ is a name-formative functor. Note: Π binds all variables simultaneously.

For simplicity we now stipulate that for each constant symbol of our extended Ontology there corresponds an object of 3-4th kind the name of which is this symbol. So we deal now simultaneously with both the extended and restricted version of Ontology, which really deserves the name of Twardowskian Formal Phenomenology (TFP) since it describes objects only in the framework of the theory of presentations.

The schema of the rule for eliminating our new general quantifier is:

OΠ1. $$\frac{x\varepsilon\Pi\psi(y_{11}, \ldots, y_{1n_1}) \ldots (y_{m1}, \ldots, y_{mn_m})}{x\varepsilon\psi(a_{11}/y_{11}, \ldots, a_{1n_1}/y_{1n_1}) \ldots (a_{m1}/y_{m1}, \ldots, a_{mn_m}/y_{mn_m})}$$

where $a_{ij}$ denotes the constant name of the object of singular direct presentation from the range of the nominal variable $y_{ij}$. It is obvious now that an expression

$$\Pi x$$

defines, in fact, an empty object Λ (see Słupecki, 1984, p. 94), and hence, we have

(Λ) $y\varepsilon\Pi \equiv y\varepsilon\Lambda$

The schema of the rule for introducing the general name quantifier is as follows:

DΠ1. $$\frac{\psi(a_{11}/y_{11}, \ldots, a_{1n_1}/y_{1n_1}) \ldots (a_{m1}/y_{m1}, \ldots, a_{mn_m}/y_{mn_m}) \subset X}{\Pi\psi(y_{11}, \ldots, y_{1m_1}) \ldots (y_{m1}, \ldots, y_{mn_m}) \subset X}$$

with the usual limitations. We remark that any object corresponding to the name-formative functor is an object of oblique presentation with the respective subordination of its nominal variables to the objects of direct presentation.

More interesting in this connection seems to be the problem of an existential name quantifier Σ. We may formally define it as follows:

$$x\varepsilon\Sigma\psi(y) \equiv x\varepsilon(\Pi\psi(y)')'$$

by means of the standard negation operator from the algebra of names (Słupecki, 1984, p. 94):

$$x\varepsilon X' \equiv x\varepsilon x \wedge \neg(x\varepsilon X)$$

Such a definition of Σ simply means that we are obliquely pointing out not a non-object of some presentation to which a general object is subordinated.

Posing the problem of apprehension of this account one can easy come to the conclusion that here we dealing with another kind of abstraction, that of limitation. Now we pass on to the general object by joining metaphysical parts (characteristic notes) of singular objects of direct presentations. If we compare, then, the first abstraction gives us the join of these. The resulting general object is subordinated to the object of direct presentations.

It is clear that apart from the rules for the general name quantifier we can introduce the schema of the rule for eliminating the existential name quantifier:

$$\text{O}\Sigma 1. \qquad \frac{\Sigma\psi(y_{11},\ldots,y_{1n_1})\ldots(y_{m1},\ldots,y_{mn_m}) \subset X}{\psi(a_{11}/y_{11},\ldots,a_{1n_1}/y_{1n_1})\ldots(a_{m1}/y_{m1},\ldots,a_{mn_m}/y_{mn_m})\subset X}$$

and the following rule for adding the existential name quantifier:

$$\text{D}\Sigma 1. \qquad \frac{x\varepsilon\psi(a_{11}/y_{11},\ldots,a_{1n_1}/y_{1n_1})\ldots(a_{m1}/y_{m1},\ldots,a_{mn_m}/y_{mn_m})}{x\varepsilon\Sigma\psi(y_{11},\ldots,y_{1n_1})\ldots(y_{m1},\ldots,y_{mn_m})}$$

where again the letter $a_{ij}$ denotes the constant name of the object of singular direct presentation from the range of the nominal variable $y_{ij}$ . The following equivalence is evident:

$$\text{(V)} \qquad y\varepsilon\Sigma x \equiv y\varepsilon \text{V}$$

where V is the term denoting the universal class of individuals (see, Słupecki, 1984, p. 82).

Our intentions become clear when given in the form of a schema of axioms:

$$(\Pi) \qquad \begin{array}{l} x\varepsilon\Pi\psi(y_{11},\ldots,y_{1n_1})\ldots(y_{m1},\ldots,y_{mn_m}) \rightarrow \\ x\varepsilon\psi(a_{11}/y_{11},\ldots,a_{1n_1}/y_{1n_1})\ldots(a_{m1}/y_{m1},\ldots,a_{mn_m}/y_{mn_m}) \end{array}$$

The dual schema of axioms for the existential name quantifier will be, obviously, as follows:

$$(\Sigma) \qquad \begin{array}{l} \psi(a_{11}/y_{11},\ldots,a_{1n_1}/y_{1n_1})\ldots(a_{m1}/y_{m1},\ldots,a_{mn_m}/y_{mn_m}) \subset \\ \Sigma\psi(y_{11},\ldots,y_{1n_1})\ldots(y_{m1},\ldots,y_{mn_m}) \end{array}$$

The intuitive meaning of these axioms relies on our previous examination of Twardowski's theory of a general object. It is, however, still highly mereological in character. In order to remove this defect we should remove the metaphysical 'contamination'. So we simply identify the given parts with the name '$x$' from (Π) but simultaneously we may say that every metaphysical part of $\Pi\psi(x_1,\ldots,x_{n_1})$ is $\psi(a_1/x_1,\ldots,a_{n_1}/x_{n_1})$, mereologically modifying the sense of the ⊂-term.

Finally, we get a system of Twardowski's Formal Phenomenology by adding to the sole axiom of Leśniewski's Ontology (Słupecki, 1984, p. 79) our axioms

(Π) (or (Σ)), (Λ) (or(V)) and the rules OΠ1, OΣ1, DΠ1, DΣ1 to the set of rules of inference.

As to the consistency of TFP we have the following theorem of weak consistency:

THEOREM 3.1 (WEAK CONSISTENCY). *If Leśniewski's Protothetic is consistent, then TFP is also consistent.*

*Proof.* By Słupecki (1984, p. 117) if in the axiom of Ontology we consider nominal variables and the functor '$\varepsilon$' as the symbol of conjunction, then this axiom becomes a theorem of protothetics. Under this interpretation, all the rules of inference of Ontology become primitive or secondary rules of Protothetics.

Our proposals now could be posed as follows: we also consider name-formative functors as proposition-formative ones and our name quantifiers Π, Σ first as the usual quantifiers, and then as the propositional ones of Protothetics. It is clear that our axioms (Π), (Σ), (Λ), (V) become theorems of Protothetics. The rules of inference of TFP also become secondary rules of Protothetics under this interpretation. This proves the theorem.

The problem of the completeness of TFP is more complicated. But it might be demonstrated that it could be proved by means of the model theory for Leśniewski's Ontology elaborated by Z. Stachniak (1981) while using, *mutatis mutandis*, the complete Boolean algebra as the basic ontological Boolean system.

## 4. TWARDOWSKIAN FORMAL ONTOLOGY

And now the time has come to shift our emphasis and answer the question: do we need to deal exclusively with objects of the 3-4th kind? A negative answer to this question allows us to expand our approach to any kind of object we find in Twardowski's teachings on presentation. And such an expansion leads us from Formal Phenomenology to Formal Ontology when our language becomes more uniform and devoid of limitation.

In order to improve our considerations we should recall the attempt to modify Ontology undertaken by K. Ajdukiewicz in (1949–1950). Loosely speaking, his proposal could be formulated as double reduplication, both of the object domain and the term '$\varepsilon$'. As a consequence we have two kinds of objects — real and intentional — and two kinds of primitive terms '$\varepsilon$' — $\varepsilon_r$ and $\varepsilon_h$ where $x\varepsilon_r y$ means 'the real object $x$ is the real object $y$' and $x\varepsilon_h y$ means 'the intentional object $x$ is the intentional object $y$', i.e. $\varepsilon_r$ deals only with real objects whereas $\varepsilon_h$ deals only with intentional ones. This extension of Ontology allows us to operate with distinct kinds of objects and this is of essential moment to our further inquiry.

The reason is that, very likely, as a first step we may reduplicate our object domain so as to comprise the ontological (real) objects of Twardowski's theory, according to Poli's classification. And then we may introduce into the language of TFP one more primitive term $\varepsilon_h$ which governs only names of objects of presentation. Hence, $x\varepsilon_h y$ will mean 'the object $x$ of presentation is the object $y$ of presentation'. Clearly, our rules and axioms for $\Pi, \Sigma$ will now contain only $\varepsilon_h$ and thus, for example, $O\Pi1$, $D\Pi1$ ought to be rewritten as follows:

$$O\Pi1. \quad \frac{x\varepsilon_h\Pi\psi(y_{11},\ldots,y_{1n_1})\ldots(y_{m1},\ldots,y_{mn_m})}{x\varepsilon_h\psi(a_{11}/y_{11},\ldots,a_{1n_1}/y_{1n_1})\ldots(a_{m1}/y_{m1},\ldots,a_{mn_m}/y_{mn_m})}$$

$$D\Pi1. \quad \frac{\psi(a_{11}/y_{11},\ldots,a_{1n_1}/y_{1n_1})\ldots(a_{m1}/y_{m1},\ldots,a_{mn_m}/y_{mn_m}) \subset_h X}{\Pi\psi(y_{11},\ldots,y_{1n_1})\ldots(y_{m1},\ldots,y_{mn_m}) \subset_h X}$$

where $X \subset_h Y$ means that every '$X$ of presentation is $Y$ of presentation'. As to ($\Lambda$),(V) then we have

($\Lambda$) $\quad y\varepsilon_h\Pi x \equiv y\varepsilon_h\Lambda$

(V) $\quad y\varepsilon_h\Sigma x \equiv y\varepsilon_h V$

positing that, notwithstanding the nature of the objects, the empty object and the universal class of objects is counted the same either for real objects or for objects of presentation.

One more way to convey this distinction can be found in free logics, where a special existence predicate is used to fix really existing objects. Most interesting for us in this connection is the metalinguistic symbol $T$ used by S. Jaśkowski in his natural deduction system which, in contrast to the most other such systems, allows for two different kinds of assumptions (or 'suppositions') (Jaśkowski, 1934). The rules of this system are as follows.

(a) *Supposition of a term:* at any point in a deduction it is possible to introduce an assumption of the form $T\tau$, where $\tau$ is a new term.
(b) *Universal Instantiation:* $\varphi[\tau/x]$ follows from $\forall x$ and $T\tau$.
(c) *Universal Generalization:* if $\varphi$ follows from $T\tau$ then it is possible to deduce $\forall\tau\varphi$, and this conclusion does not depend on the assumption $T\tau$.

The idea of a supposition itself seems very attractive when aiming to distinguish kinds of objects: it is sufficient to regard supposition as presentation and thus the introduction of an assumption of the form $T\tau$ is regarded as an act of presentation when $\tau$ is not simply a new term but the name of an object of presentation. So, let us introduce a predicate of presentation $Ph$ and, as a consequence, the following modification of rules and axioms:

$$D\Pi1. \quad \frac{\psi(\overline{a}_{1n_1}/\overline{y}_{1n_1})\ldots(\overline{a}_{mn_m}/\overline{y}_{mn_m}) \subset X \wedge Ph(\psi_{mn_m})}{\Pi\psi(y_{11},\ldots,y_{1n_1})\ldots(y_{m1},\ldots,y_{mn_m}) \subset X}$$

$$D\Sigma 1. \quad \frac{x\varepsilon\psi(\overline{a}_{1n_1}/\overline{y}_{1n_1})\dots(\overline{a}_{mn_m}/\overline{y}_{mn_m}) \wedge Ph(\psi_{mn_m})}{x\varepsilon\Sigma\psi(y_{11},\dots,y_{1n_1})\dots(y_{m1},\dots,y_{mn_m})}$$

$$(\Sigma) \quad x\varepsilon\psi(\overline{a}_{1n_1}/\overline{y}_{1n_1})\dots(\overline{a}_{mn_m}/\overline{y}_{mn_m}) \wedge Ph(\psi_{mn_m}) \rightarrow x\varepsilon\Sigma\psi(y_{11},\dots,y_{1n_1})\dots(y_{m1},\dots,y_{mn_m})$$

where $\psi(\overline{a}_{1n_1}/\overline{y}_{1n_1})\dots(\overline{a}_{mn_m}/\overline{y}_{mn_m})$ means $\psi(a_{11}/y_{11},\dots,a_{1n_1}/y_{1n_1})\dots \dots(a_{m1}/y_{m1},\dots,a_{mn_m}/y_{mn_m})$ and $Ph(\psi_{mn_m})$ means $Ph(\psi(a_{11}/y_{11},\dots \dots,a_{1n_1}/y_{1n_1})\dots(a_{m1}/y_{m1},\dots,a_{mn_m}/y_{mn_m}))$, because $\Pi$ and $\Sigma$ deal only with objects of presentation. Also it does not come as a surprise that the lack of $Ph(x)$ in $D\Sigma 1$ and $(\Sigma)$ is not of an occasional character: doing so we, in effect, propose to count the nature of '$x$' as universal, i.e. $x$ may be an object of any kind.

The same peculiarity eventually appears to reside in our previous modification of TFP if we extend the meaning of $x\varepsilon_h y$ in the following way: 'an object $x$ is the object $y$ of presentation'. This slight modification will have strong consequences because it destroys the rigid borderline between objects of presentations and real ones.

Finally, what is most essential is that we now, in fact, obtain the desired systems of Twardowskian Formal Ontology (TFO) instead of TFP.

*Section of Logic, Institute of Philosophy Russian Academy of Science*

## REFERENCES

Cocchiarella, N.: 1974, 'Formal Ontology and the Foundations of Mathematics', in G. Nakhikian (ed.), *Bertrand Russell's Philosophy*, Duckworth, London, pp. 29–46.

Husserl, E.: 1913, *Ideen zu einer reinen Phänomenologie und phänomenologischen Philosophie*, Husserliana **III–V**.

Husserl, E.: 1929, *Logische Unterzuchungen*, Husserliana **XVIII 1& 2, XIX**.

Husserl, E.: 1929a, *Formale und transzendentale Logik*, Husserliana **17**.

Husserl, E.: 1950, *Cartesianische Meditationes*, Husserliana **I**.

Husserl, E.: 1952, *Ideen zu einer reiner Phänomenologie und phänomenologischen Philosophie*, II und III Buch, Husserliana **IV, V**.

Husserl, E.: 1954, *Erfahrung und Urteil*, Claassen, Hamburg.

Husserl, E.: 1956–1959, *Erste Philosophie, I Teil: Kritische Ideengeschichte (1923)*, Husserliana **VII**, *II Teil: Theorie der Phänomenologischen Reduktionen (1923/4)*, Husserliana **VIII**.

Husserl, E.: 1966, *Zur Phänomenologie des inneren Zeitbewusstseins (1893–1917)*, Husserliana **X**.

Iwanuś, B.: 1984, 'On Leśniewski's Elementary Ontology', in J.T.J. Srzednicki, V.F. Rickey (eds.), (1984), pp. 165–215.

Jaśkowski, S.: 1934, 'On the rules of supposition in formal logic', *Studia Logica* **1**, pp. 5–32.

Leśniewski, S.: 1927–31, 'O podstawach matematyki', *Przegląd Filozoficzny* **30**, pp. 164–206; **31**, pp. 261–291; **32**, pp. 60–101; **33**, pp. 77–105; **34**, pp. 142–170. Eng. trans. as 'On the Foundations of Mathematics' in Leśniewski, 1991, *Collected Works*, 2 vol., ed. by S. Surma,

J. Srzednicki, D.I. Barnett, V.F. Rickey, Kluwer Academic Publishers, Dordrecht/PWN, Warsaw, vol. I, pp. 181–382.

Poli, R.: 1992, *Ontologia Formale*, Marietti, Genova.

Słupecki, J.: 1984, 'S. Leśniewski's Calculus of Names', in J.T.J. Srzednicki and V.F. Rickey (eds.), (1984), pp. 59–122.

Srzednicki, J.T.J., Rickey, V.F., Czelakowski, J. (eds.): 1984, *Leśniewski's Systems. Ontology and Mereology*, Martinus Nijhoff Publishers, The Hague, Ossolineum, Wroclaw.

Stachniak, Z.: 1981, *Introduction to model theory for Leśniewski's Ontology*, Wydawnictwo Uniwersitetu Wrocławskiego, Wroclaw.

Twardowski, K.: 1894, *Zur Lehre vom Inhalt und Gegenstand der Vorstellungen*, Hölder, Vienna.

Twardowski, K.: 1991, 'A theory of judgments' (in Russian), in B. Dombrowski, *Positive metaphysics. The philosophy of Lvov-Warsaw School*, Preprint of Institute of Appl. Problems of Mathemat. Mech., UkrSSR Acad. Sci., N 5–91, Lvov, pp. 71–92.

Vasyukov, V.L.: 1993, 'A Leśniewskian Guide to Husserl's and Meinong's Jungle', *Axiomates* **1**, pp. 59–74.

Vasyukov, V.L.: 1993b, 'Antidiodorean logics and the Brentano-Husserl conception of time', *Axiomates* 3, pp. 373–388.

TOSHIHARU WARAGAI

# ON SOME ESSENTIAL SUBSYSTEMS OF LEŚNIEWSKI'S ONTOLOGY AND THE EQUIVALENCE BETWEEN THE SINGULAR BARBARA AND THE LAW OF LEIBNIZ IN ONTOLOGY

## 1. INTRODUCTION

The main aim of this paper is to show that in Leśniewski's Ontology, the law of Leibniz and a special case of *Barbara* which we refer to as the *singular Barbara* are equivalent to each other. To show this, we choose a group of theses of Ontology which are sufficient to establish the intended equivalence. From the result it follows that the problematic characters concerning the law of Leibniz reduce either to the validity of the singular *Barbara* or that of the theses used in establishing the intended equivalence.

We suggest that the most dubious thesis is the one which correctly expresses the operation of comprehension. According to this result, we claim 1) that the dubiousness of the law of Leibniz reduces to that of the notion of comprehension and 2) that not every property is convertible to a name. With respect to these results, the doctrine of limitation of size in set theories will be criticized.

Since I presented and used an extended version of Ontology to show the main results at the conference held in Lvov, I will give a sketch of the system I made use of at the conference. The logical relation between the sole axiom of Leśniewski's Ontology and Russell's theory of description will be made clear, too.

## 2. THREE SUBSYSTEMS OF LEŚNIEWSKI'S ONTOLOGY

We consider here three subsystems of Leśniewski's Ontology. The first of them is the most fundamental. The other two pay heed to the degree of extensionality.

### 2.1. *The Essential Subsystem of Ontology*

The first one has as its axioms the following theses:

AO $[ab](a \varepsilon b. \equiv .[\exists x](x \varepsilon a) \wedge [xy](x \varepsilon a \wedge y \varepsilon a. \supset .x \varepsilon y) \wedge$
$[x](x \varepsilon a. \supset .x \varepsilon b))$

AC $[x\phi](x \varepsilon \mathrm{trm}{<}\phi{>} . \equiv .x \varepsilon x \wedge \phi(x))$



*K. Kijania-Placek and J. Woleński (eds.),*
*The Lvov–Warsaw School and Contemporary Philosophy,* 169–180.
© 1998 *Kluwer Academic Publishers.*

AI $[ab](a = b. \equiv .a \varepsilon b \wedge b \varepsilon a)$
AP $[ab](\varepsilon* \text{⦃}a\text{⦄}(b). \equiv .a \varepsilon b)$

where the semantic categories and the intended meaning of the symbols are the following:

| | the semantic category of the symbol | intended meaning |
|---|---|---|
| 1 $\varepsilon$ | $s/n, n$ | singular-proposition formative copula |
| 2 $\varepsilon*$ | $(s/n)/n$ | verbalizer |
| 3 = | $s/n, n$ | singular non-reflexive identity |
| 4 trm | $n/(s/n)$ | predicate-to-name converter |

In the original system of Ontology, AO is the sole axiom, while AC is a thesis introduced into Ontology on the basis of the rule of definition which he called *ontological* definition. AI and AP are theses introduced into Ontology on the basis of the rule which he called *protothetical* definition that is of combinatorial character. Hereafter the system the axioms of which are AO, AC, AI and AP will be referred to as OL. OL is a subsystem of Leśniewski's original Ontology.

### 2.2. *A Naive Semantics of a Sentence of the Form* $[a \varepsilon b]$

Taking into consideration the difficulty in understanding the above axioms, it is desirable to give a naive semantics concerning $[\varepsilon]$ to facilitate the understanding of OL.[1] The naive semantics is the following:

TC '$a \varepsilon b$' is true iff $a$ is an object and $a$ is (a) $b$.

We deduce from TC some results:

TC1 if $a \varepsilon b$ then $a$ is an object.

Under the (apparently) natural supposition that if $a$ is an object, then $a$ is $a$, we have:

TC2 $a \varepsilon a$ iff $a$ is an object.

From TC1 and TC2, we have:

TC3 if $a \varepsilon b$, then $a \varepsilon a$

so that we have:

TC4 if $a \varepsilon b$, then for some $x$: $x \varepsilon a$

We have from TC and TC2:

TC5 $a$ is-identical-with $b$ iff $a \varepsilon b$ and $b \varepsilon b$

As a definition of identity, it suffices to accept the following instead of TC5:

TC6 $a$ is-identical-with $b$ iff $a \varepsilon b$ and $b \varepsilon a$

Now we pass on to the following thesis:

TC7 if $a \varepsilon b$, $x \varepsilon a$ and $y \varepsilon a$ then $x \varepsilon y$

*Proof.* suppose $a \varepsilon b$, $x \varepsilon a$ and $y \varepsilon a$. In the light of TC and *Barbara*, we have that $x$ is $a$. By noticing that $x$ is taken to be an object, we have that $x$ is an

object such that it is $a$. In the same way, $y$ is an object such that it is $a$. Since $a$ is an object, we have that $x$ is-identical-with $y$.
By TC6, we obtain the expected result:

TC8 if $a \varepsilon b$ and $x \varepsilon a$ then $x \varepsilon b$

*Proof.* suppose that $a \varepsilon b$ and $x \varepsilon a$, then by TC, $x$ is an object and $x$ is $a$ and $a$ is $b$. Using *Barbara*, we have that $x$ is an object and $x$ is $b$, i.e. by TC $x \varepsilon b$.
From TC4, TC7 and TC8, we have:

TC9 Under TC, $a \varepsilon b$ implies the right-hand formula of AO.

Now let us examine the reverse direction. The right-hand of AO is the conjunction of the following sentences:

(1) $[\exists x](x \varepsilon a)$
(2) $[xy](x \varepsilon a \wedge y \varepsilon a. \supset .x \varepsilon y)$
(3) $[x](x \varepsilon a. \supset .x \varepsilon b)$

From (1) we conclude that there is an object that is $a$. Let us call it $c$. From (2) we have that there is at most one object that is $a$. Hence $c$ is-identical-with $a$. From this we have $a \varepsilon a$, so that we have $a \varepsilon b$. Thus we have:

TC10 Under TC, the right-hand of AO implies $a \varepsilon b$.

Combining TC9 and TC10, we have that under TC, AO must hold. Thus under TC, AO is a sentence that is true.

### 2.3. *An Extremely Extensional Version of Ontology*

Here in this section we consider an extension of Ontology. Let us define the functor $[\circ]$ in the following way:

DE $a \circ b. \equiv .[x](x \varepsilon a. \equiv .x \varepsilon b)$

If we add the following formula as a thesis of Ontology:

AE $[ab](a \circ b. \supset .[\phi](\phi(a). \equiv .\phi(b)))$

AO becomes equivalent to the following very short thesis:[2]

T1 $[ab](a \varepsilon b. \equiv .[\exists c](a \varepsilon c \wedge c \varepsilon b))$

which is a well known logical fact about Ontology.

A point to be noticed here is that AE is the ontological version of the axiom of extensionality in the axiomatic set theories in the sense that AE treats the names in a completely extensional way, i.e. if two names have the same extension, then they are logically indistinguishable.

## 3. A MODERATELY EXTENSIONAL VERSION OF ONTOLOGY

Since the above claim concerning extensionality is a very strong one, when seen from the point of knowledge representation,[3] we shall not take this position. Instead we shall take a weaker but more plausible version of AE. Since we have good reason to suppose that $[a]$ and $[\mathrm{trm}{<}\{a\}{>}$ (that which is-$a$)$]$ have the same value from a logical point of view, i.e. they are logically indistinguishable in any context, we put the following as a thesis to be added to OL:

$$\text{AEW } [a](\phi(a). \equiv .\phi(\mathrm{trm}{<}\varepsilon{*}\{a\}{>}))$$

and we shall refer to the system thus resulting, i.e. the system ⟨AO, AC, AI, AP, AEW⟩ as OLW. Let us refer to the system ⟨AO, AC, AI, AP, AE⟩ as OLE. OLE is the strongest system, and OL is the weakest among these three systems. OLW is in between. OLW is very suitable as a representational tool for our everyday knowledge network in which it is required to treat the *copula* as a logical relation.[4] What is more, OLW provides the basic logical framework which pays heed to the degree of extensionality.

## 4. A SYSTEM WHICH IS INFERENTIALLY EQUIVALENT TO OLW

Though the main purpose of the present paper is to establish the equivalence between the singular *Barbara* and the law of Leibniz, and to show it it suffices to consider a group of theses of OL, I will present in this section a system that is a natural extension of OL. The system will be referred to as $\mathrm{LI}^+$. The merit of $\mathrm{LI}^+$ is that the axioms of it are intuitively quite easy to capture,[5] and it has the law of Leibniz, with which the main theme of the present paper is concerned, as an axiom. Another reason for presenting this system is that it is a nice natural extension of OL and can be modified in a natural way. Moreover, the real logical relation between Russell's theory of definite description and the sole axiom of Leśniewski's Ontology is easy to see in this system.

### 4.1. *The Axioms of $LI^+$*

Its primitive symbols are just those of OLW. The axioms of $\mathrm{LI}^+$ are the following:

A1 $[ab](a = b. \supset .b = a)$
A2 $[abc](a = b \wedge b = c. \supset .a = c)$
A3 $[xy](x = y. \supset .[\phi](\phi(x). \equiv .\phi(y)))$
A4 $[\phi a](\phi(a). \equiv .\phi(\mathrm{trm}{<}\varepsilon{*}\{a\}{>}))$
A5 $[xa](x\,\varepsilon\,a. \equiv .\varepsilon{*}\{a\}(x))$
A6 $[a](a = a. \equiv .a\,\varepsilon\,a)$
A7 $[\phi\psi](\mathrm{trm}{<}\phi{>}\ \varepsilon\ \mathrm{trm}{<}\psi{>}\ . \equiv .[\exists x](x\,\varepsilon\,x \wedge \phi(x))\wedge$
$[xy](x\,\varepsilon\,x \wedge y\,\varepsilon\,y \wedge \phi(x) \wedge \phi(y). \supset .x = y)\wedge [x](x\,\varepsilon\,x \wedge \phi(x). \supset .\psi(x)))$

### 4.2. *An Explanation of the Axioms of $LI^+$*

A1, A2 and A3 are concerned with the notion of identity. Notice that this identity is non-reflexive, i.e. $[x](x = x)$ does not hold universally.[6] A4 is a kind of extensionality axiom which states that $[a]$ and [that which is-an-$a$] are logically indistinguishable. A5 states that $[x$ is-a $a]$ if and only if $[x$ is-a-$a]$. A6 states that $[a = a]$ if and only if $[a$ is-a $a]$. A7 is an ontological version of Russell's theory of definite descriptions.[7]

### 4.3. *On the Real Logical Relation between Russell's Theory of Description and the Axiom AO of Ontology*

In order to see the logical relation between Russell's theory of description and the sole axiom of Leśniewski's Ontology in a clear manner, we may introduce the functor $[ob(.)]$ the reading of which is [(.) is-an-object] on the basis of TC2.

DO $ob(a). \equiv .a \,\varepsilon\, a$

and rewrite A7, getting:

A7.1 $[\phi\psi](\mathrm{trm}{<}\phi{>} \;\varepsilon\, \mathrm{trm}{<}\psi{>}\,. \equiv .[\exists x](\mathrm{ob}(x) \wedge \phi(x))\wedge$
$[xy](\mathrm{ob}(x)\wedge \mathrm{ob}(y) \wedge \phi(x) \wedge \phi(y). \supset .x = y)\wedge$
$[x](\mathrm{ob}(x) \wedge \phi(x). \supset .\psi(x)))$

Now let us take the so-called object-oriented[8] reading of quantifiers.[9] Then A7.1 becomes the following scheme, $[\mathrm{ob}(x)]$ and $[\mathrm{ob}(y)]$ being absorbed into the quantifiers:[10]

A7.2 $\mathrm{trm}{<}\phi{>} \;\varepsilon\, \mathrm{trm}{<}\psi{>}\,. \equiv .(\exists x)(\phi(x))\wedge$
$(x)(y)(\phi(x) \wedge \phi(y). \supset .x = y) \wedge (x)(\phi(x). \supset .\psi(x))$

According to TC and AC, $[\mathrm{trm}{<}\phi{>} \;\varepsilon\, \mathrm{trm}{<}\psi{>}]$ is read [the unique object such that $\phi$'s $\psi$'s]. Now let us write [the unique object such that $\phi$'s], after Russell, as $[\iota x\phi(x)]$. Then the whole sentence becomes $[\psi(\iota x\phi(x)]$. Hence we have:

A7.3 $\psi(\iota x\phi(x)). \equiv .(\exists x)(\phi(x))\wedge$
$(x)(y)(\phi(x) \wedge \phi(y). \supset .x = y)\wedge (x)(\phi(x). \supset .\psi(x))$

This is the thesis Russell proposed to cope with sentences containing definite descriptions.

Now let us see the reverse direction. If we 1) shift the criterion of ontological commitment from the object-oriented one to TC, [11] and 2) express explicitly the objecthood at the syntactic level by introducing the functor *ob*[12] and 3) accept AC, we obtain A7 from A7.3 simply by quantifying the predicates. Thus:

AO and Russell's theory of definite description are equivalent to each other modulo TC and AC.

This is the real logical relation between these two ingenious theories.

### 4.4. *Basic Results of LI$^+$*

The most important result concerning this axiomatic system is that

LI$^+$ is equivalent to OLW.[13]

so that:

AO, AC, AI are theses in LI$^+$.

A merit of considering this system is that we can see the logical relation between Russell's theory of definite descriptions and the sole axiom of Leśniewski's Ontology in a clear manner. Another merit of LI$^+$ is that it has the law of Leibniz in an explicit way, i.e. as an axiom.[14] As is widely known, this is a law which is viewed to cause some philosophical-logical troubles,[15] so that it is at times regarded as dubious. But as we shall soon see, this is equivalent to the singular *Barbara* under some (apparently) plausible theses of Ontology, so that those who doubt the validity of this law are forced to examine carefully the validity of the special case of *Barbara* or that of the theses used for the establishment of the intended equivalence. In a word, the dubiousness of the law of Leibniz reduces to that of the singular *Barbara* or that of the theses that will be discussed below.

## 5. SINGULAR BARBARA, THE LAW OF SINGULAR EXTENSIONALITY AND ITS SYLLOGISTIC FORM

The singular *Barbara* is the following thesis which we hereafter will refer to as L1:

L1 $[abc](a \varepsilon b \wedge b \varepsilon c. \supset .a \varepsilon c)$

The following formula is the law of Leibniz, which in LI$^+$ has the status of an axiom:

L2 $[ab](a = b. \supset .[\phi](\phi(a). \supset .\phi(b)))$

The following proposition L3 is a thesis in OL:

L3 $[abc](a = b. \supset .(a \varepsilon c. \supset .b \varepsilon c))$

It should be noticed that L3 is a syllogistic form of the law of Leibniz.

## 6. EQUIVALENCE BETWEEN THE LAW OF LEIBNIZ AND SINGULAR BARBARA

We will examine the relationship between L1, L2 and L3 with respect to other theses of Ontology. For this purpose, we shall be concerned with a group of theses in Ontology:

TL1 $[ab](a = b. \equiv .a\,\varepsilon\, b \wedge b\,\varepsilon\, a)$
TL2 $[ab](a = b. \supset .a = a)$
TL3 $[abc](a\,\varepsilon\, b \wedge b\,\varepsilon\, c. \supset .b\,\varepsilon\, a)$
TL4 $[ab](\varepsilon * \not(b\not)(a). \equiv .a\,\varepsilon\, b)$
TL5 $[a\phi](a\,\varepsilon\, \mathrm{trm}{<}\phi{>}\,. \equiv .a\,\varepsilon\, a \wedge \phi(a))$

The following follow directly from TL1 and T2:

TTL1 $[ab](a = b. \supset .b = a)$
TTL2 $[ab](a = b. \supset .a\,\varepsilon\, a)$

Now it is time to cope with the main purpose of the present paper:

L4 TL1$\wedge$TL3. $\supset$:
$$[abc](a = b. \supset .(a\,\varepsilon\, c. \supset .b\,\varepsilon\, c))). \supset .[abc](a\,\varepsilon\, b \wedge b\,\varepsilon\, c. \supset .a\,\varepsilon\, c)$$

| | | |
|---|---|---|
| Pr. | 1 TL1 | [sup.] |
| | 2 TL3 | [sup.] |
| | 3 $[abc](a = b. \supset .(a\,\varepsilon\, c. \supset .b\,\varepsilon\, c))$ | [sup.] |
| | 4 $a\,\varepsilon\, b$ | [sup.] |
| | 5 $b\,\varepsilon\, c$ | [sup.] |
| | 6 $b\,\varepsilon\, a$ | [4,5,2] |
| | 7 $a = b$ | [4,6] |
| | 8 $b = a$ | [7,TTL1] |
| | 9 $b\,\varepsilon\, c. \supset .a\,\varepsilon\, c$ | [3;*a*/*b*,*b*/*a*,8] |
| | 10 $a\,\varepsilon\, c$ | [5,9] |

Thus in a system in which TL1 and TL3 hold, L3 implies L1.
Now we pass on to examining the reverse direction. This holds under TL1.

L5 TL1. $\supset$:
$$[abc](a\,\varepsilon\, b \wedge b\,\varepsilon\, c. \supset .a\,\varepsilon\, c). \supset .[abc](a = b. \supset .(a\,\varepsilon\, c. \supset .b\,\varepsilon\, c))$$

| | | |
|---|---|---|
| Pr. | 1 TL1 | [sup.] |
| | 2 $[abc](a\,\varepsilon\, b \wedge b\,\varepsilon\, c. \supset .a\,\varepsilon\, c)$ | [sup.] |
| | 3 $a = b$ | [sup.] |
| | 4 $a\,\varepsilon\, c$ | [sup.] |
| | 5 $b\,\varepsilon\, a$ | [1,3] |
| | 6 $b\,\varepsilon\, c$ | [4,5,2;*a*/*b*,*b*/*c*,*c*/*c*] |

Thus in a system in which TL1 holds, L1 implies L3. Putting together L4 and L5, we have:

L6 TL1$\wedge$TL3. $\supset$:
$$[abc](a\,\varepsilon\, b \wedge b\,\varepsilon\, c. \supset .a\,\varepsilon\, c). \equiv .[abc](a = b. \supset .(a\,\varepsilon\, c. \supset .b\,\varepsilon\, c))$$

We now pass on to a proof of an equivalence between L2 and L3.

L7 TL4 . $\supset$:
$$[ab](a=b.\supset.[\phi](\phi(a).\supset.\phi(b))).\supset.[abc](a=b.\supset.(a\,\varepsilon\,c.\supset.b\,\varepsilon\,c))$$

| | | | |
|---|---|---|---|
| Pr. | 1 | TL4 | [sup.] |
| | 2 | $[ab](a=b.\supset.[\phi](\phi(a).\supset.\phi(b)))$ | [sup.] |
| | 3 | $a=b$ | [sup.] |
| | 4 | $a\,\varepsilon\,c$ | [sup] |
| | 5 | $[ab](a=b.\supset.(\varepsilon*\langle c\rangle(a).\supset.\varepsilon*\langle c\rangle(b))))$ | [2;$\phi/\varepsilon*\langle c\rangle$] |
| | 6 | $[ab](a=b.\supset.(a\,\varepsilon\,c.\supset.b\,\varepsilon\,c))$ | [5,1] |
| | 7 | $a=b.\supset.(a\,\varepsilon\,c.\supset.b\,\varepsilon\,c)$ | [6;$a/a,b/b$] |
| | 8 | $a\,\varepsilon\,c.\supset.b\,\varepsilon\,c$ | [3,7] |
| | 9 | $b\,\varepsilon\,c$ | [4,8] |

Thus in a system in which TL4 holds, L2 implies L3. Now we prove the reverse direction of L7.

L8 TL1$\wedge$TL2$\wedge$TL5. $\supset$:
$$[abc](a=b.\supset.(a\,\varepsilon\,c.\supset.b\,\varepsilon\,c)).\supset.[ab](a=b.\supset.[\phi](\phi(a).\supset.\phi(b))$$

| | | | |
|---|---|---|---|
| Pr. | 1 | TL1 | [sup.] |
| | 2 | TL2 | [sup.] |
| | 3 | TL5 | [sup.] |
| | 4 | $[abc](a=b.\supset.(a\,\varepsilon\,c.\supset.b\,\varepsilon\,c))$ | [sup.] |
| | 5 | $a=b$ | [sup.] |
| | 6 | $\phi(a)$ | [sup.] |
| | 7 | $a\,\varepsilon\,a$ | [5,2,1(or 5,TTL2)] |
| | 8 | $a\,\varepsilon\,a\wedge\phi(a)$ | [6,7] |
| | 9 | a $\varepsilon$ trm$<\phi>$ | [8,2] |

Hence we have:

L9 TL1$\wedge$TL2$\wedge$T4$\wedge$TL5. $\supset$:
$$[abc](a=b.\supset.(a\,\varepsilon\,c.\supset.b\,\varepsilon\,c)).\equiv.[ab](a=b.\supset.[\phi](\phi(a).\supset.\phi(b)))$$

Combining L6 and L9, we obtain:

L10 TL1$\wedge$TL2$\wedge$TL3$\wedge$TL4$\wedge$TL5.$\supset$.
$$[abc](a\,\varepsilon\,b\wedge b\,\varepsilon\,c.\supset.a\,\varepsilon\,c).\equiv.[ab](a=b.\supset.[\phi](\phi(a).\supset.\phi(b)))$$

Thus we established the equivalence between the singular *Barbara* and the law of Leibniz.

## 7. THEOREMS CONCERNING THE EQUIVALENCE BETWEEN THE SINGULAR BARBARA AND THE LAW OF LEIBNIZ

According to the result obtained in the previous section, we have the following theorem:

THEOREM 1: The law of singular *Barbara* is equivalent to the law of Leibniz under TL1, TL2, TL3, TL4, TL5

If we allow *protothetical* definitions, TL1 can be viewed as a definition of [=]. TL4 is also to be viewed as a definition which introduces *verbalizer.* Then all we need to prove the equivalence between the singular *Barbara* and the law of Leibniz are TL2, TL3 and TL5. Notice that TL5 is the correct formulation of the operation of separation.[16] Thus we proved that singular *Barbara* is equivalent to the law of Leibniz under TL2, TL3 and TL5. Since this result seems to be worth a notice in several respects, we record it as a theorem:

THEOREM 2: The law of singular *Barbara* is equivalent to the law of Leibniz under TL2, TL3 and TL5 and the rule of protothetical definition.

## 8. A REFLECTION ON THE EQUIVALENCE OF THE SINGULAR SYLLOGISM AND THE LAW OF LEIBNIZ

Now we are in a position to state some propositions concerning the dubiousness usually linked to the law of Leibniz:

PROPOSITION 1: The dubiousness of the law of Leibniz reduces to that of the law of singular *Barbara* or that of TL1, TL2, TL3, TL4, TL5.

If we follow the truth condition TC, then TL1, TL2, TL3, TL4 seem to be logically safe. This implies that the dubiousness of the singular *Barbara* reduces to that of TL5, i.e. AC, which is a correct formulation of the operation of comprehension.[17] From this consideration we have the following proposition:

PROPOSITION 2: The dubiousness of the law of Leibniz reduces to that of the law of singular *Barbara* or that of AC.

from which we have:

PROPOSITION 3: If the law of singular *Barbara* is valid, the dubiousness of the law of Leibniz reduces to that of AC.

Insofar as the law of singular *Barbara* is regarded to be a safe logical law, we are forced to conclude:

PROPOSITION 4: The dubiousness of the law of Leibniz reduces to that of AC.

Hence, if the law of Leibniz is dubious, then:

PROPOSITION 5: AC is not universally acceptable.

This has a very interesting implication with respect to the so-called doctrine of limitation of size in set theories. Proposition 5 directly implies:

PROPOSITION 6: There are some predicates which cannot be converted into a name.

## 9. A REMARK ON THE DOCTRINE OF LIMITATION OF SIZE

At this point it should be noticed that the notion of *proper* class we encounter in the axiomatic set theories possibly stems from the fact that not every predicate is convertible to a name already at the level of the logical device we use in everyday knowledge processing. Thus seen, it seems that there is a good reason to doubt the logical validity of the doctrine of size limitation which is used in explaining the notion such as a proper class or a set which is *too big in size.*[18] The doctrine of limitation of size as explained in the textbooks of set theories is far from our (at least my) intuition, though it is very strongly entrenched.[19] Thus we conclude that the validity of the doctrine of size is to be looked for in the logical systems which are suitable for knowledge representation, and not in the axiomatic theories of class/set.

*Tokyo Institute of Technology, O-okayama, Meguro-ku, Tokyo*

## NOTES

[1] Cf. Lejewski (1958). The reader will notice soon that the following TC corresponds to his Ontological Table.

[2] Cf. Sobociński (1933).

[3] We distinguish [the round square] from [Pegasus], [human being] from [featherless biped]. The general position about the names in everyday logical activities is *not* extensional.

[4] It seems that Ontology is the most suitable and consistent theory to treat what is called IS-A relation in the field of artificial intelligence. For this reason it is desirable to enrich Ontology with AEW, and not AE. For IS-A relation, cf. e.g. Brachman (1983).

[5] It does not contain as an axiom AO whose intention is hard to capture. Notice that this difficulty has been the basic obstacle for popularizing Ontology. Concerning the relation of Leśniewski's Ontology to Aristotelian Ontology, cf. Waragai (199a).

[6] On the notion of non-reflexive identity, cf. Lejewski (1967) and Waragai (1987).

[7] As for a detailed discussion of the system, the reader is asked to confer Waragai (1990) and Hiż (1979).

[8] Or, referential.

[9] There are some ways of reading of quantifiers, and the claim that the ontological commitment is immediately linked with quantifiers is just one of the possible readings. There has been a great amount of discussion on this problem. On this topic, cf. e.g. Orenstein (1978), Gochet (1982) and (1985). The so-called object-oriented/referential reading of quantifiers is just a candidate, though it is the most popular one. To maintain that a candidate is the true one is a logical mistake, and we call it at times *dogma.*

[10] We use [(.)] for the general quantifier with the object-oriented reading and [(∃.)] for the particular quantifier with the object-oriented reading.

[11] TC implies directly that we do not follow the so-called object-oriented reading of quantifiers. The category and the place of expressions convey ontological information. In this case, [$a$] appearing in [$a \varepsilon b$] carries the ontological burden.

[12] This is carried out on TC in this case.

[13] For a proof, the reader is asked to consult Waragai (1990).

[14] In OL, it is fairly easy to prove this law.

[15] As an example, cf. Heijenoort (1974a).

[16] Cf. Waragai (1996).

[17] On this topic, cf. Waragai (1996).

[18] On a more detailed discussion, cf. Waragai (1996).

[19] I confess that I have never been able to understand what is meant by the phrase [*too* big in size].

## REFERENCES


Brachman, R.J.: 1983, 'What IS-A is and isn't: An Analysis of Taxonomic Links in Semantic Networks', *IEEE Computer* **16**, pp. 30–36.

Fraenkel, A., Bar-Hillel, Y., Levy, I.: 1973, *Foundations of Set Theory*, second revised edition, North-Holland Publishing Company, Amsterdam.

Gochet, P.: 1982, 'The Syncategorematic treatment of Predicates', in Matilal, B.K. (ed.), *Analytical Philosophy in Comparative Perspective*, D. Reidel Publishing Company, Dordrecht.

Gochet, P.: 1985, *The Ascent to Truth*, Philosophia Verlag, Munich.

Hallet, M.: 1982, *Cantorian theory of set and the limitation of size*, Clarendon Press, Oxford.

Heijenoort, J. Van: 1974, 'On the number of planets', reprinted in *Selected Essays*, Bibliopolis, Naples, 1985, p. 35.

Heijenoort, J. Van: 1974a, 'On Kripke's puzzle', reprinted in *Selected Essays*, Bibliopolis, Naples, 1985, pp. 37–38.

Hintikka, J.: 1982, 'Semantics: A Revolt Against Frege', *Contemporary philosophy. A new survey* **1**, Martinus Nijhoff Publishers, Hague, pp. 57–82.

Hiż, H.: 1977, 'Descriptions in Russell's Theory and in Ontology', *Studia Logica* **36**, 4, pp. 271–283.

Lejewski, C.: 1958, 'On Leśniewski's Ontology', *Ratio* **1**, No. 2, pp. 150–176. Reprinted in Srzednicki, T.J., Rickey, V.F., Czelakowski, J. (eds.) (1984), pp. 123–164.

Lejewski, C.: 1967, 'A Theory of Non-Reflexive Identity and its Ontological Ramification', *Grundfragen der Wissenschaften und ihre Wurzeln in der Metaphysik*, pp. 65–102.

Orenstein, A.: 1978, *Existence and the Particular Quantifier*, Philosophical Monographs, Second Annual Series, Temple University Press.

Orenstein, A.: 1984, 'Referential and Non-referential Substitutional Quantifiers', *Synthese* **60**, pp. 145–157.

Quine, W.V.O.: 1948, 'On What There is', reprinted in Quine (1953).

Quine, W.V.O.: 1953, *From a Logical Point of View*, Harvard UP, Cambridge.

Słupecki, J.: 1955, 'S. Leśniewski's Calculus of Names', *Studia Logica* **1**, pp. 44–111. Reprinted in Srzednicki, T.J., Rickey, V.F., Czelakowski, J. (eds.) (1984), pp. 59–122.

Sobociński, B.: 1930, 'Successive Simplification of the Axiom Systems of Leśniewski's Ontology', reprinted in McCall, S., (ed.), *Polish Logic 1920–1939*, Clarendon Press, Oxford, pp. 188–200.

Srzednicki, T.J., Rickey, V.F., Czelakowski, J., (eds.): 1984, *Leśniewski's Systems*, Martinus Nijhoff Publishers, Hague.

Waragai, T.: 1987, 'Ontology Based on Non-Reflexive Identity and Product Name Functor', *The Annals of Japan Association for Philosophy of Science* **7**, No. 2, pp. 73–84.

Waragai, T.: 1990, 'Ontology as a Natural Extension of Predicate Calculus Equipped with Description', *Annals of the Japan Association for Philosophy of Science* **7**, No. 5, pp. 23–40.

Waragai, T.: 1995, 'IS-A Relation, the Principle of Comprehension, and the Doctrine of Limitation of Size', *The Annals of Japan Association for Philosophy of Science* **9**, No. 9, pp. 23–34.
Waragai, T.: 199a, 'Aristotle's Master Argument about Primary Substance and Leśniewski's Logical Ontology: A Formal Character of Metaphysics', to appear.

# PART III

# PHILOSOPHY OF LANGUAGE

ANDRZEJ GRZEGORCZYK

# THE PARADOX OF GRELLING AND NELSON PRESENTED AS A VERIDICAL OBSERVATION CONCERNING NAMING

Antipsychologism was adopted in semantical analysis at the turn of the 20th century. This might be seen as an advance in the development of European rationalism: it signified an isolation of semantic phenomena and helped analytic philosophers capture the specific structure of semantic relations; their work was done by the end of the first half of our century. However, already at the beginning of the century some undesirable consequences of the antipsychologistic tendency became apparent. I dare to conjecture that it was because of the antipsychologistic paradigm that new semantical antinomies appeared: Antipsychologistic semantics could not find a solution for the new as well as for the old semantical paradoxes, demonstrating that antipsychologism is an idealistic simplification.

I am inclined to say that a psychological analysis enables us to capture the very meaning of fundamental semantic concepts. This analysis also provides philosophers with notions basic to any reasonable philosophical theory of human beings.

In the following I would like to sketch the rudiments of semantics on the basis of a philosophical anthropology which may be regarded as psychologistic. I would like also to proclaim something like a general, and rather trivial, psychologism. The evidence for psychologism (or mentalism) is, roughly speaking, the following:

> Semantics and logic deal with human utterances. Human utterance stems from mental states of human beings and is an object of human mental activity. The basic human mental action towards an utterance is human intentional reference of the utterance to reality.
>
> One can distinguish two fundamental references to reality: the first one consists in *denoting* particular real objects by corresponding expressions, whereas the second is a *narration* about reality.
>
> Assertion and denial are fundamental human cognitive states (attitudes) towards one's own narration or towards somebody else's narrative.
>
> Logic is a set of rules which are tools of the mind. The essential feature of the rules of logic is that they preserve the assertions of the majority of educated (and not deviant) people.



*K. Kijania-Placek and J. Woleński (eds.),*
*The Lvov–Warsaw School and Contemporary Philosophy,* 183–190.
© 1998 *Kluwer Academic Publishers.*

I would like to present a solution to the Grelling-Nelson antinomy from this point of view. A similar solution of the antinomy of the liar is published in my book *Logic a Human Affair* (Wydawnictwo Naukowe Scholar, Warsaw, 1997).

In the following I shall first make a psychological analysis of the action of naming. I shall then concentrate on a property of naming which I will call 'accuracy.' The Grelling-Nelson paradox will turn out to concern the accuracy of words.

## 1. NAMING

The meaning of the verb 'to name' or 'to call' may be defined for a large (and for many purposes sufficiently large) set of objects as follows:

An expression $E$ is used by a person $P$ (or by any person of a group $G$) as the name of the thing $X$ (or: is used for calling the thing $X$) if and only if the person $P$ (any person $P$ of the group $G$) uses the expression $E$ while describing what happens with the thing $X$ in situations which are observed or performed by the person $P$.

For example:

- when $P$ is touching $X$, $P$ says or is ready to say: *I am touching E*,
- when $P$ is pointing to $X$, $P$ is ready to say: *I am pointing to E*,
- when $P$ is eating $X$, $P$ is ready to say: *I am eating E*,
- when $P$ is looking at $X$, $P$ is ready to say: *I am looking at E*. And so on.

Hence the 'meaning' of the verb 'to name' (or 'to call'), may be understood as an account of the use of a given expression by a group of people under consideration in some situations. This concerns individual names as well as collective names.

An expression $E$ is an individual name in the usage of person $P$ iff $E$ is used by person $P$ only to call one thing.
E.g. the expression 'la tour Eiffel' is an individual name of a tower in Paris in the usage of all Frenchmen. The French use the expression 'la tour Eiffel' only when they are pointing to this tower in Paris, or when they are speaking about it.

An expression $E$ is a collective name in the usage of a person $P$ iff there are two different things $X$ and $Y$ such that person $P$ uses the expression $E$ for calling the thing $X$ and also for calling the thing $Y$. E.g. many people say: 'I am pointing to a cat' when they are pointing to one cat or another.

According to the definition above, the same expression may be used in different situations as a name of different concrete things which belong to the same set. Hence one can say that in the case of collective names the use of the verb 'to name' concerns sets and may be also defined as follows:

People of group $G$ call the objects of set $X$ by name $E$ (or: Expression $E$ names the objects of set $X$ in the usage of people of group $G$) if and only if for any member $P$ of group $G$:

- when $P$ is touching an object belonging to $X$, $P$ is ready to say: *I am touching an E*;
- when $P$ is pointing to an object belonging to $X$, then $P$ is ready to say: *I am pointing to an E*. And so on.

Naming is simply the using of names in some situations. Names of things are names which are used by people dealing with things. The use may be more or less consistent. Generally we do not suppose anything about regularity or consistency of the use of names by people. A kind of regularity is be defined as *accuracy*. The accuracy of the use of some names by some people is an interesting semantic property which deserves further study.

(It should be noted that the above definitions of naming (or calling) may be thought insufficiently exact from a formal-methodological point of view. The use of italics in the above definitions cannot be consistently defined. A careful reader will guess that the use of italics in the above definitions plays a role similar to the role of the quotation mark. On the other hand the variables $P$ and $E$ are also written in italics and they are not quotations of the corresponding characters. They are ad-hoc names of some entities, as variables usually are. $P$ stands for a person and $E$ stands for an expression.

Hence to be absolutely correct we should reformulate the above definitions of naming in the following way:
Instead of the phrase:

> $P$ is ready to say: *I am touching E*

it should be written:

> $P$ is ready to say the expression composed of two expressions: first the expression '*I am touching* immediately followed by second which is the expression $E$.

And the analogical change should be done for the expression: *I am pointing* ... and so on.

Such a composition of an expression is called concatenation.)

## 2. ACCURACY

An essential (or at least very important) feature of naming as referring to reality will be called in this paper the *accuracy* of the use of a given name. The 'meaning' of the notion of accuracy will now be explained step by step. First let me note that accuracy should not be confused with the conformity of an individual's use to the standard meaning of a language considered. A user of a given linguistic system may employ a given word either in conformity with the use adopted by the majority of users of the system or not. However, the use of a word by an individual human being or by a small group of people may be more accurate than the use of a large population. Frequently the use of a word established by an individual specialist, or by a very narrow specialised circle of people, is more precise (more accurate) than the use of a larger population.

In a systematic exposition of semantics, naming and its accuracy should be studied for different categories of entities in separate chapters. I would like to mention several possible titles (with a few words about their content) of the consecutive chapters of a systemic exposition of this part of semantics.

### 2.1. *Naming of individual observable things*

The simplest case which should be considered is the naming of individual observable things. This is a very safe semantical operation. We can indicate a thing by a universally accepted gesture and say: 'this is what I call ... '. If the thing is easily distinguishable, no confusion should result. In many situations we may be sure that people accurately use a given individual name of an individual thing.

The intuition of accuracy is in this case very simple. For example: we can say that some parents accurately use the utterance 'our son John' if and only if they use this utterance for and only for calling their son John.

Our normal every-day language allows us to construct names for many individual things, e.g. 'the little finger of my right hand' is a good name of the little finger of my right hand. It is generally assumed that a univocal description of an object may stand for the name of the object. Hence every individual object has many names in every language.

### 2.2. *Naming similarity classes*

The second much more general case is naming similarity classes. Names of similarity classes are the most important class of names. Many natural sets are defined using a relation of similarity. The linguist A. Wierzbicka wrote: "Human being are things like we are." Sometimes we may point out a prototype of a gender. Scientists find a basis for some spectacular similarities of living beings. Biologists say that similarity of genes is responsible for similarity between horses. (They say 'identity' of genes but, then, they are not logicians.)

I hope that my friend zoologist $Z$ accurately uses the name 'horse'. This means that if $Z$ encounters an object $X$ then:

(1) $Z$ names $X$ by the word 'horse' if and only if $X$ is horse.

There is a very important class of similarity classes with which we shall deal below. This is the class of expressions. Expressions are fundamental linguistic entities. Before formulating a general definition of accuracy we ought to first describe the range of linguistic entities which may stand for names, that is, the class of expressions.

The most perceptible objects of linguistic reality are called *individual occurrences of expressions*. They are physical events if they are spoken and physical entities if they are written. An occurrence of an expression occurs in an individual place and time. It is a material thing different from another occurrence written in another place or said at another time. However, we can say that sometimes two occurrences have the same *linguistic shape* which is distinguishable by the users of a given language. Two occurrences are similar or discernible by their linguistic shape. Users of a given language recognise which physical linguistic entity has the same linguistic shape as another. Hence, if we consider the relation between physical linguistic entities expressed by the saying:

a *has the same linguistic shape as* b

then, for a given set of users of that language, this is an empirical relation.

If, for example, we confine ourselves, for a moment, to written text then the physical linguistic entities are individual inscriptions. If $a$ and $b$ are two individual occurrences of some letters then we easily recognise whether $a$ is an occurrence of the same letter as $b$ or not. This means that we recognise whether or not one occurrence of a letter has the same linguistic shape as another occurrence of this letter. We can see that from the perspective of the logical theory of relations: *The relation of having the same linguistic shape is an equivalence relation.* Such a relation may be called a similarity relation. *Expressions are classes of abstraction of the similarity relation of having the same linguistic shape.*

I introduce into our analysis a part of the machinery of set theory to discuss classes of abstraction. (Some linguists try to make use only of words of everyday language in their scientific vocabulary, but this rests on a misunderstanding. The language of scientific inquiry may be different from a language of people who are the objects of our investigation.)

Expressions are often denoted by use of the quotation mark. Putting an inscription of an expression between quotation marks is a general method of making names for expressions which are written. The use of quotation marks seems to be a very safe method of naming things. Employing quotation marks we state the following familiar general rule:

An inscription with a quotation mark at the beginning and at the end is the name of the class of all inscriptions which have the same linguistic shape as the inscription between the quotation marks.

The inscription in quotation marks is the prototype of a class. Naming expressions by the use of quotation marks may be thought of as a kind of iconic naming, that is, naming by showing an icon or a picture of the object of a given class.

We may conclude this part of our analysis concerning the expressions defined as similarity classes by saying that every expression is a similarity class, and is well defined — as well as, for example, the species of horses.

Before I formulate the general definition of accuracy let me mention some other topics that should be considered in a systematic exposition of this part of semantics. Such an exposition would embrace (for example) *names of coherent individual sets* like: London, General Motor Company, my own family, or the set of things in this room. ('My own family' and 'the set of things in this room' are, of course, occasional names, e.g. the first denotes the family of the speaker. It denotes my own family only if I am the speaker of this expression.) Also as a separate category we should consider *names of similarity classes of relations*. (It is remarkable how easily small children pass from understanding that Mother is only the Mother of myself to understanding that every child has its mother. The second understanding is the understanding of a much more abstract situation: a name for a relation.) And at the very end of this sequence of topics we should consider: *names of sets (relations, properties) defined by arbitrary formulas which are well formed in a theoretical language*. This last category is, of course, the most dangerous. The Grelling-Nelson paradox finds its proper place here and demonstrates that naming in this class is not safe at all.

Let us now introduce a general definition of accuracy:

Definition 1. A human being $A$ accurately uses the expression $E$ for denoting the objects of set $S$ if and only if $E$ is an expression and for every object $X$:

(2) $A$ uses $E$ as the name of $X$ if and only if $X$ belongs to $S$.

In the definition above the variable $E$ ranges over expressions. Hence one may substitute for $E$ an expression defined by the use of quotation marks. Thus making the substitutions: $A$ / my friend zoologist $Z$, $E$ / horse, $S$ / the set of horses, we get example (1). Formula (2) is then a generalisation of example (1).

Another example: make the substitutions: $A$ / an anthropologist — $Anth$, $E$ / human being, $S$ / the species *homo sapiens*. We then get:

$Anth$ accurately uses the expression 'human being' for denoting the objects of the species *homo sapiens* iff for every object $X$:

$Anth$ uses 'human being' as the name of $X$ if and only if $X$ belongs to the species *homo sapiens*.

## 3. THE ANALYSIS OF THE PARADOX

Every argument reveals a truth. Logical paradoxes reveal a truth about human beings: about a deficiency of our thinking. The paradoxes show the limits of our possibilities, and the infeasibility of some intellectual dreams. Hence the paradoxes are veridical using W.V. Quine's terminology in 'The Ways of Paradox' (in *The Ways of Paradox*, Random House, New York, 1966).

It is well known that the paradox of the crocodile simply proves that we may give a promise which seems to be realistic if we do not take into account a small fact which makes it infeasible. The paradox of the barber shows the same: no barber can carry out the decision to shave those and only those men of his village who do not shave themselves. In making such a decision he simply forgets that he is also a man of his village.

I would like to interpret the so-called semantical paradoxes as veridical as well, contrary to Quine's claim in the book mentioned above. The Grelling-Nelson paradox dashes the natural human ambition always to behave accurately. Especially if we are intellectuals, we are convinced that we can always use our language accurately. Having the above definition of accuracy we can prove the following theorem without any additional assumptions, using only pure logic, methodology and empirical evidence.

THEOREM 1. There is a methodologically correctly defined empirical notion such that no human being accurately uses this notion.

*Proof.* Let us take an arbitrary human being $A$. Then we may introduce the notion Heterological (Het) as follows:

Definition 2. $X$ is Het if and only if $X$ is an expression and $A$ does not name $X$ by $X$.

Of course, by the convention governing the use of the quotation mark we have the following empirical premise:

(3) 'Het' is an expression.

The set Het is methodologically correctly defined and has a clear empirical meaning. This is always displayed with perverse satisfaction by logicians when they present the antinomies to the profane. Let me repeat two examples. First: no expression is a horse. Hence also the expression 'horse' is not a horse. Then our arbitrary human $A$ does not need to be very clever in order to be sufficiently accurate and not to name the expression 'horse' by the name 'horse'. Hence accordingly to Definition 2:

'horse' is Het.

The second example deals with the expression 'expression of Latin origin', recognised by philologists as an expression of Latin origin. Then we (and our $A$, if sufficiently educated) are entitled to name the expression 'expression of Latin origin' by the expression 'expression of Latin origin'. Hence accordingly to Definition 2:

'expression of Latin origin' is not Het.

The examples are, of course, not necessary for the proof of the Theorem. The formal proof is rather simple. From Definition 1 (accuracy) by putting: $E$ / 'Het', $X$ / 'Het' and $S$ / Het, we get:

If $A$ accurately uses the expression 'Het' for denoting the set Het then:

(4) $A$ names 'Het' by 'Het' if and only if 'Het' belongs to Het.

From the Definition 2 (Het) by putting $X$ / 'Het' we get that:

(5) 'Het' is Het if and only if 'Het' is an expression and $A$ does not name 'Het' by 'Het'.

From (3) and (5) we get:

(6) 'Het' is Het if and only if $A$ does not name 'Het' by 'Het'.

From (4) and (6) we get:

(7) If $A$ accurately uses the expression 'Het', then

'Het' is Het if and only if 'Het' is not Het.

Thus we get the conclusion: $A$ does not accurately use the expression 'Het', irrespective of how clever $A$ may be.

We can state that philosophers are in a position similar to that of barbers. Representatives of both professions cannot realise their dreamed-of design. Barbers cannot shave all inhabitants of their villages who do not shave themselves. Philosophers cannot accurately operate with every notion that they have very correctly defined. (Perhaps the main difference between the two professions is that barbers are paid for shaving (which is easily verifiable), while philosophers have managed to be paid not only for enlightening but also for obscuring minds.)

Theorem 1 is in some sense paradoxical. But in pure mathematics there are much more paradoxical theorems. This theorem may also inspire an interesting methodological endeavor: the search for rules of introducing new empirical notions that preserve the accuracy of introduced terms. Set-theoretical antinomies have contributed to formulations of rules of set-theoretical constructions. Semantical antinomies may play a similar role in the domain of empirical cognition.

*Polish Academy of Science, Warsaw*

MICHELE MARSONET

# DĄMBSKA, QUINE, AND THE SO-CALLED EMPTY NAMES

In a short but important article,[1] the Polish philosopher Izydora Dąmbska criticised the thesis — endorsed by Tadeusz Kotarbiński[2] — that there are 'empty' terms which denote no objects at all, besides the usual general and singular terms. Dąmbska remarked that "we usually find cited as examples of empty names such self-contradictory names as 'square circle' or 'son of a childless mother', or names of mythical deities — fictitious figures that exist only in legends, poems, novels, etc."[3] She also pointed out, however, that the basic semantic function of names consists in denoting. This means that, if we admit the concept of a name which denotes nothing, either we arbitrarily change the meaning of the term 'name' or run into contradiction. In fact, she observed, "for the everyday interpretation of the term 'name', a name which does not denote anything is practically a non-name".[4]

In order to solve this problem, Kotarbiński modified the definition of a name, replacing the usual semantic definition with a syntactic one, so that being a term amounts to being usable as a 'predicative word' within a sentence of the type '*A* is *B*'. By adopting such a strategy, we are allowed to use empty names but, at the same time, there is no true statement in which an empty name might occur as a predicative word. Kotarbiński's proposal, however, is not very helpful from the philosophical viewpoint. No doubt the interesting problem in this context is philosophical and not formal, since it regards primarily the relationships between logic and ontology. We can claim that Kotarbiński's proposal solves the problem only by adopting a view endorsed by logical positivism and some representatives of analytic philosophy, i.e., by identifying logic and ontology. But, if one refuses to adopt such a stance, clearly the problem retains its full philosophical significance.

Dąmbska continued, saying that if "we substitute empty names for both *A* and *B*, or an empty name for *A* and a non-empty one for *B*, we obtain statements which may be true or false. The statement 'Erato is a Muse' is true; 'Erato is one of the Fates' is false; 'Zeus was a deity in Greek mythology' is true; and 'Zeus was a deity in Chinese mythology' is false. The latter case (a statement in which an empty name occurs only as subject) is not an argument against the definition of empty names, but the former is. For to questions about who Erato and Atropos were, I can give the true answer: Erato was a Muse, and Atropos was one of the Fates, even though supposedly neither the Muses nor the Fates exist."[5] Dąmbska cleverly underlined that, here, we face a problem which is first of all metaphysical, even though it obviously has an important



*K. Kijania-Placek and J. Woleński (eds.),*
*The Lvov–Warsaw School and Contemporary Philosophy*, 191–198.
© 1998 *Kluwer Academic Publishers.*

logical side too. One is entitled to claim that the metaphysical dimension of this same problem is not important only if (1) he endorses a materialistic viewpoint (by claiming that only physical, perceptible entities exist), or (2) he rejects any distinction between the logical dimension and the ontological-metaphysical one.

It is interesting to note that in Quine's works we find a position with the same problem mentioned above when dealing with Dąmbska's criticisms of Kotarbiński's stance. Throughout Quine's writings, in fact, we constantly find the question 'What kinds of objects exist?', and the so-called 'ontic decision', that is to say the decision concerning what objects should be admitted into ontology, is the cornerstone of his philosophical research. A central position has been given to these problems in Quine's speculations since the late 1930s, and important formulations may be found in such writings as 'On What There Is', *Word and Object*, and 'Ontological Relativity'.[6] Quine believes that ontological problems can be addressed (and eventually solved) only by means of logic. Obviously, no one means to deny that formal logic is an essential tool when ontological questions are dealt with, but in Quine's view the above mentioned statement is much stronger, since he claims that ontological problems are, at bottom, logical matters: standard quantification theory, and in particular existentially quantified sentences, are in his opinion the ontological idiom *par excellence*. This position leads us to take into account the relationship between language and the world. Why does Quine assume such a position?

The answer is provided by three interconnected stances which can be detected in his works: (A) Only formal logic can adequately describe reality; (B) It is impossible to make a distinction between logic, language and reality; and (C) It is difficult to ascertain whether there is a world independent of the language we use for talking about the world. Quine underlines that the latter — i.e., (C) — is precisely the problem we have to solve in order to get a satisfactory comprehension of what ontology is and what it means to formulate ontological questions. And in fact, when posing existential problems in his works, Quine always adopts this formulation: (D) Are we allowed to use a term $x$ without presuming, by that same fact, the existence of the entity (object) named by $x$? In other words, what worries him is the possibility that a speaker uses non-denoting or general terms just believing that there is something real named by them. But — we may ask — are those really ontological problems?

For example, we may find someone who believes in the existence of winged horses. If he tells us that Pegasus exists, in order to show him that this statement is incorrect we will not use quantification theory but, rather, we will invite him to pursue an empirical recognition. It may turn out, at some point, that he believes in the existence of mental entities, thus rendering a direct empirical recognition quite hard. Even in that case, though, we may agree on the fact that mythological texts report many stories about Pegasus and winged horses.

Provided we are ready to accept an extended ontology including mythological objects, we may also concede that the sentence 'Pegasus is the winged horse of Bellerophon' is true, while the sentence 'Pegasus is the winged horse of Hercules' is false. We may subsequently formalise those two sentences using first order logic, but this move will neither solve nor add anything to the question of whether there are winged horses or not. The solution, in fact, depends on two facts: (1) Our interlocutor's ability to actually find winged horses; and (2) His background ontological assumptions. If he is an empiricist, he will make one claim, and if he is a mentalist he will make yet another claim.

If we pose the previous questions in a Quinean style, we can find out that non-denoting singular terms like 'Pegasus' or general terms like 'centaur' do not work in language as usual names do.[7] No doubt this is an important result, but its consequences concern formal logic and the philosophical analysis of language. It is thus difficult to understand how one may determine the existence (or the non-existence) of certain objects by adopting Quine's point of view. We may find a possible solution to this puzzling problem if, and only if, the existence of a reality independent of knowing subjects is questioned. It would therefore be necessary to adopt a different notion of reality, inextricably connected with language and such that its meaningfulness coincides with our capacity to express it at the linguistic level. And in fact he often gives hints in this direction, claiming that "ontology is internal to language." This move, however, does not justify the elimination of so-called empty names.

Let us note now that Quine always claims that he is a convinced empiricist, and this is the reason why he is so often a prisoner of two alternatives which seem incompatible. (1) On the one hand he supports a strong version of empiricism, according to which reality is simply (and only) what we can have experience of from the sensory viewpoint; but, (2) on the other, he supports a sort of logical and linguistic realism, according to which reality is simply (and only) what we are able to express within our language adequately modified by having recourse to logical formalization. His neopositivist legacy takes him to a form of radical empiricism, while his logical realism leads him in quite a different direction.

Only by taking this inner tension into account is it possible to make sense of Quine's claim that Pegasus does not exist because the term 'Pegasus' does not work as a real name when it is encapsulated in a sentence. It follows — in his opinion — that we must resist the temptation to adopt an overcrowded ontology (an "overpopulated universe").[8] We would like to stress that the attempt to base ontology on this sort of linguistic/aesthetic preference looks unjustified: If reality were truly overcrowded from the ontological viewpoint, would it follow — according to Quine — that we should give priority to his (or even our) preferences rather than to the ontological structure of reality? Maybe Ockham's razor, after all, is not such a good ontological criterion: Why should we decide to simplify, following our personal taste and opinions, a reality which is itself complex?

It is worth noting at this point that Quine himself gives us some good hints for facing the aforementioned problem. Let us ask again: 'What objects are we allowed to admit into our ontology?'. Since according to Quine we cannot admit possible objects, meanings, intentional and abstract entities, what remains to be done is to determine the scope of our ontological commitment. Quine's answer is the following: the existence of any object is relative to the language we speak, and is therefore based on such a language. If this is the situation, it would seem to follow that there are no objects independent of the language we use to talk about them and that, furthermore, there are no objects independent of the linguistic terms which denote them. Perhaps it would be better to substitute the word 'ontology' with the expression 'logical universe of discourse', but Quine never makes such a move, and so we face the following alternative: Either language and reality are the same thing, or language and reality, even though they do not coincide completely, are so inextricably connected that it is impossible to distinguish one from the other. No wonder, then, that when Quine writes about ontology, he always refers to language. Since the existence of objects is determined by the language one speaks, any decision concerning ontological problems implies the clarification of which ontological commitments are embedded in a particular language, and this, in turn, means that speakers of different languages assume (or may assume) different ontological commitments, so that ontology is internal to language as we said before.

At this point we think it is better to abandon talk about logic and logical quantification in order to shift our attention, instead, to conceptual schemes. Nothing prevents us from claiming that there indeed is a reality independent of human subjects. But, at the same time, we should admit that the description of this same reality involves reference to minds. In other words, we should recognize that we get in touch with reality — including ourselves as part of it — through conceptualisation which, in turn, can be expressed by means of thought (plus language). And, having granted this fact, we must now try to verify what kind of consequences may be drawn from it. Since mind is responsible for nature as we understand it, and not for nature itself, we eliminate from the onset one of the major critical points of idealistic philosophies.

On the one hand mind does not causally produce nature but, on the other, we always need the mind for conceiving nature itself: Our reference to the so-called external world is always mind-correlative. Note, however, that we are not compelled to have recourse to apriorism in this context. We can take this creative human capacity as fitting in the order of nature, stemming from a long process of evolution which, although not completely known thus far, can nevertheless be described with sufficient precision. In this process man comes on the stage relatively late, and gradually acquires a predominant position thanks to his capacities which are in turn a product of natural evolution. If we follow this path, the human capacity of conceptualising becomes something natural and explainable, although our comprehension of its historic origin and concrete development is by no means complete.

Here, however, a further step is mandatory. Let us think of (A) reality-as-we-conceive-it. We can grant that it cannot be identified with (B) nature itself which, as such, is mind-independent. What are we supposed to do with reality-as-we-conceive-it? A naive realist would most likely answer that we do not need it: Being a product of our conceptual apparatus, we can just rule out the possibility that it has any ontological value whatsoever. We do not find this position acceptable, and for a very simple reason: How can we deny the existence of entities like dreams or mythological characters which are clearly there, although not in the same sense as physical objects? As Aristotle claims, 'being' can be talked about in many ways. We live in an extremely complex reality. Besides facing a world which is not produced by us, we also build up a reality which is the product of our mind.

On the one hand (i) we cannot identify those two kinds of reality, but on the other (ii) we can neither deny existence to one of the two, because it is a fact that they both exist, nor (iii) can we divide them with too rigid a boundary. There is in fact *continuity* between (A) and (B): A dream cannot be separated from the subject who dreams it. But note — and this is very important for our discourse — that there is continuity even between us and such typical fictional objects as Pegasus. To deny that Pegasus exists despite the fact that (a) mankind created this fictional character and (b) it can be found in our mythology texts is, in our view, untenable. And it should also be noted that there is a difference between dreams and mythological characters. In the case of mythological characters, in fact, we have operative, intersubjective criteria that allow us to say when we pronounce true or false statements about them.

Quine, once again, would disagree with such an 'overcrowded' ontology but, as previously noted, we think that Ockham's razor, which he so often applies, poses too many problems to be adopted uncritically. Quine is still prisoner of the positivist dogma according to which 'real' is only what is material and empirically observable, a dogma that contemporary science has shown to be untenable (think, for example, of quantum mechanics).[9] Moreover, this prompts us to be rather sceptical about Kotarbiński's reism. If reality is complex, ontology is bound to be complex as well: We cannot oversimplify reality according to our personal opinions. So we have different kinds of existence. First of all a conceptual existence which is predicated of conceptual reality: It is tied to our conceptual apparatus and is bound to vanish along with it. Secondly we have a wider concept of existence, which may be predicated of external reality too. It should be clear, however, that both kinds of existence are real, the only difference being that the first is tied to worlds in which men with their minds are present, while the second has a broader range, since it is relative to any kind of world, whether or not it is inhabited by mind-equipped entities.

By granting the mind-independent existence of reality as such, we can admit without problem not only that (1) mind shapes reality as-we-conceive-it, but also that: (2) only through mind can we get in touch with reality itself. Unlike the other entities that form reality itself, in fact, we are mind-equipped, which means that we get in touch with external reality through conceptualisation. However, since this faculty need not to be assimilated to the a priori faculties envisioned by Kant, but may instead be explicable as the outcome of a natural process of evolution, we need not assume, either, that reality itself exists as long as it gets in touch with our perceptual apparatus. We are indeed creators of some type of reality, i.e. reality-as-we-conceive-it, and in this sense literature, philosophy, mythology, etc., are all products of the human mind. *Our* world (i.e., nature-as-we-conceive-it) would not even exist without mind's capacity of conceptualising and, as a matter of fact, we cannot even imagine a different way of getting in touch with reality itself. Certainly no 'vision of the world' would then be possible.

Going back now to Dąmbska's article on empty names, we find there a clear understanding (as early as 1948) of the untenability of the neopositivist and purely linguistic approach, along with some clear anticipations that are today widespread following the post-empiricist turn. She recognized, for example, that the extension of the term 'empty name' varies according to our *Weltanschaung*, so that "For a pious ancient Greek the word 'Zeus' was not an empty name, but for an 18th century atheist the word 'God' was".[10] What follows from such remarks? Resuming our previous distinction between (A) reality-as-we-understand-it, and (B) reality itself, one may wonder whether it is ontological or epistemological. To answer this question, we might say that ontology aims at discovering what kind of entities make up reality ('what there is', to use the Quinean expression), while epistemology's task is to discover the principles by which we get to know reality.

As a matter of fact, however, men understand reality itself (i.e., B) in a great number of different ways, and the history of science gives us a clear picture of this situation. So reality-as-we-understand-it truly changes. Our vision of it is quite different from the vision held by the ancient Greeks or the men living in the Middle Ages. We may admit that a distinction can be drawn between the natural world on the one hand, and the socio-linguistic world on the other. But it should also be specified that, from an historical viewpoint, we began to identify ourselves and the objects that surround us only when the socio-linguistic world emerged from the natural one, and this in turn means that our criteria of identification are essentially social and linguistic. We need an intersubjective criterion giving rise to the notion of a world which is both objective and mind-independent. This means that the subject/object distinction cannot be found *in* nature: It arises when men have such an intersubjective criterion, i.e., within a social world created by men themselves. It is important to note that these remarks do not entail the total identification of the aforementioned two worlds. The conclusion is rather that, of the natural world as such, little can be said. We can suppose that a border-line between ontology and epistemology really exists but, at the same time, as long as *we* are concerned, such a distinction looks less important today than it was usually thought to be.

There are two main reasons which explain why things are so. On the one hand conceptualisation gives us access to the world, while, on the other, it is the most important feature of our cultural evolution (which is distinct from — although not totally separated from — biological evolution).[11] This does not mean to diminish the importance of the latter, which is specifically geared to the natural world and, after all, is supposed to precede our cultural development from the chronological viewpoint. However, it is cultural evolution that distinguishes us from all other living beings that happen to share our planet with us. While the thesis according to which the mind produces natural reality looks hardly tenable, it is reasonable to claim that we perceive this same reality by having recourse to the filter of a conceptual apparatus whose presence is, in turn, geared to the development of language and social organisation.

This is the reason preventing any neat distinction between ontology and epistemology. It is obvious that if our conceptual apparatus is at work even when we try to pave our way towards unconceptualised reality, our access to it entails anyhow the involvement of the mind. Resorting to a paradox, it might be said that the unconceptualised reality turns out to be an image of the mind (even though, it is worth repeating, this recognition does not force us to deny the mind-independent existence of an unconceptualised reality).

So, we must distinguish between 'son of a childless mother' and 'Pegasus'. The latter belongs to mythology, which is part of the social-linguistic world which forms our shared cultural heritage (and this means, in turn, that 'Pegasus' is anything but an empty name). As far as we are concerned, socio-linguistic reality is as real as natural reality, the only difference being that the former is strictly connected to us. We cannot claim that only material-empirical reality exists, or that it is the only reality worth taking into account: Socio-linguistic reality exists as well, and we must make room for it in our ontology.

One should always take into account the broader models (cultural traditions, conceptual schemes) by means of which we judge our sentences — including the mythological ones — to be true or false. They are part of the 'framework of conceptual thinking'[12] and, as long as men are concerned, they can think because they are able to measure their thoughts by having recourse to standards of correctness and of relevance. The aforementioned 'framework of conceptual thinking' somehow transcends the individual thought of individual thinkers.

This explains why there is truth and error with respect to it, even though we may talk of entities which do not exist in the physical world. There is indeed a correct and an incorrect way to describe this framework.

Eventually, we may wonder to what extent Izydora Dąmbska can be deemed to have endorsed such a view. The answer is that this view is somehow implicit in her own words. She realised that "The claim that empty names do not denote anything is [ . . . ] just an unattractive formulation which poisons human souls",[13] but she also kept her analysis at the level of logic and the philosophy of language. Jan Woleński noted in this regard that "Dąmbska's argumentation [ . . . ] raises an important issue, which particularly now is a subject matter of lively discussion among logicians in connection with so-called free logics, i.e., logics without existential assumptions".[14] To this we would like to add, as our previous analysis has hopefully shown, that her short 1948 article is a very interesting anticipation of the current debate on conceptual schemes as well.

*Department of Philosophy, University of Geneva*

NOTES

[1] I. Dąmbska, 'Concerning the So-Called Empty Names', in J. Pelc (ed.): *Semiotics in Poland: 1894–1969*, D. Reidel Publishing Company, Dordrecht-Boston, 1979, pp. 126–130. The paper was originally published in Polish in 1948.

[2] T. Kotarbiński, *Gnosiology: The Scientific Approach to the Theory of Knowledge*, Pergamon Press, Oxford-London, 1966, pp. 389–409.

[3] I. Dąmbska, op. cit., p. 126.

[4] ibid., p. 127.

[5] ibid.

[6] W.V. Quine, 'On What There Is', in W.V. Quine, *From a Logical Point of View*, Harvard University Press, Cambridge (MA)-London, 1980, 4th pr., pp. 1–19; *Word and Object*, The MIT Press, Cambridge (MA), 1960; 'Ontological Relativity', in W.V. Quine, *Ontological Relativity and Other Essays*, Columbia University Press, New York-London, 1969, pp. 26–68.

[7] For a more detailed analysis of Quine's stance see M. Marsonet, *Science, Reality, and Language*, State University of New York Press, Albany (NY), 1995.

[8] W.V. Quine, 'On What There Is', cit., p. 4.

[9] See P. Davies, J. Gribbin, *The Matter Myth*, Simon & Schuster, New York-London, 1992.

[10] I. Dąmbska, ibid.

[11] The distinction biological/cultural evolution is clearly stated in N. Rescher, *A Useful Inheritance: Evolutionary Aspects of the Theory of Knowledge*, Rowman & Littlefield, Savage (MD), 1990. See also M. Marsonet, *The Primacy of Practical Reason. An Essay on Nicholas Rescher's Philosophy*, University Press of America, Lanham-New York-London, 1996.

[12] For a definition of this expression see W. Sellars, 'Philosophy and the Scientific Image of Man', in W. Sellars, *Science, Perception and Reality*, Routledge & Kegan Paul, London-New York, 1963, pp. 1–40.

[13] I. Dąmbska, ibid., p. 130.

[14] J. Woleński, *Logic and Philosophy in the Lvov-Warsaw School*, Kluwer Academic Publishers, Dordrecht, 1989, p. 270.

KRYSTYNA MISIUNA

# TRUTH AND TIME*

## I

Polish philosophers such as Twardowski, Kotarbiński, Łukasiewicz and Leśniewski had been concerned with the concept of truth long before the famous *Wahrheitsbegriff in den formalisierten Sprachen* was published. Each of them made an important contribution to the discussion of the concept of truth relativised to time. Twardowski's paper concerning so called relative truths appeared in 1900. In 1913 Kotarbiński published one of his most impressive philosophical articles 'The Problem of the Existence of the Future'. In the same year two of Leśniewski's papers appeared: the polemical essay 'Is all truth only true eternally or is it also true without a beginning?', and 'A Critique of the Logical Principle of Excluded Middle'. Łukasiewicz was occupied with the concept of truth while he was examining many-valued logics. The first of Łukasiewicz's notes on truth which are important for us may be found in his study *On the Principle of Contradiction in Aristotle* and in his short article 'On the Law of the Excluded Middle' which appeared in 1910. Next came the papers 'On three valued logic' (1920), 'On Determinism' — an address delivered in 1922 — and 'Philosophische Bemerkungen zur mehrwetigen Systemen des Aussagenkalküls' (1930). I do not intend to give a detailed historical account of the discussion of the concept of truth in the Lvov-Warsaw School before Tarski. I would like rather to give a coherent interpretation of some ideas which may be found in the works mentioned above.

## II

Kotarbiński argues that there are sentences which became true at a given moment of time. Although their being true has a beginning, once they became true, they are true eternally. Leśniewski argues, on the other hand, that all truths are true not only eternally, but also sempiternally, that is without a beginning. The controversy is a good starting point to consider more general questions.

1. Are there complete tensed sentences which are neither true nor false?
2. Are there natural tensed sentences which have a third truth-value?
3. Is the law of the excluded middle applicable to declarative sentences without any restrictions?
4. What arguments for the discussion of realistic and epistemic conceptions of truth are provided by our analysis of tensed sentences?
5. What is a tensed interpretation of possibility?



*K. Kijania-Placek and J. Woleński (eds.),*
*The Lvov–Warsaw School and Contemporary Philosophy*, 199–208.
© 1998 *Kluwer Academic Publishers.*

One category of tensed sentences is formed by those containing indexical expressions like 'now', 'tomorrow', 'yesterday', and so on. Even if some sentences do not explicitly include any temporal indexical, like for instance the utterance: 'It is raining', their truth-conditions are essentially dependent on a context of use. This dependence usually reveals a context's parameters appearing in truth conditions.

Sentences with temporal indexicals may be also regarded as incomplete.[1] They become complete when appropriate context parameters are attached to them. It is easy to give an example in question. The sentence

(1) It is now raining in Warsaw

is incomplete and ambiguous. Adding its time datum makes it a complete one, like the following utterance:

(2) On 28 October 1995 at 10.30 a.m. it is raining in Warsaw.

Sentences with indexicals behave semantically like ambiguous expressions. The ambiguity of sentence (1) is a result of the semantic behaviour of the indexical 'now', which may be adequately used at indefinitely many moments of time when this sentence is uttered. Thus, it is obvious that at some moments of time sentence (1) is true, and at others false. That is why such sentences are misleadingly called relative truths. However, this kind of relativity disappears in the case of sentences like (2). What is important is that attaching a time datum creates a difference as to content of respective sentences, which are true or false depending on their content and relevant empirical facts. It is not the case that the same sentence changes its truth-value when uttered at different moments of time. In fact, the different moments of utterance change the content of the sentence in question, and produce different propositions whose truth-value remains unchanged.

Both complete tensed sentences and incomplete ones may be used for forming more complex tensed sentences by means of the so called tense operators, that is, the following kinds of phrases: 'It will be the case that (at least once in the future)', 'It was the case that (at least once in the past)'; or their duals: 'It is always going to be the case', 'It has always been the case'. The analysis of tensed sentences in terms of the tense operators goes back to Prior who wrote:

> I want to suggest that putting a verb into the past or future tense is exactly the same sort of thing as adding an adverb to the sentence. 'I was having my breakfast' is related to 'I am having my breakfast' in exactly the same way as 'I am allegedly having my breakfast' is related to it, and it is only an historical accident that we generally form the past tense by modifying the present tense, e.g. by changing 'am' to 'was', rather than by tacking on an adverb. In a rationalised language with uniform constructions for similar functions we could form the past tense by prefixing to a given sentence the phrase 'It was the case that' or 'It has been the case that' (depending on what sort of past we meant), and the future tense by prefixing 'It will be the case that'.[2]

## III

The distinctions made above enable us to describe the discussion between Kotarbiński and Leśniewski with greater precision. Kotarbiński claims that there is a class of true sentences which became true at a given moment of time. Sentences describing results of human creative activity belong to this class.[3] Obviously, some sentences of this kind will become true at a certain moment of time in the future. I share this conviction, although the argument given by Kotarbiński in support of this view seems to me very misleading.[4] He argues that the same sentence had been neither true nor false before that moment of time at which it became true. To make the matter more perspicuous, let us take an extremely simple example. Suppose that the sentence

(3) This piece of paper is white at $t$

is true at $t$. Hence the sentence

(4) This piece of paper is red at $t$

where 'this piece of paper' denotes in both sentences the same object, is false at $t$. Suppose again that the very piece of paper was painted and became red at $t'$. Thus the sentence

(5) This piece of paper is red at $t'$

is true at $t'$, where $t'$ is later than $t$. The question is: How is it possible to utter seriously the same complete tensed sentence at different moments of time? (4) and (5) are different sentences, because they have different content. Thus it is no wonder that (4) is false at $t$ and (5) is true at $t'$. Every attempt to utter the same complete tensed sentence at different moments of time leads to different utterances each of which has different time datum or is compounded with the past or future tense operator.

Another question arises *when* we evaluate (5) as to its truth-value. If we evaluate it at $t$, where $t$ is an earlier moment of time than $t'$, in fact we evaluate the sentence

(6) It will be the case that this piece of paper is red at $t'$

Sentences (5) and (6) are equivalent — however the latter conforms to common usage. Both sentences are neither true nor false when evaluated at $t$. This example shows that an adequate analysis of tensed sentences requires

(i) a time datum
(ii) a time of evaluation

for each sentence explicitly given. In the case of tensed sentences which do not contain tense operators, the time datum is the same moment of time as that of utterance, as in examples (4) and (5). The moment of utterance differs from the time datum in the case of sentences containing tense operators, as in example (6). The time of evaluation is either the same as the time of utterance, being also the same as the time datum (example (4)), or is earlier than the time datum (examples (5) and (6)). In the case of sentences with a past tense operator, the time of evaluation is later than the time datum. It needs to be emphasised that evaluated at every later moment of time, tensed complete sentences do not change their truth-value, if they have any. Thus (3), being true at $t$, remains true at every moment of time later than $t$, and (4), being false at $t$, remains false at every moment of time later than $t$. Therefore, sentence (4) is false at $t'$, although (5) is true at $t'$.

## IV

Leśniewski[5] holds that there are no sentences which are neither true nor false, and consequently all truths are true without a beginning. His argument goes as follows. Suppose that the sentence of the form '$A$ is $B$' is true evaluated at $t$, and the same sentence is false evaluated at $t'$, where $t'$ is an earlier moment of time than $t$. Then '$A$ is not-$B$' is true at $t'$. However, on the basis of the law of contradiction, if '$A$ is $B$' is true at $t$, then '$A$ is not-$B$' is always false. Hence, '$A$ is not-$B$' is false at $t'$, which contradicts that '$A$ is not-$B$' is true at $t'$. Therefore Leśniewski concludes that the supposition that there are truths which became true at a given moment of time leads to contradiction. However it seems that the argument contains the false assumption that the same sentence evaluated at different moments of time changes its truth-value: it is false at an earlier moment of time, and true at a later moment of time. It seems that the proper candidates for sentences which are neither true nor false, which Leśniewski did not consider, are complete tensed utterances with the future tense operator such as, for instance:

(7) It will be the case that $A$ is $B$ at $t'$

evaluated at $t$, where $t$ is an earlier moment of time than $t'$. However, '$A$ is $B$ at $t'$' evaluated at $t'$ is true or false depending on the relevant facts. The question which arises in connection with future tense sentences is whether all sentences beginning with the future tense operator are neither true nor false. Kotarbiński and Łukasiewicz answer the question in the negative. They hold that not all sentences about future facts are indeterminate today. Łukasiewicz wrote:

I do not doubt at all that some future facts have their causes already in existence today and have had them from eternity. By means of observations and the laws of motion of the heavenly bodies astronomers predict eclipses of the moon and sun with great precision many years in advance. But nobody is able to predict today that a fly which does not yet exist will buzz into my ear at noon on 7 September of next year.[6]

Instead of causes I prefer to speak about temporal properties. Let us consider two different groups of those properties. Suppose that to one of them belongs:

(i) the temporal property of being this white piece of paper painted red,
(ii) the temporal property of being John in Warsaw,

(iii) the temporal property of being John an author of this book.

Let us call such properties contingent temporal properties. Examples of the properties which belong to the second group may be

(i) the temporal property of John's headache during viral infection,
(ii) the temporal property of my increased blood pressure immediately after my drinking a cup of coffee,
(iii) the temporal property of the sun's eclipse.

Let us call such properties permanent temporal properties. By contrast to contingent temporal properties, permanent temporal properties exist today, although they exist intermittently.[7] A sentence with the future tense operator describing a property belonging to the first group is neither true nor false evaluated at the present moment of time. On the other hand, such sentences describing properties belonging to the second group are true or false evaluated today. Contingent temporal properties may (or may not) come into existence in the future. That is why the sentence

(8) It will be the case that John is in Warsaw on 1 November 1996

is indeterminate as to its truth-value on 1 November 1995, but the sentence

(9) It will be the case that my blood pressure is increased on 1 November 1996 immediately after my drinking a cup of coffee

is true evaluated on 1 November 1995.

Maybe someone could argue that a sentence like (8) is not known to be true on 1 November 1995, but in fact it is true, because what is not knowable may nevertheless be true. I do not think that this reasoning is quite correct. What is known about sentence (8) on 1 November 1995 is exactly that it is indeterminate, because the contingent temporal property described by this sentence has not come into existence yet, and its coming into existence later on is not excluded. In this respect, contingent temporal properties differ from permanent and non-temporal ones. The indeterminacy of sentences with future tense operators is not a consequence of an epistemic theory of truth, that is, explaining truth by the concept of verification. Quite the contrary, it is in agreement with a realistic conception of truth, understood as a correspondence between a sentence and a part of the world. Łukasiewicz, who uses the concept of cause explained in terms of 'known laws',[8] seems to be unclear on this matter. It is not clear whether the relevant temporal property does not exist, or the property in question is not known to us.

## V

Łukasiewicz and Kotarbiński, contrary to Leśniewski, hold that there are tensed sentences which are neither true nor false. While Kotarbiński does not provide us with the grammatical structure of sentences which are neither true nor false, Łukasiewicz's examples indicate that those sentences have the form

$$F\alpha(t')$$

where $F$ symbolizes the future tense operator 'It will be the case that', and $(t')$ indicates that a time datum is explicitly stated in $\alpha$. Our analysis shows that sentences of this form evaluated at the time of utterance are neither true nor false, even if we accept metaphysical realism concerning the concept of truth. The acceptance of sentences with truth-value gaps leads to the question of the validity of the principle of the excluded middle. The principle has at least three major versions. Łukasiewicz accepts the following statement: "two contradictory sentences are not false together, that is that one of them must be true."[9] The principle formulated in that way is also called the metalogical principle of the excluded middle. This version of the principle should be discerned from its ontological version which takes rather the following form: "there is so-and-so or there is not so-and-so".[10] In its logical version, the principle says that the formula

$$\alpha \vee \neg\alpha$$

is a tautology, where the logical constants have their classical meanings.

In his paper 'On Determinism', Łukasiewicz in fact rejects the metalogical version of this principle. However, in his three-valued logic, where 1 is the only designated truth-value, the principle in its logical form is rejected. It seems to me that rejection of the metalogical version of the principle need not lead to the rejection of its logical version.

If logical constants have their classical meanings, then the rejection of the metalogical principle of the excluded middle means that it may be the case that neither $\alpha$ nor $\neg\alpha$ is true. Turning to example (8), it is the case that neither $F\alpha(t')$ nor $F\neg\alpha(t')$ is true. Consequently, the class of all sentences of a given language is divided into three subclasses:

(i) sentences which are true,
(ii) sentences which are false,
(iii) sentences which are neither true nor false.

It is worth noticing that in such a language truth conditions for $\neg\alpha$ are not equivalent to the conditions asserting that $\alpha$ does not hold. In other words, in this case we cannot use the standard model-theoretic clause:

$$M \models \neg\alpha \iff M \not\models \alpha.$$

As far as the validity of the logical principle of the excluded middle is concerned, an answer depends on an interpretation of logical constants. If we do not accept the metalogical version of this principle, then we have atomic sentences which are neither true nor false. However even then we may retain the classical interpretation of logical constants and the law of the excluded middle in its logical version. Indeed, it is the case in the semantics of supervaluation, where a formula is called true (false) if and only if it is true (false) by the ordinary classical valuation under every interpretation. Let us consider an example showing how this procedure can be adopted for tensed atomic sentences with the future tense operators. Let us assume that the sentence $F\alpha(t')$, evaluated at the time of utterance $t_u$, is neither true nor false. However, the sentence $F\neg\alpha(t')$ is also neither true nor false when it is evaluated at the same moment of time $t_u$. We shall use brackets for the time of evaluation, so that $[t_u]$ attached to a formula plays a role of a metalanguage parameter. We should examine the validity of the following formula:

(EM) $F\alpha(t')[t_u] \vee F\neg\alpha(t')[t_u]$.

In the semantics of supervaluation the formula (EM) turns out always true, because it is true on every classical valuation under every interpretation, that is when

$$F\alpha(t')[t_u] = 0 \quad \text{and} \quad F\alpha(t')[t_u] = 1.$$

It is worth mentioning that (EM) has the truth-value $u$ ('undefined') when it is evaluated by the Kleene strong rules.[11] In the Łukasiewicz system of three-valued logic, $L_3$, the formula (EM) has the third truth-value, because every contingent sentence evaluated at an earlier moment of time has the third truth-value. Łukasiewicz comes to this conclusion by analysing contingent sentences about the future.

## VI

In his paper 'On Determinism' Łukasiewicz wrote:

> I maintain that there are propositions which are neither true nor false but indeterminate. All sentences about future facts which are not yet decided belong to this category. Such sentences are neither true at the present moment, for they have no real correlate, nor are they false, for their denials too have no real correlate. If we make use of philosophical terminology which is not particularly clear, we could say that ontologically there corresponds to these sentences neither being nor non-being but possibility. Indeterminate sentences, which ontologically have possibility as their correlate, take the third truth-value.[12]

This third truth-value has been interpreted in many philosophically different ways, but I do not intend to review this issue. The question which needs to be considered is the tensed interpretation of possibility. It seems that there is a concept of possibility which does not involve the third truth-value. It is interesting that such an interpretation is suggested by Łukasiewicz who wrote:

If John's presence at or absence from home tomorrow noon is not yet decided, we say 'it is possible that John will be at home tomorrow noon, but also it is possible that John will not be at home tomorrow noon'.[13]

On this interpretation, possibility may be expressed in the following way:

(P1) $\Diamond\alpha$ iff $F\alpha(t')[t_r] \vee F\neg\alpha(t')[t_r]$,

where $\Diamond$ symbolizes the one-place operator 'It is possible that', and logical constants have their classical meanings. Adopting Reichenbach's terminology,[14] we call $t_r$ a time of reference. It is the moment of time included in the time datum in those cases when a time datum is explicitly given. Evaluated at $[t_r]$, $F\alpha(t')$ is true and $F\neg\alpha(t')$ is false, or $F\alpha(t')$ is false and $F\neg\alpha(t')$ is true.

What remains to be considered is the question of which postulates concerning the concept of possibility could suggest the introduction of the third truth-value. Some explanations concerning this issue may be found in Łukasiewicz.[15] One of those postulates could take the following form:

(P2) $\Diamond\alpha$ iff $F\alpha(t')[t_u] \equiv F\neg\alpha(t')[t_u]$,

The equivalence is true in the Łukasiewicz system of three-valued logic, $L_3$, just in case $F\alpha(t')[t_u]$ has the third truth-value, because in that case $F\neg\alpha(t')[t_u]$ also has the third truth-value. To see how this concept of possibility works, let us consider the following inference:

$$\Diamond\alpha, \Diamond\beta \models \Diamond(\alpha\&\beta)$$

It is easy to give a counterexample to this inference. It is possible that it is raining everywhere and possible that it is dry everywhere, but it is not possible to have both these states of affairs obtaining together.[16] This incorrect inference should also remain incorrect on our definition of possibility (P2). Let us notice first that in the three-valued logic $L_3$, each formula: $F\alpha(t')[t_u]$, $F\neg\alpha(t')[t_u]$,$F\beta(t')[t_u]$, $F\neg\beta(t')[t_u]$ has the third truth-value, and each formula $\Diamond\alpha$, $\Diamond\beta$, defined by (P2), has the truth-value 1.[17] The conclusion of our inference takes the following form:

$$F(\alpha\&\beta)(t')[t_u] \equiv F\neg(\alpha\&\beta)(t')[t_u].$$

It is easy to see that if $\alpha$ is replaced by an elementary sentence like 'It is raining everywhere' and $\beta$ is replaced by the sentence 'It is dry everywhere', then $F(\alpha\&\beta)(t')[t_u] = 0$, and $F\neg(\alpha\&\beta)(t')[t_u] = 1$. Thus the above equivalence is false in $L_3$.

What about the Principle of Possibility under definition (P2)? The Principle asserts that if a proposition is true, then it is also possible.[18] Let us assume that $\alpha$ is replaced by a true elementary proposition, like 'Snow is white'. Then $F\alpha(t')[t_u]$ is true, but $F\neg\alpha(t')[t_u]$ is false, therefore $\Diamond\alpha$ is false. Thus the Principle of Possibility is false for (P2). (It is easy to show that the principle is true for (P1))

The discussion shows that the proposals (P2) is unsatisfactory. However, it is easy to modify (P2) so that the new concept of possibility satisfies the Principle of Possibility and invalidates the inference: $\Diamond\alpha, \Diamond\beta \models \Diamond(\alpha\&\beta)$. The modification in question is the following:

(P3) $\Diamond\alpha$ iff $F\neg\alpha(t')[t_u] \rightarrow F\alpha(t')[t_u]$.

If $F\alpha(t')[t_u]$ takes the third truth-value, $\Diamond\alpha = 1$. We cannot so easily obtain a truth-functional concept of possibility in classical two-valued logic. A separate question is that of which formal system of many-valued logic is the best one to provide us with an adequate concept of possibility.

Intuitively, the third truth-value plays the same role as not yet being true nor false. The formula $F\alpha(t')[t_u]$ evaluated at $t_u$, where $t_u$ is the time of utterance, is indeterminate, because the temporal property described by $\alpha$ does not exist at $t_u$. For similar reasons, the formula $F\neg\alpha(t')[t_u]$ is neither true nor false when it is evaluated at $t_u$. The negated property with respect to that described by $\alpha$ also does not exist at $t_u$. Tensed sentences which are neither true nor false at a given moment of evaluation tell us something about future properties of the world. On the other hand, sentences which tell us something about properties which exist or do not exist at the time of evaluation are true or false. A particular property of sentences with the future tense operator, like $F\alpha(t')$, describing future contingent facts, is that being neither true nor false at some time of evaluation they have a determinate truth-value when evaluated at a time sufficiently close to the time datum $t'$. This property of those sentences leads to a serious trouble, if we want to give their truth-conditions.[19]

*Institute of Philosophy, University of Warsaw*

## NOTES

* During the preparation of this paper the author was supported by the KBN grant: GR 511.

[1] This view is characteristic of Twardowski, cf. Twardowski (1900). Similar ideas may be found in Frege, see the discussion in Evans (1985).

[2] Prior (1967), pp. 7–8.

[3] The same idea may be found in Łukasiewicz when he wrote: "We can ourselves somehow shape the future of the world in accordance with our design", see Łukasiewicz (1922), p. 127.

[4] Kotarbiński (1913), pp. 79–81.

[5] Leśniewski (1913).

[6] Łukasiewicz (1922), p. 121.

[7] A discussion of the concept of intermittent existence may be found in Simons (1987), pp. 195 ff.

[8] Łukasiewicz (1922), p. 118.

[9] Łukasiewicz (1922), p. 114.

[10] Reichenbach (1947), p. 164.

[11] Kleene (1952), p. 334.

[12] Łukasiewicz (1922), p. 126.

[13] Łukasiewicz (1922), p. 125.
[14] Reichenbach (1947), pp. 288 ff.
[15] Łukasiewicz (1930).
[16] The example is borrowed from Forbes (1985), p. 5.
[17] It is worth noting that (P2), evaluated by Kleene's rules, is not 1, but 'undefined'. This is a consequence of the fact that Kleene's sentential operations are monotonic, while Łukasiewicz's for $\equiv$ and $\rightarrow$ are not.
[18] von Wright (1951), pp. 21–22 and 32.
[19] The problem is discussed in my paper 'Future Contingencies', forthcoming.

## REFERENCES


Evans, G.: 1985, 'Does Tense Logic Rest upon a Mistake?', in G. Evans, *Collected Papers*, Clarendon Press, Oxford, pp. 343–363.

Forbes, G.: 1985, *The Metaphysics of Modality*, Clarendon Press, Oxford.

Kleene, S.C.: 1952, *Introduction to Metamathematics*, North-Holland Publishing Company, Amsterdam.

Kotarbiński, T.: 1913, 'Zagadnienie istnienia przyszłości' (The Problem of the Existence of the Future), *Przegląd Filozoficzny* **16**, pp. 74–92.

Leśniewski, S.: 1913, 'Czy prawda jest tylko wieczna, czy też i wieczna i odwieczna?', *Nowe Tory* **VIII, 10**, pp. 493–528; Eng. trans. 'Is all Truth only True Eternally or is it also True without a Beginning', in S.J. Surma, J.T. Srzednicki, D.J. Barnett, V.F. Rickey (eds.): *Stanisław Leśniewski: Collected Works*, vol. I , PWN, Warsaw, Kluwer Academic Publishers, Dordrecht, 1992, pp. 86–114.

Łukasiewicz, J.: 1922, 'O determinizmie', in *Z zagadnień logiki i filozofii* (On Logic and Philosophy), PWN, Warsaw, 1961, pp. 114–126. Eng. trans. in *Jan Łukasiewicz: Selected Works*, North-Holland Publishing Company, Amsterdam, PWN, Warsaw, 1970, pp. 110–128.

Łukasiewicz, J.: 1930, 'Philosophische Bemerkungen zur mehrwertigen Systemen des Aussagenkalküls' (Philosophical remarks on many-valued systems of propositional logic), *Comptes rendus de la Société des Sciences at des Lettres de Varsovie*, Cl. III, 23, pp. 51–77, Eng. trans. in S. McCall, (ed.), *Polish Logic 1920–1939*, Clarendon Press, Oxford, 1967, pp. 19–39.

Prior, A.: 1967, *Past, Present, Future*, Oxford University Press, Oxford.

Reichenbach, H.: 1947, *Elements of Symbolic Logic*, Macmillan, New York.

Simons, P.: 1987, *Parts. A Study in Ontology*, Clarendon Press, Oxford.

Twardowski, K.: 1900, 'O tak zwanych prawdach względnych', reprinted in Kazimierz Twardowski: *Wybrane Pisma Filozoficzne* (Selected Philosophical Writings), PWN, Warsaw, 1965, pp. 315–340. Eng. trans. as 'On So-called Relative Truths' in Brandl, J., Woleński, J., *Kasimir Twardowski. Selected Writings*, Rodopi, Amsterdam (forthcoming).

Wright von, G.H.: 1951, *An Essay in Modal Logic*, North-Holland Publishing Company, Amsterdam.

MARIAN PRZEŁĘCKI

# THE POSTULATE OF PRECISION: ITS SENSE AND ITS LIMITS

The postulate of precision is generally considered to be an essential element of philosophical rationalism — in any of the forms it has taken in the history of Western philosophy. It is characteristic, in particular, of that kind of rationalistic philosophy which was proclaimed and practiced by the Lvov-Warsaw School. According to its representatives, precision and sufficient justification are two conditions which any rationally accepted statement must satisfy. The postulate of precision so conceived raises a number of questions. One of them concerns the rationale of the postulate. Why should we require precision of our statements? What reasons may be given for such a requirement? A common opinion maintains that the postulate of precision does not need any justification, since it is self-evident. Its validity becomes obvious if it is couched in suitable terminology. Precision may be identified with clarity, imprecision with obscurity, and these are words with a non-neutral, value-laden meaning. What is clear is good, what is obscure is bad. It seems evident then that we should aim at clarity and avoid obscurity. Such an approach, however, begs the question. It simply assumes that precision is something valuable rather than providing any reasons.

Usually, attempts to provide such reasons treat the precision of expressions as a condition of their intelligibility. Only precise statements are said to be intelligible. So, if we want to be understood, we should be precise in what we say or write. It seems, however, that this argument cannot be accepted without qualification. As is often stressed, our language has different functions, and it is only with respect to some of them that the precision of expressions seems to play an essential role. The functions of language may be divided roughly into cognitive and non-cognitive, the latter including emotive, i.e. expressive or impressive, uses of language. Now, it is only the cognitive use of linguistic expressions that can be shown to require precision. A purely emotive function of language seems to be independent of its precision. We do not need precise statements in order to express our feelings or to impress other minds. On the contrary, we can sometimes do this better by means of enigmatic enunciations and vague slogans. Precision seems to be needed in the cognitive use of language only. This kind of use may be defined as a use concerned with the truth-value of the statements in question. The cognitive contexts of language are those in which the speaker and his audience are interested in the truth-value of the speaker's statement, i.e. in the question whether it is in fact so as



*K. Kijania-Placek and J. Woleński (eds.),*
*The Lvov–Warsaw School and Contemporary Philosophy,* 209–218.
© 1998 *Kluwer Academic Publishers.*

the statement says it is. To decide the question one has to know what exactly the given statement says. Its meaning must be such as to endow the statement with a definite truth-value, to guarantee its truth or falsity. Only then can the question of its validity be decided in one way or other. Now, to satisfy this condition, the meaning of the statement must be sufficiently precise. Its precision is a necessary condition of its decidability. In this one can see a rationale for the postulate of precision.

The kind of reason given in support of the postulate determines its content and its scope. Generally speaking, lack of precision may be said to be important in so far as it affects the truth-value of the given statement. The mechanism of this semantic phenomenon varies with different kinds of imprecision. Two such kinds are usually distinguished: vagueness and ambiguity in its many varieties. Though distinct, both phenomena can be characterised in a common way by means of the general notion of interpretation. An imprecise expression of whatever kind may be defined as expression that admits of different interpretations. Let us call such an expression a semantically indefinite one. Its meaning then may be said to be defined by a class of its admissible interpretations (containing more than one member). In the view of what has been said about the postulate of precision, it is semantically indefinite statements that are the proper object of the postulate. The main problem is how the indefiniteness of a given statement bears on its truth-value, since its possession of a definite truth-value is a necessary condition of its decidability. Only if this condition is fulfilled does it make sense to ask whether a given statement is true or false. From this point of view, two main kinds of indefinite statements may be distinguished: determinate and indeterminate ones. A semantically indefinite statement is determinate if all its admissible interpretations have the same truth-value: all are true or all are false; it is indeterminate if its admissible interpretations differ as to their truth-value: some are true and some are false. It is the occurrence of indeterminate statements that is incompatible with the postulate of precision. If under one interpretation a statement turns out to be true and under another false it cannot be the object of rational discussion: it does not make sense to ask whether it is so as the statement says.

However, it is not only indeterminate statements that give rise to philosophical difficulties. Some fundamental semantic problems are raised by indefinite statements, whether determinate or indeterminate. The main problem concerns the notion of truth with regard to that class of statements. Is the notion of truth applicable to such statements, and in what sense, if any? We can, of course, say that an indefinite statement is true, or false, on a given interpretation, but can we say that it is true, or false, *tout court*? What would this mean?

It is to be noticed that there is a straightforward answer to this question which refers to Tarski's famous schema. It is based on the assumption that the object-language for which the concept of truth is being defined is a part of the meta-language in which the definition is couched. Its interpretation is thus assumed to be given. What any statement of the object-language means may be rendered by the very statement itself. Hence Tarski's schema:

sentence '$p$' is true if and only if $p$,

may be formulated for all statements of the object language, definite as well as indefinite. In this way then, all indefinite statements, including indeterminate ones, may be said to be true or false.

It is obvious, however, that such an approach does not solve the above mentioned problem. It does not provide any explanation of the meaning and truth of indefinite statements; it simply takes them for granted. In order to account for these phenomena, one has to resort to a meta-language which is free from any kind of indefiniteness — which, in contradistinction to the object-language, is fully precise. Within such a language, the interpretation of an indefinite statement cannot be identified with a single sentence; it is to be represented by a class of sentences, in accordance with what we have assumed above. The problem of truth with regard to indefinite statements so interpreted has no straightforward solution. Different ways of coping with it have been proposed and discussed. In order to consider the main ones we have to resort to a more definite conceptual framework.

Let it be the framework of standard model-theoretic semantics. Being generally known, it is in no need of an introduction. Restricted to languages formalisable in standard way, it identifies the interpretation of such a language with a set-theoretic entity, called its model (or possible model). Providing denotations for its constant expressions and ranges for its variables, a model determines a possible interpretation of the given language. The fundamental concept of model-theoretic semantics is the concept of truth which may be termed relative; it is relativised to a given interpretation of the language in question, that is to say, to a given model of it. Its definition explains what it is to say that:

sentence $S$ is true in model $M$.

The definition of the absolute concept of truth — truth *tout court* — is based on the assumption that one of the possible models of the language in question is distinguished as its intended (or actual) model — model which determines its intended (or actual) interpretation. Truth *tout court* is then defined as truth in the intended model.

This semantic approach appears to be restricted to languages free from any kind of indefiniteness. The intended interpretation of an indefinite — vague or ambiguous — expression cannot be identified with its standard model-theoretic denotation. Various ways of solving that difficulty have been suggested. The simplest (and much simplified) solution consists in assigning to an indefinite expression as its intended interpretation a class of its standard model-theoretic denotations — a class which corresponds to all admissible ways of making the expression precise. This applies to vague expressions as well as to ambiguous ones. What, *inter alia*, differentiates these two cases is the formal structure of the corresponding classes of denotations. A vague expression, in contrast with an ambiguous one, seems to be characterised by the fact that the class of its denotations is, as a rule, ordered by the relation of inclusion and contains the greatest and the least element. According to the present approach then, the intended interpretation of a semantically indefinite language is represented by a class of its standard models rather than by a single one. This makes it impossible to define the absolute concept of truth for such a language in the standard way: as truth in the intended model. Is then the absolute concept of truth devoid of any meaning with regard to indefinite languages? Or may it be introduced in some other way? Three main types of answer to this question have been propounded and discussed. According to the first of them, no indefinite statement is either true or false; according to the second, some indefinite statements are either true or false and some are neither; according to the third, all indefinite statements are either true or false.

The first solution sticks with the standard concept of truth defined as truth in a given model, conceived of as the intended model of the language in question. The concept so defined does not apply to languages for which not one intended model, but a class of such models is given. With regard to such languages it makes no sense to ask whether its sentences are simply true or false. All one may ask is whether they are true or false in such-and-such a model belonging to the given class, that is to say on such-and-such an admissible interpretation. It is my contention (which I cannot substantiate here) that in most problems concerning semantically indefinite languages that relative concept of truth proves to be quite sufficient.

The remaining solutions introduce some special concepts of absolute truth for semantically indefinite languages. The second solution defines a notion of truth called supertruth. What is essential for this notion is the distinction between determinate and indeterminate statements. Statements which have the same truth-value on all admissible interpretations are said to be either true or false, statements which do not satisfy this conditions are said to be devoid of an absolute truth-value. If, according to the model-theoretic approach, the intended interpretation of an indefinite language is identified with a certain class of models, called the class of intended models, the definition of the concept of supertruth will read as follows: sentence $S$ is true if and only if it is true in all the intended models; it is false if and only if it is false in all these models. Thus all sentences which in some intended models turn out to be true and in others false are said to be neither true nor false in the absolute sense. These truth-value gaps result in a rejection of some classical meta-logical laws such as the meta-logical Law of Excluded Middle. If sentence $S$ is devoid of any truth-value, so is its negation, not-$S$, while their disjunction, $S$ or not-$S$, is true and their conjunction, $S$ and not-$S$, false. In spite of that, this conception of truth may be said to preserve classical logic. It guarantees that all classical logical laws (including the logical Law of Excluded Middle) are true and the classical logical consequences are valid.

Supertruth theory is well known and widely discussed. In what follows then, I shall concentrate upon the last of the standpoints mentioned — less known and more controversial than the former. It presents an attempt to define the absolute concept of truth for all indefinite statements, including indeterminate ones. The definition refers to different ideas, drawn from philosophy of science as well as from philosophy of language. Philosophers of science appeal to the idea of approximate truth, philosophers of language to the idea of information conveyed by a given statement. It is the last suggestion that seems to be of special importance in the present context.

The fundamental concept of the theories of language which I have in mind[1] is the concept of intralinguistic meaning of a given statement conceived of as its interpretation taken in abstraction from the context of its use. Meaning so understood is assumed to be determined by information conveyed by the statement in isolation from its context; statements which convey the same information are assumed to possess the same meaning. Now, according to the view discussed, information conveyed by a statement which admits of different interpretations corresponds to its weakest interpretation understood as an interpretation equivalent to the disjunction of all its admissible interpretations. In effect, the meaning of an indefinite statement is claimed to be the same as the meaning of its weakest interpretation. And so, taken in isolation from the context of its use, the statement 'John has a pen' informs us only that John has an enclosure for farm animals or an instrument for writing. Its intralinguistic meaning is then claimed to be the same as the meaning of that disjunction. In this way the ambiguity of the original statement has, in fact, been eliminated. This approach to indefiniteness treats that phenomenon as superficial, concerning only the way of expressing a given statement; its proper meaning turns out to be definite, contrary to appearances. Since all sentences obtain thereby a definite interpretation, there is no problem in applying to them the standard concept of truth; they all turn out to be true or false statements.

I think, however, that the above approach should be looked upon in a different way. The approach departs in fact, from the standard concept of truth. A non-standard interpretation of indefinite statements results, in effect, in a non-standard concept of truth as applied to such statements. The assumption that the meaning of an indefinite statement reduces to the meaning of its weakest interpretation defines the truth-conditions for the statement in question: it is said to be true if and only if the disjunction of its admissible interpretation is true, or — in other words — if and only if at least one of the interpretations turns out to be true. Identifying the class of those interpretations with the class of intended models, one arrives at the following model-theoretic definition of the absolute concept of truth for the language in question: sentence $S$ is true if and only if it is true in at least some of the intended models; it is false if and only if it is false in all these models. This is surely a non-standard notion of truth (which, in comparison with supertruth, might be termed subtruth). Though, as the former, it preserves the truth of all classical logical laws, it violates some classical meta-logical laws in a drastic way. If $S$ is an indeterminate sentence, $S$ as well as its negation not-$S$, are said to be true, and their conjunction, $S$ and not-$S$, false. From the logical point of view this seems hardly acceptable, whatever reasons might be adduced for the above conception.

But are the reasons convincing enough? The argument outlined above consists of two steps. The first assumes that an indefinite statement conveys the same information as its weakest interpretation. The second concludes that, in virtue of this, the meaning of the statement is the same as the meaning of its weakest interpretation; the same then also must be their truth-conditions. The authors of the argument do not provide any explication of the notion of information, but refer to its intuitive sense. In the light of it, the initial assumption of the argument seems quite plausible. If a given statement admits of several interpretations, the information that one gets from it seems to be the same as the information conveyed by the disjunction of those interpretations. Let sentence $S$ admit of two interpretations: $S_1$ and $S_2$. We seem warranted in identifying the information conveyed by $S$ with that conveyed by: $S_1$ or $S_2$. But are we thereby warranted to identify the meaning of $S$ with that of $S_1$ or $S_2$? The answer seems to be in the negative. This is especially obvious in cases where the disjunction $S_1$ or $S_2$ does not belong to the admissible interpretations of $S$. Situations like this seem to be quite common. As any English dictionary confirms, 'pen' may mean an enclosure for animals or may mean an instrument for writing, but it cannot mean whatever is either an enclosure for animals or an instrument for writing. The word may refer to either of these sets of objects but not to their union. Generally speaking then, there are cases where the weakest interpretation of an indefinite statement — interpretation equivalent to the disjunction of all its admissible interpretations — does not itself belong to its admissible interpretations. In such cases the meaning of an indefinite statement can hardly be taken to be the same as the meaning of its weakest interpretation. There are, of course, cases which do not raise that difficulty. They concern expressions whose admissible interpretations include the weakest interpretation. This is the case of typical vague expressions, due to the characteristic structure of the class of their denotations. In cases like this we could identify the meaning of an indefinite statement with its weakest interpretation. Such an identification, however, though possible, seems quite unwarranted, in view of the counterexamples mentioned. They cast doubt on the analogy between information and meaning. An indefinite statement may be claimed to convey the same information as its weakest interpretation, but it cannot be said to have the same meaning as the latter. Replacing the statement by its weakest interpretation results in sharpening its meaning and thus modifying it to some extent. Since the original statement is indefinite and the modified one definite, their meaning can hardly be identical. The meaning of a statement seems to be determined by the class of its admissible interpretations; changing that class, we change the statement's meaning.

In view of this, the attempt to define the absolute concept of truth for indefinite statements in the way proposed cannot be considered successful. No acceptable sense of truth as applied to such statements appears to be available. This pertains, first of all, to indeterminate statements. With regard to indefinite but determinate statements, the concept of supertruth seems a plausible suggestion. But it is not that kind of statement that creates the main difficulty. Whether considered true or false or denied any truth value, they do not conflict with the postulate of precision understood as a condition of decidability, since they are, in fact, decidable. The real problem is with indeterminate statements. Being essentially undecidable, they cannot be an object of rational discussion. The postulate of precision requires their elimination, or suitable modification. Generally speaking, they should be sharpened so as to become determinate: admitting only true or only false interpretations.

The view discussed above indicates a way of accomplishing that task which seems free from certain typical objections directed against this kind of procedure. The objections concern the question whether we are warranted to modify our statements in the required way. A sharpening of the meaning of an indefinite statement results, as a rule, in strengthening the statement, i.e. in replacing the original statement by a stronger one. Now, are we entitled to do this? Does the evidence available to us justify the statement so modified? The answer in typical cases seems to be: no. The objection, however, does not pertain to the way of sharpening an indefinite statement which has been considered above. Replacing the statement by its weakest interpretation does not strengthen the original statement. The assumption that an indefinite statement conveys the same information as its weakest interpretation (which we have found plausible) seems to warrant the claim that whatever justifies one of them justifies the other. If we have sufficient reasons for accepting a statement which admits of different interpretations, we have thereby such reasons for accepting any statement equivalent to the disjunction of all those interpretations. We may conclude then that this way of obtaining precision is not vulnerable to the objections mentioned and presents an acceptable general procedure of implementing the postulate of precision.

Let me note, however, that there are various reasons which prevent its realisation in practice. These include linguistic as well as epistemic reasons. The main linguistic reason — well known and widely discussed — refers to the phenomenon of vagueness, which appears to be characteristic of any empirical discourse. I shall not dwell on this point on the present occasion, since it has been an object of detailed examination, especially with regard to the language of empirical theory. It has been convincingly argued that the language of such a theory is linked to empirical reality in an exceedingly loose way, and successful attempts have been made to show how, in spite of that, it is able to fulfill its cognitive functions.

The problem which in the present context calls for some explanation concerns what I have referred to as the epistemic reasons for semantical indefiniteness. In his classic essay 'On clear and obscure styles of philosophical writing',[2] Kazimierz Twardowski puts forward a supposition that "the obscurity of the style in which some philosophers write ... has its source in the vagueness and obscurity of the way they think." I agree with that supposition, but — contrary to the author's opinion — I am inclined to think that the obscurity of philosophical thought may be a "consequence of the factors inherent in (its) subject matter." Because of its 'intricacy', the view a philosopher can get of it reminds a dim outline rather than a clear image. If he is to be true to his insight, he must content himself with some more or less indefinite statements. Since I cannot substantiate that view in general terms I shall restrict myself to considering some examples, which illustrate the point.

One type of example concerns certain moral intuitions, which have been the object of notorious controversies. What seems characteristic of them is their conceptual dependence upon our common-sense notion of 'man' (or of 'human being' — to use a more literary style). The vagueness of this notion results in the indefiniteness of the corresponding intuitions. Vagueness of our common concept of 'man' manifests itself in the existence of various borderline cases. No sharp boundary line can be drawn between human and non-human beings, either in the phylogenetic or in the ontogenetic development of man. According to the common concept of man, a process of becoming (and of ceasing to be) a man is a gradual one. This pertains to the evolution of man from some species of ape as well as to the development of human beings from human zygotes. In some transitional stages of these processes no criteria of application of the concept are available. There are contexts in which vagueness of the concept of man turns out to be highly troublesome; hence various attempts — biological, theological, legal — to get rid of it, or at least to reduce it considerably. The biological definition of man includes in this concept any organism belonging to the human species in any phase of its development, the theological any being endowed with a human soul, etc. All such postulates result in sharpening the common concept and diminishing its original vagueness. Now, my contention is that our fundamental moral intuitions concerning human beings are dependent upon our common-sense concept of man, however vague and enigmatic it may be. They cannot just be transferred to refined and sharpened notions. Whenever we postulate some 'human rights', condemn some acts of 'homicide', the concept of 'human being' that we — explicitly or implicitly — appeal to is the concept present in our everyday discourse. It is 'human being' thus understood that prompts our moral emotions and supports our moral intuitions. To express these intuitions in an adequate way we have to refer in our moral statements to the common concept of man and to accept the fact that its vagueness makes the statements semantically indefinite.

Another example of indefinite philosophical knowledge which I have in mind concerns intuitions of a metaphysical rather than moral nature. Their indefiniteness comes not so much from vagueness as from the ambiguity of some expressions involved in their statement — expressions which, as a rule, have the status of metaphor. To illustrate the point, let me refer to an idea which constitutes an essential element of the philosophical theory presented by Andrzej Grzegorczyk in his treatise 'Life as a challenge'.[3] The title refers directly to the idea in question. Seeing in life — in all that happens to me — a challenge addressed to me is claimed to be an important metaphysical intuition, based on some fundamental existential experience. But what exactly is the meaning of this belief? The statement which says that life is a challenge cannot be taken literally. Expressing it, the author himself provides it with qualifications such as 'as if' or 'like'. It is evident that the word 'challenge' is used here in a metaphorical way. What then is its proper meaning in the context given? Taken literally, 'challenge' means the same as 'summons, call or invitation to trial or contest.' Thus it assumes someone who is summoning, calling or inviting. If what constitutes the challenge is all that happens to me in my life, he who challenges must be identified with God (taken in a personal sense). Such a religious interpretation of that metaphysical belief has been explicitly advanced by Grzegorczyk. May it be regarded as a literal paraphrase of the original metaphorical statement? This is rather doubtful, since with respect to God the notions such as summoning, calling or inviting cannot be understood in a strictly literal way. To explain the notions, the author resorts to the concept of analogical meaning, but in the absence of a satisfactory explication of the latter it can hardly be said to differ from the concept of metaphorical meaning.

The religious interpretation of our metaphysical belief is not the only admissible one. The author himself postulates a more liberal interpretation, which — free from any religious assumptions — might play the role of the weakest interpretation of the given statement. What would it amount to? The metaphorical meaning of a word is said to be typically weaker than its literal meaning; it attributes to the referent only some of the characteristics attributed to it by the literal meaning. In the case of 'challenge' these will be some traits connected with those which characterise a challenge literally understood. Being designed by someone, that which presents a challenge is always something purposeful. If its designer is God, it is not only purposeful, but also providential and helpful. In a metaphorical sense then, we may call something a challenge, if we assume that it is meaningful and acceptable. When I believe that life — all that happens to me — is a challenge, I am convinced that it has a kind of positive meaning for me, that it is not purely accidental, absurd or destructive. The problem is what is meant by 'positive meaning' in this context, in which we cannot appeal to any aim or intention. One cannot escape the conclusion that what we are dealing with is but another metaphor. So our attempts to provide a literal paraphrase of the metaphorical statement considering life as a challenge result in paraphrases which do not seem free from metaphorical expressions and which, like the original statement, suffer from considerable semantical indefiniteness. The reason is that the very metaphysical intuition which the statement tries to convey is — as Twardowski would say — obscure. At the same time, the intuition seems to be — as Grzegorczyk contends — theoretically and practically important, being based on some deep existential experience. Thus, if we do not want to give up that kind of philosophical reflection, we are doomed to some degree of indefiniteness in our philosophical thinking.

The postulate of precision requires that we strive for precision in so far as it is a necessary condition of the decidability of the problems under discussion. But, as our examples suggest, there are some limits to such an endeavour, because there are some limits of precision — of a linguistic and epistemic nature. Absolute precision is unattainable. We must then learn to live with imprecision — to think and to speak rationally in spite of its all-pervasive presence.

*Institute of Philosophy, University of Warsaw*

NOTES

[1] See, e.g., B. Stanosz, 'Meaning and Interpretation', *Semiotica* **11:4**, 1974, pp. 347–358.

[2] See *Semiotics in Poland 1894–1969*, ed. by J. Pelc, Kluwer Academic Publishers, Dordrecht 1979, pp. 1–2.

[3] A. Grzegorczyk, *Życie jako wyzwanie* (Life as a Challenge), Wydawnictwo IFiS PAN, Warsaw, 1993, pp. 103, 243–244.

RYSZARD ZUBER

# POLISH LOGIC, LANGUAGE AND PHILOSOPHY OF LANGUAGE

## 1. INTRODUCTION

In spite of the strong development of logic and logical methods there is still a level of reflection which presupposes a tight interdependence between logic and language. Many members of the Lvov-Warsaw School, who for short I will call Polish logicians,[1] contributed not only to the techniques of logic but also to various domains more properly considered as belonging to the philosophy of language. The obvious observation concerning their work is that although all of them published in various foreign languages, surely all of them were *thinking* in Polish. This paper is an attempt to present some results belonging, roughly speaking, to the philosophy of language and to the theoretical linguistics developed by Polish logicians in the linguistic context of the Polish language. As has sometimes been pointed out, some of Leśniewski's logical systems have some flavour of Polish. Although it was never made quite clear, the main influence of the structure of the Polish language that one thinks about immediately concerns the structure and semantics of Polish, or more generally Slavic noun phrases. Indeed, in Polish there are no definite or indefinite articles, and there are many more bare noun phrases, than, say, in English or French.[2] Probably for this reason Polish logicians in general were much less interested in the problem of definite descriptions than scholars interested in the logical form of natural languages whose native language was English. In fact, the division of names into sub-classes of definite and indefinite descriptions, although very popular among Anglo-Saxon logicians, was not so common among Polish logicians and apparently neither Ajdukiewicz nor Kotarbiński make such a distinction (cf. Pelc, 1971). Furthermore, Polish logicians usually distinguished (at least) two categories of names and this distinction had rather semantic origins, being 'justified' by the number of objects denoted by name. In particular Ajdukiewicz, who conceived his categorial grammar mainly for the purpose of analysing *logical* (thus probably formal) languages apparently felt a need for two such distinct categories of names when analysing 'ordinary' language. I will present some other features of Polish and suggest that in some sense they are also 'responsible' for the final state of some results of Ajdukiewicz and Leśniewski, among others, in the domain related to natural language studies. More specifically, I will try to relate some difficulties Ajdukiewicz had with the complete categorisation of linguistic expressions, in particular quantifiers,



*K. Kijania-Placek and J. Woleński (eds.),*
*The Lvov–Warsaw School and Contemporary Philosophy*, 219–238.
© 1998 *Kluwer Academic Publishers.*

to the difficulties one has with the syntactic and semantic status of noun phrases in Polish. This will lead to some speculations about the difficulties encountered by members of the Polish tradition in logic with the semantic analyses of noun phrases and in particular the absence of the notion of the generalised quantifier at that period.

The paper is organised as follows. First, I will mention some results and ideas of Polish logicians from the Lvov-Warsaw School which I think are relevant to contemporary research in the philosophy of language, in a broad sense. In the next section I will briefly present Ajdukiewicz's categorial grammar, and oppose it to richer contemporary versions of extended categorial grammars, mentioning in particular specific constructions which were made on the basis of those various extensions. Next I will discuss the problem of noun phrases in Polish as opposed to languages having basic determiners. Finally I will speculate as to why the structure of Polish noun phrases prevented Ajdukiewicz and logicians of that period from the proper categorisation of quantifiers and quantified noun phrases and consequently from the necessary introduction of additional rules of category composition.

## 2. SOME POLISH CONTRIBUTIONS TO THE STUDY OF THE STRUCTURE OF NATURAL LANGUAGE

The work of various logicians and philosophers who were members of the Lvov-Warsaw School or their followers are frequently quoted in many introductory or historically oriented papers dealing with various aspects of the contemporary formal studies of natural languages. Their work and ideas are usually mentioned as examples of results which considerably influenced, in a not always specified sense, the development of such studies and their contemporary state. One can notice in this connection, however, that the reasons for which some work was mentioned are not always obvious since the importance of this work can vary and its influence when not clearly established should probably be relativised. Furthermore, in addition to the work often mentioned in the literature and which had some real influence, it is also possible to find various ideas in the work of Polish logicians which preceded some independently developed ideas which seem to be less known, according to the number of their quotations in appropriate contexts. Thus the article of Ajdukiewicz from 1935, usually known in its English translation (Ajdukiewicz 1967), is considered as a pioneering work which influenced research on formal grammars and in particular laid the basis for the various versions of formal grammar known as *extended categorial grammars* (cf. Geach, 1971; Oehrle *et al.*, 1988; Steedman, 1990). In the next section I will present in some detail the main ideas of Ajdukiewicz in this respect. I shall recall why his grammar is empirically inadequate and then show how it has been extended in order to capture more syntactic and semantic phenomena found in natural languages.

Of course Ajdukiewicz's work on the categorisation of linguistic expressions parallels Leśniewski's development of his particular logical systems. I will indicate how some problems in ontology can be related to my considerations on the categorisation of noun phrases in Polish. Here I would like to add that Leśniewski's mereology has also been used as a model for the description of some specific phenomena in natural languages such as the mass-count distinction (Pelletier, 1979). It seems also that one can find in Ajdukiewicz's contribution to the philosophy of language many other important ideas related to some current trends. As is well-known, Ajdukiewicz was very concerned with the notion of meaning, with different senses of this term when applied to various linguistic categories, and many of his papers deal with various aspects of this complex topic. Concerning some of his ideas about the meaning of sentences (or propositions) it has been mentioned for instance that these ideas on meaning can be related to some basic ideas developed in so-called *situation semantics* (cf. Cooper, 1988). In particular the (modern) notion of *situation* might have been very close to the notion of state of affairs as used by Ajdukiewicz.

Another precursory idea can be discerned in Ajdukiewicz's work concerning the problem of *intensionality* of sentential operators formed from verbs of propositional attitudes. These operators give rise to a stronger intensionality (*hyperintensionality*, cf. Cresswell, 1975) than the classical modal operators, for instance. Thus they are sensitive not only to the extension but even to the intension of their arguments, which means that the substitutivity of logical equivalents is not possible in their context. This fact constitutes a serious challenge to formal semantics of constructions involving verbs of propositional attitudes, even if the models used in such semantics are enriched with the concept of possible worlds. In recent semantic literature various proposals have been made to solve this problem. The dominant proposal (Cresswell, 1984; Larson and Ludlow, 1993) is apparently to consider propositional attitude verbs as expressing relations between an agent (referred to by the grammatical subject) and *interpreted logical forms* of their sentential arguments. Interpreted logical forms correspond to *structured meanings* of the sentential arguments since they represent not only semantic values of the arguments but also their 'linguistic structure' or *linguistic mode of presentation*. Technically they can be phrase markers whose nodes pair terminal and non-terminal categorial description with semantic values. In other words, strong intensionality is dealt with by reinforcing the semantic values of arguments which are in the scope of intensional operators by their categorial description.

Now, if one looks at the proposal presented in Ajdukiewicz (1967), a paper based on a talk on intensional expressions given by Ajdukiewicz in 1959, one notices that the resolution of the problem of intensionality proposed there very much resembles those involving *structured meanings* presented above. Indeed, Ajdukiewicz explicitly proposes that we consider objects to which intensional operators can be applied as composed not only of the semantic values of the linguistic components but also of the information concerning the syntactic position of these components. Thus, according to Ajdukiewicz, if we consider two expressions to be (strongly) equivalent if they are not only semantically equivalent but also that their syntactic components have the same syntactic position then the substitution of equivalents is possible in the context of verbs of propositional attitude and consequently they are not intensional (relative to the strong equivalence).

The last result of Polish logicians that I would like to mention is Tarski's famous theorem (or *criterion*) of the finite axiomatisability of deductive systems. Although the importance of results on finite axiomatisability for our understanding of the nature of natural languages was never, as far as I know, discussed in the literature on theoretical linguistics, it seems to me that they may be important for at least two reasons: first, the axiomatisability of extended categorial grammars and, second, in the psychological context of learnability. Tarski's criterion concerns the deductive systems in which in addition to the rule of substitution only the rule *modus ponens* is valid. Now if we consider categorial grammars as particular deductive systems in which the rule of *modus ponens* corresponds to the composition function, then extended categorial grammars are deductive systems with additional rules of category composition, and thus with additional deductive rules. So the axiomatisability of extended categorial grammars is a problem on its own.

The problem of *learnability* is a particular problem which arose fully in the philosophy of language inspired by Chomsky's work in theoretical linguistics. According to Chomsky, one of the major goals of linguistic theory is to give an account of how language learning and mastery is possible (Chomsky, 1986). A normal child is capable of learning any natural language when brought up in the appropriate linguistic environment: a child can learn Polish if brought up in a Polish speaking environment and similarly for any other language and linguistic environment. Given the way a child learns a language this means that the mechanism responsible for the (first) language acquisition must function in any learning situation determined by the pair of a language and corresponding linguistic environment. Consequently natural languages must be learnable and any linguistic theory must indicate what constraints apply to the the ability to learn a language. The part of linguistic study concerned with this problem is *learnability theory*.

One of the formal approaches used in the study of learnability (cf. Wexler and Culicover, 1980) originates in the work of Gold (1967) on the identifiability of language in the limit. According to Gold's proposal a class of languages is said to be learnable if there exists a learner who successfully constructs (learns) a correct grammar from any infinite sequence of sentences corresponding to a language in that class. Language learning is a process in which the learner is presented with a sequence of sentences, one at a time. After he encounters a new sentence he makes a guess as to the form of the grammar of the language, on the basis of finitely many sentences encountered so far. Successive guesses may give different grammars but if after a certain time the learner always indicates one grammar, which is a correct grammar of the language, then the learning is considered successful. Using these ideas of learnability Kanazawa (1996) shows that some specific classes of categorial grammars are identifiable in the limit and thus learnable in the Gold sense. Now, although it has to be shown precisely, it seems to me that results related to the use of the Tarskian criterion of finite axiomatisability (cf. Lemmon, 1965; Czelakowski and Malinowski, 1984; Zuber, 1992) could be used to interpret the results inspired by Gold's work concerning identification in the limit.

As another result, this time clearly from the domain of the philosophy of language which explicitly uses Tarski's criterion, one can mention the work of Tokarz (1994) on the finite non-axiomatisability of Grice's implicatures. Tokarz shows that a deductive system in which conversational implication (*implicature*) is formalised as an operation in a simple propositional language cannot be finitely axiomatised.

## 3. POLISH LOGIC AND EXTENDED CATEGORIAL GRAMMARS

I will consider the syntactic aspects of Polish logic by looking at the concept of categorial grammars and its essential component, the syntactic category. Indeed, one can say that the problem of the syntax of logical expressions, and thus of the categorisation of their parts, is a recurrent problem in the work of Leśniewski, Ajdukiewicz, Tarski and many others. Given the fact that various notions used in categorial grammars have important ramifications in logic in general (cf. van Benthem, 1991) this step seems to me justified.

Although some classes of categorial grammars are formally equivalent to phrase structure grammars, all of them differ from the latter by the philosophy which they impose by way of considering linguistic expressions. Categorial grammars impose a *computational* approach to linguistic constitution, whereas phrase structure grammars correspond to a purely positional and configurational treatment of the syntactic structures. Consequently, categorial grammars categorise any expression as function-expression or as argument-expression, functor and argument, respectively, in Ajdukiewicz's terminology, which I will adopt in what follows. Of course functors need not be lexically complex. In fact, the fundamental idea of categorial grammar is that lexically simple items may be assigned to complex categories corresponding to function expressions. As we will see later on, in the case of extended flexible categorial grammars, an expression can have multiple categorisation and, in particular, in different contexts it can be categorised as a functor and as an argument. It follows, however, from this paradigm, that each complex expression is the result of applying a certain functor to some or all of its admissible arguments.

This way of categorising well-formed expressions has immediate consequences for the (compositional) semantics of the expressions so treated: an expression categorised as a functor $F$ denotes the function $f$, in the algebraic sense, whereas the expressions categorised as arguments $< a_1, \ldots, a_k >$ of $F$ denote objects $< \alpha_1, \ldots, \alpha_k >$ which are in the domain of $f$. Furthermore, the complex expression denotes the value $f(\alpha_1, \ldots, \alpha_k)$.

Given this correlation with the semantic interpretation any functor can be categorised according to the number of arguments it 'takes', the category of the arguments and the category of the resulting complex expression. In Ajdukiewicz's original notation, categories of functors are written with a vertical stroke, the category of the result over the category (or categories) of argument. I will not follow his clear but space consuming notation and adopt the notation in which the horizontal stroke is replaced by the slash and the parenthesis, allowing us to to avoid horizontal stacks of symbols. Thus the symbol A/B designates the category of a functor which takes an expression of the category B and gives as an output the (complex) expression of the category A. Similarly A/$(B_1, \ldots B_k)$ designates a complex category of the functor which takes $k$ argument expressions (each of the category $B_1, \ldots, B_k$ respectively) and gives as output the expression of the category A.

This part concerning the set of categories is usually defined more formally by a recursive definition to the effect that a set of basic categories is first fixed and from it the set of functor categories is defined by various functor creating operations similar to the operation of slash '/'. Then the set of categories is the least set containing the set of basic categories and closed with respect to the functor creating operations. In Ajdukiewicz's categorial grammar the set of basic categories contains only the category *n*–nouns and *s*–sentences. There is only one functor creating operation, the slash operation.

The second part of any categorial grammar is composed of rules of category reduction (or cancellation) which make it possible to categorise a sequence of expressions each of which is already categorised. This categorisation of sequence of expressions is done by the successive application of rules of category reduction to appropriate sub-sequences. Ajdukiewicz uses only one category reduction rule, sometimes called the rule of *function application*: using the conventions of Polish notation, the sequence composed of the expression of the category $c/(c_1, \ldots, c_k)$ followed by $k$ expressions, respectively of the category $c_1, \ldots, c_k$, yields, by the rule of function application, the category $c$.

It is important to realise that the procedure introduced by Ajdukiewicz serves not only for string categorisation. What is more important, it gives an algorithm for deciding for any string of symbols (of a given language) whether this string is 'grammatical' or not, relative to a given basic category of the language. In other words, the set of rules proposed by Ajdukiewicz constitutes a formal grammar. It seems to me that in order to show how these rules can be considered as a formal grammar it is preferable to present first various extensions of Ajdukiewicz's proposal.

Notice first that Ajdukiewicz, being mainly interested in formal languages, uses Polish notation. In order to avoid the use of this purely formal notation, which does not in general correspond to the natural language's syntax, various extensions of Ajdukiewicz's system have been proposed. In many cases these extensions appeared necessary for empirical reasons –in order to account for the regularity and richness of syntactic constructions found in natural languages.

There are various directions in which Ajdukiewicz grammar can be extended. In addition to the possibility of having more basic categories it is possible to introduce new ways of constructing derived or functor categories. In particular, concerning natural languages, it is obvious that the order of functor with respect to its argument should be taken into consideration. Thus the unidirectional slash operation can be split into two, bi-directional operations (cf. Bar-Hillel 1953): '$/_L$' and '$/_R$'. Given this extension the notation $A/_LB$ designates the category of the functor which has on its left the argument of the category $B$ with the resulting complex expression of the category $A$. Similarly $A/_RB$ denotes the category of the functor which takes its argument (immediately) on the right. This extension of the operation, creating a new *backward functor*, necessitates, of course, the addition of the rule of reduction corresponding to these new backward functors.

Of course, the distinction between right and left side arguments of a functor, although in agreement with the syntax of natural languages, is in conflict with some principles of Polish notation and with the existence of $n$–ary functors. Since Ajdukiewicz was interested rather in the description of formal languages, he used Polish notation for sequences of categories, since he thought that by using this notation one could allow for linear, left-to-right (incremental) processing. In fact the situation is more complicated and, as Buszkowski (1995) shows, such left-to-right processing is not possible in general: it is possible, in conjunction with Polish notation, if only unary functors are used and impossible if $n$–ary functors are used which have other functors as arguments. I think that some other cases concerning in particular $n$–ary functors in which all arguments are of the same category — the usual case in natural languages — have yet to be investigated.

It should be added that modern versions of categorial grammars do not use $n$–ary functors but only unary ones. This is formally possible due to the following Schönfinkel-Curry result: any $n$–ary function can be broken down into a series of unary functions. For instance, the binary propositional connective of conjunction which Ajdukiewicz would categorise as $s/(s, s)$ can be categorised as $s/((s/s)/s)$, or, in the bi-directional symbolism, either as $(S/_RS)/_LS$ or as $(S/_LS)/_RS$.

Let me come back to the ways in which categorial grammars can be extended. The new systems of categorial grammars can differ from those proposed by Ajdukiewicz, and also Bar-Hillel, by having category changing rules: some expressions can have many categories, usually related in some systematic way. Since, as we have seen, expressions can be considered as functors or arguments, it is quite possible that given a structure functor-argument, one can invert the role of functor and argument and consider what was argument as functor and other way around. This move, called *type-lifting*, has the following semantic justification: any element $\alpha$ in the domain of some function $f$ can always be identified with a (higher order) function $g$ such that $f(\alpha) = g(f)$. Following this move we obtain the rule of *subject raising* which allows us to associate with the subject noun, which primarily is of the category $n$, the higher order category $s/_R(s/_Ln)$. Thus the subject becomes the functor which takes as its argument the predicate (which is of the category $s/_Ln$).

Another natural rule used in extended categorial grammars concerns items which clearly have a pluri-categorial character. This is the case for instance with many functors corresponding to the Boolean operations. Thus we have not only sentential negation but also predicative negation, nominal negation, etc. Similarly the conjunct can conjoin not only sentences but also nouns, adjectives, etc. In order to account for this phenomenon Geach (1971) proposes the following rule of type shifting (written here without the distinction of directionality): any expression which has the category $A/B$ can also have the category $(A/C)/(B/C)$. Thus a one place sentential functor, like negation, for instance, which is of the category $s/s$, is also, according to the Geach rule, of the category $(s/n)/(s/n)$, which is just a one place functor taking predicates as arguments.

The new rules which have been added have mainly an empirical justification since grammars containing them can generate and accept many more strings than classical categorial grammars. In particular they account, with additional rules of category reduction, for sentences with non-standard constituents like the following: 'Most philosophers hate but Kasia loves logic' where the conjunction *but* conjoins two non-standard constituents 'Most philosophers hate' and 'Kasia loves'. Similarly, sentences with ellipses like, for example, 'Logicians think quickly and philosophers slowly' need additional rules of category reduction.

Usually, additional rules of category reduction are necessary to account for constructions with non-standard constituents, as well as for many other syntactically complicated constructions. For instance, one generally accepts a rule according to which a conjunction of two strings is of the same category as the category of the conjoined expressions (which both have to be of the same category). Another common rule is the rule of *function composition*: the string of two consecutive expressions of the category $A/B$ and $B/C$ respectively is itself of the category $A/C$.

I will not analyse here any specific linguistic construction. In the next section I will make some remarks concerning the categorial structure of noun phrases. Various linguistic applications of categorial grammar can be found in Oehrle *et al.* (1988) and Steedman (1990), for instance. I would like, however, to stress that in general the additional rules of category shifting and category reduction are not given in any arbitrary way: they are logically interrelated and form all together logical systems, very close to the systems of propositional logic we know of (cf. van Benthem, 1991; Zuber, 1994). For instance, as I have already indicated, the only rule of category reduction that Ajdukiewicz uses, the rule of function application, corresponds to *modus ponens*, if the slash is interpreted as the sign of logical implication. Similarly, one notices that the Geach rule corresponds to one of the hypothetical syllogisms and the rule of function composition is equivalent to the transitivity of implication. In fact, the rules which can be added are those which can be derived, in the technical sense, in the grammar considered as a deductive system. In such systems one can also derive rules which, although not having direct linguistic applications, show that systems are 'logically sound'. For instance in any system with the Geach rule one can derive the rule of extensionality (cf. Zuber, 1991).

We can now more formally define a categorial grammar and the notion of the string generated or recognised by it. First we have a finite set $BCAT$ of (basic) category symbols. This set is a subset of the set $CAT$ of all category symbols which is the smallest set containing $BCAT$ and closed with respect to the operations '$/_L$' and '$/_R$'. This means that if $A, B \in CAT$ then $A/_L B \in CAT$ and $A/_R B \in CAT$. The next step is to define the binary relation *reddir*, 'reduces directly to' on pairs of category sequences. This relation essentially uses the rules of category reduction and category shifting presented above. For simplicity let us define this relation using only the rule FA of function application introduced by Ajdukiewicz and which is used by all extended categorial grammars, and the rule FC of function composition. The relation *reddir* (relative to the rule FA) is defined as follows:

For any categories $X, Y$ in $CAT$ and any finite sequence of categories $\alpha$ and $\beta$,

$$\alpha X/_R Y Y\beta \textit{ reddir } \alpha X\beta \textit{ and } \alpha Y X/_L Y\beta \textit{ reddir } \alpha X\beta$$

The relation *reddir* relative to the rule FC is defined as follows, where $Z$ is also a category in $CAT$:

$$X/_RYY/_RZ \text{ } reddir \text{ } X/_RZ$$

The relation *red*, reduces (directly or indirectly), is the reflexive and transitive closure of *reddir*.

Now we have the following definition: a categorial grammar G is a triple

$$\text{G} = < V_T, CAT, F >$$

where $V_T$ is a finite set of terminal symbols (words of the language), $S$ is a designated member of $CAT$ and $F$ is a function assigning to each terminal symbol a set of categories from $CAT$ (thus it is a function from $V_T$ to the power set of $CAT$).

A sequence $w$ of elements of $V_T$ is a string of category $C$ *generated* by G iff $w = w_1, \ldots, w_k$ and there is some sequence of categories $C_1, \ldots, C_k$ such that $F(w_i) = C_i$ for all $1 \leq i \leq k$, and $C_1, \ldots, C_k$ *red* $C$. Thus, $w_1, \ldots, w_k$ is a string of the category $C$ generated by G iff there is some choice of categories assigned by $F$ to the symbols in the sequence $w_1, \ldots, w_k$ which reduces to $C$.

In general, due to the influence of phrase structure grammars, one considers only strings of one, designated, category, usually the category of sentences. The above formulation is more general and allows us to speak about strings as being grammatical in any given category. Of course they are usually grammatical in one category and not grammatical in another category.

## 4. NOUNS AND NOUN PHRASES

An important way of extending categorial grammars is by enriching the inventory of basic categories. The choice can be motivated in various ways, empirical considerations being important but only one among others. In addition different criteria can lead to different choices. Notice, for instance, that the number of basic categories corresponds to the number of propositional variables and thus systems with different number of variables can have different formal properties. Ajdukiewicz, and for that matter Montague, who had mainly a semantic motivation in mind, followed the logical tradition by choosing nouns (roughly proper nouns or expressions referring to individuals) and sentences as basic categories. Lewis (1972) proposes a system in which there is in addition the category of the common noun among the basic categories. In some cases a purely linguistic motivation leads to the choice of other categories (cf. McGee Wood, 1993) — in particular the category of embedded sentences (*that S*, which is a sentence with a sentential determiner).

Notice that on the semantic side, where syntactic categories correspond to logical *types* (used in the corresponding logical language), some other decisions can be made. For instance given the existence of the phenomenon of intensionality Montague proposed the type corresponding to a possible world, which need not have a syntactic correlate. On the other hand it is not necessary to posit some types as basic given the fact that some of them are interdefinable. For instance the type of individuals can be defined with the help of the type of properties as a filter, in the algebraic sense, in the algebra of properties. Similarly the type of properties can be defined with the help of the type of individuals and the type interpreting sentences as a function from the set of individuals to the truth values. In fact it has been suggested that properties and truth values, enriched by the Boolean structure, may be the only basic types (cf. Keenan and Faltz, 1985).

So we are in the following situation: all proposed systems include sentences as a basic category and their interpretation as a basic type. Divergences begin with respect to nouns (i.e. common nouns denoting properties) and what linguists call noun phrases, whose principal subclass are proper nouns (denoting individuals). Although semantically the two types corresponding to these two categories need not be considered as basic, an important empirical problem is whether they should be distinguished at the syntactic level in a language like Polish. In the remaining part of this section I would like to develop this problem and relate it to some historical observations concerning the category of quantifiers in the work of Polish logicians.

It might be useful to recall the criteria used to distinguish expressions such as nouns (*nazwa* in Polish) and other expressions. This classification, also presented in various manuals of general logic, sometimes uses an auxiliary notion of the *atomic sentence of the form A is B* (*A jest B*). Then an expression is qualified as a noun (*nazwa* in Polish) if it can occur as a subject (*podmiot*) or as a nominal predicate (*orzecznik*) in an atomic sentence (Kotarbiński, 1961). A much more restricted definition is given by Czeżowski (1949) who considers only the expressions which can occur as the subject in this construction as nouns.

Let me make some remarks about the above definitions. First of all notice that the notion of subject is hopelessly ambiguous and that to explain the notion of 'nominal predicate' one probably needs to use the notion of the noun itself. In addition for the definition of Kotarbiński we have another problem, specific to the Polish language: the nouns in Polish are often case marked and thus one should decide whether case marking plays any role in the definition of a noun. This question is not trivial since the nouns in the predicate should in principle be in the instrumental case. One cannot answer this question by saying that the case marking is just an accidental linguistic feature without importance for the logical status of nouns. The reason is that the case marking can change the status of the noun and in particular switch a nominal expression from its role as argument into that of a functor. This is in particular the case with the genitive case's marking of the possessive clauses: in the expression *problem logika* (logician's problem) the noun *logika* is in the genitive case and should be considered as a functor which takes the common noun *problem* and gives the noun phrase *problem logika*. Thus case marking can be of importance when the categorial status of nouns is considered.

Moreover, Polish is a language with a relatively flexible word order. In general the role of word order in any language is far from trivial (cf. Zuber, 1995). In the case at hand one notices that in Polish the order of the basic sentence '*A* is *B*' can be reversed and consequently, according to the proposed criteria, any noun which was in the predicative position (and marked by the instrumental case) can take the subject position. Thus in parallel with *Kazio jest logikiem* (Kazio is a logician) we have in Polish *Logikiem jest Kazio*. But still it does not follow from this that the noun in the instrumental case is the (logical) subject. It only means that some other criteria for subjecthood are needed. And indeed it is known that not all syntactic operations apply equally to the subject noun and to the predicate noun. For instance in Polish as in many other languages the cleft operation is not possible on the predicate noun: we have *To Kazio jest logikiem* (It is Kazio who is a logician) but not **To logikiem jest Kazio*.

Finally, another difficulty with the application of the '*A* is *B*' test to Polish is the fact that strictly speaking the position of *A* cannot be easily occupied by a bare common noun (in the nominative case): sequences like *Logik jest filozofem* (Logician is a philosopher) or *Matematyk jest szaloną osobą* (*Mathematician is a mad person) are not very acceptable even with generic interpretations.[3] If, however, the common nouns in these sequences are completed by a determiner the sequences become acceptable: thus *Ten/każdy/nasz logik jest filozofem* (This/every/ our logician is a philosopher) are grammatical and easily interpretable. Given the fact that similar sequences with a proper noun in the subject position are also acceptable, this means that proper names and common names with a determiner, and not just proper nouns and common nouns, are of the same category.

One of the most important tests used in theoretical linguistics to establish that two expressions belong to the same category is based on conjunction: if the conjunction of two expressions is grammatical then these two expressions belong to the same category and in addition their category is the same as the category of the whole conjunction (cf. Lyons, 1968). We have already seen the general form of this principle in the case of non-standard constituents. Its application to Polish nouns shows that we should again distinguish between noun phrases and common nouns. For indeed, according to the principle of conjunction we have to consider proper nouns as being different from common nouns: their conjunction is in principle not possible, as the non-acceptability of, or difficulty with, the non-contextual interpretation of the following sequences shows: **Jan i student są logikami* (*Jan and student are logicians) or ?*Ani Jan ani student nie jest logikiem* (*Neither Jan nor student is a logician). These sequences become much better from the point of view of grammar if we place in the corresponding conjunctions nominal determiners such as *ten/Kazia/każdy* (this, Kazio's, every) etc. before bare common nouns. Furthermore, the conjunction of bare common nouns with (complex) nouns phrases (i.e. common nouns proceeded by a determiner) is in general not easily acceptable in Polish either. This means that in general bare common nouns and common nouns with a determiner (i.e. noun phrases) do not belong to the same category.

The two categories, the category CN, common noun, and the category NP, noun phrases, differ with respect to the type of objects they denote. I will indicate the type of NPs later on, but for the moment let me notice that they cannot both denote the same type. For suppose, for instance, that the NP *at least five philosophers* denotes the set containing at least five philosophers. Then in general there are many such sets and in a model with an infinite universe of discourse in which there are at least five philosophers there would be an infinite number of such sets. This would mean that the indicated NP denotes an infinite number of objects and so has an infinite number of 'meanings', which is absurd. For similar reasons the NP *all logicians* cannot denote the set containing all logicians and the NP *most students* cannot denote the set containing most students. Consequently the NPs do not denote sets (or properties) as do CNs.

So what is the status of NPs as opposed to CNs? First consider the syntactic structure of NPs: they are usually complex expressions, even in Polish, and in the simplest case they are of the form $DetCN$, where $Det$ is the so-called *nominal determiner* which has the category $NP/_{L}CN$. CNs, on the other hand, need not be complex and this fact was used to show another important difference between NPs and CNs. Keenan (1987) showed that CNs, and not NPs are *lexically free* in the sense that anything which is denotable by a complex expression of that category is available as a denotation for a lexically simple expression in that category.

Determiners can also be binary and, more generally $n$–ary (cf. Keenan and Moss, 1985). As an example of a binary determiner one can quote the determiner *more ... than ...* as in *more logicians than philosophers*. Unary determiners can be discontinuous in the same way as binary ones, like the determiner *most ... including the saddest one* which forms the NP *most philosophers including the saddest one*. Now, the point which is important for us is, on the one hand, that the so-called articles, which Polish lacks, are nominal determiners, and, on the other hand, the traditional quantifiers like *every, all, some* are also determiners. Various results concerning the determiners in other languages, English in particular (Keenan and Stavi, 1986), can be applied to Polish as well and they show without doubt that the lack of articles in no way implies the poverty of the system of determiners in Polish. Among these determiners, in addition to the 'classical' quantifiers, one finds many which have also clearly quantificational meaning, such as numerals, proportional determiners, the determiner *most*, etc. They all have the same category: a functor which applies to one or more common nouns and gives an NP.

Notice that I am not claiming that in Polish all NPs explicitly contain a determiner. As in many languages in Polish there are bare NPs without overt determiners and very often they are of the same nature as in other languages. This is in particular the case with NPs corresponding to mass terms and with plurals. I want only to stress that in some cases nominal determiners are necessary components of NPs and are fundamental to the difference between CNs and NPs. Similarly in Polish as in other languages there are other noun-like elements. In particular there are *that S* and *whether S*, nominalisation clauses and infinitives which behave in many respects as NPs. As far as I know they were not systematically studied by Polish logicians.

Concerning the mass/count distinction I would like to mention a specific semantic effect due to the nominal diminutives when applied to mass terms. In Polish there exists a so-called *diminutive suffix* which applies to a common noun and gives another common noun: *pies-piesek*, *dom-domek*, etc. Although the exact nature of this operation is far from clear, one can notice the following two points: the CNs in the diminutive form cannot be used in the predicate position and very often the application of the diminutive suffix changes a mass term into a count term. Thus the sentence *Sokrates/to zwierzę jest psem* (Socrates/this animal is a dog) is quite normal but when the CN *psem* (a dog, instrumental case) is replaced by its diminutive form, the sentence ceases to be acceptable. The second effect of the diminutive suffix can be illustrated by the following pairs of CNs where the second member, which has the meaning of a count term, is obtained from the first, having the meaning of the mass term, by the application of the diminutive suffix: *ziarno-ziarenko* (grain/seed), *słoma-słomka* (straw), etc.[4] Although mereology was invoked in this connection, the count-mass distinction was not discussed by Polish logicians in spite of its apparent philosophical and logical implications (cf. Pelletier, 1979).

To conclude this section I shall say a few words about the semantics of NPs. A popular and important approach to the semantic aspects of determiners and NPs is so-called *generalised quantifier theory* (Barwise and Cooper, 1981; Keenan and Faltz, 1985; Keenan and Stavi, 1986). In this theory, which originated with the work of Mostowski (1957) and Lindström (1966), nominal determiners and NPs are interpreted by so-called *generalised quantifiers* which are higher order objects: sets of sets, relations between sets, etc. NPs in the subject position are functions from the set of properties (which are denotations of the predicate of the intransitive verb phrase) into the set of truth values. Thus they are sets of sets. The unary determiners are binary relations between sets (the denotations of the common noun phrases which are arguments of the determiner) and denotations of the verb phrases; binary determiners are relations between Cartesian products (of the denotations of their two arguments) and sets (denotations of the verb phrases), etc. In this higher order logic the use of variables is not necessary, roughly speaking, and thus the logical representation is closer to the surface form of natural language sentences. Moreover, truth conditions can be easily expressed in a set-theoretical language enriched by the $\lambda$-operator.

## 5. A SOMEWHAT SPECULATIVE CONCLUSION

Although Polish, probably like most languages, is a language where the distinction between common nouns and noun phrases is less apparent, this distinction clearly exists in Polish. The fact that this is a purely syntactic distinction was apparently not perceived by Polish logicians. They were trying to continue the Aristotelian tradition by classifying nouns based on the type of the denoted set. This base of classification is not sound from the point of view of modern linguistics given the fact that all (at least major) categories are Boolean and thus can be conjoined and disjoined and the semantic interpretation of Boolean conjunction makes it impossible to preserve the cardinality of the denoted set. For this reason, for instance, in Leśniewski's ontology the conjunction of nouns in the subject position is impossible since all conjoined nouns in true sentences would denote the same object, contrary to what happens in natural languages, including Polish.

As we have seen, NPs are in general syntactically complex since they are composed of determiners and common nouns. Although lexical items like *every* and *some* are part of traditional lexical vocabulary, syntactically they are nominal determiners in the same way as *most*, numerals, possessives, etc. and their Boolean compositions (*most but not all, most or at least three)*. This means that they are primarily functors (applying to common nouns) and as such should be analysed with their arguments. Polish logicians analysed them as variable binders, with unrestricted variables, because traditional logical quantifiers always have unrestricted forms: *All A are P* is equivalent to *All existing objects are such that if they are A they are B*. This move made it possible to omit the argument of *all, every or some* corresponding to the presupposed universe and consider quantifiers in isolation from their argument (in the form of a common noun). But it is not true that all quantifiers can be decomposed in this way: the so-called generalised universal (for instance *All..but n*) and generalised existential have such equivalent forms but many others are *inherently sortal* (cf. Keenan, 1993) and do not have the corresponding unrestricted form. Thus *Most A are B* does not have an equivalent sentence in the form *Most existing objects are such that* ... . Similarly the quantifier corresponding to the definite article *the* is inherently sortal. But since it is lacking in their language, Polish logicians did not have to worry about it.

The difficulty Ajdukiewicz had with the categorisation of quantifying expressions may be due to the inadequacy of his logical treatment of quantifiers. Being interested, at least in the context of his work on categorial grammar, in the language of first order logic, Ajdukiewicz had to use as a target language a rather inadequate language, from the syntactic point of view, to represent sentences of natural languages with quantified expressions in which quantifier-like expressions were associated with (non-existing on the surface) variables and not with the corresponding common nouns which are their syntactic arguments. In first order language the only noun phrases are terms, which are syntactically simple objects, like proper nouns. At the same time the quantifier expressions or determiners did not have their natural arguments since the basic expressions in first order logic are expressions with unrestricted quantifiers. Ajdukiewicz's attempts to find the appropriate category of quantifier expressions failed because he did not realise that their argument has to have a category of CN which had to be different from the category of the resulting expression, which, in addition, could not be a sentence. And this is probably the consequence of the fact that in Polish there are noun phrases which are different from proper nouns and from common nouns.

The problem of the appropriate categorisation of quantifier-expressions is related to the problem of the flexibility of some categories. It is known that noun phrases can have three functions: referential, quantificational and predicative. In Polish some of these functions are correlated with the form of the NP and consequently some NP can occur in the form of a bare CN. This entails their categorial flexibility. Partee (1986) shows that one can account for the three functions of NPs by considering that they can be interpreted as denoting three different types, the type of individuals, the type corresponding to the set of sets (or generalised quantifier) and the type corresponding to sets (see also Williams, 1983). This multiple categorisation (or typing) is allowed by type-shifting (or category changing) rules which form a component of extended categorial grammars. The problem of the categorisation of quantifier expressions (of determiners) was solved by Suszko (cf. Suszko, 1958, 1960, 1964). Although Suszko did important work on the deductive rules it is not clear whether he realised that categorial grammars with reduction rules correspond to deductive systems. It became obvious in the early 60s after the publication of the important paper by Lambek (Lambek, 1958). His system, which also aimed at the syntactic description of natural languages, differed from the system of Ajdukiewicz precisely by the component of category changing rules. Their introduction was motivated in particular by the difference between a proper noun, which in general is a syntactically simple noun phrase, and a complex noun phrase. As has been indicated, these rules correspond to derived rules in deductive systems. Ajdukiewicz, who already had the notion of a derived rule (see Ajdukiewicz, 1928), apparently did not think about the possible relationship between deductive systems, derived rules and categorial grammars and even less so about the solutions these notion could bring to the difficulties he experienced with the classification of nouns.

There are at present about three thousand spoken languages in the world. In some sense they must obviously be very similar. However, serious empirical research shows that there can be important differences among various languages. It appears that there are significant differences between languages and the constructions they use which are related to all aspects of the 'quantificational module'. In particular it is not obvious that NPs in all languages can be analysed as generalised quantifiers (cf. various articles in Bach *et al.*, 1995). So it is possible that underlying languages of semantic interpretation may vary from language to language. There is no reason that this possible consequence should be surprising. So in some still to be specified sense, *logics underlying languages* can be different according to the language. It has sometimes been asserted by philosophers that Aristotle's distinction between *substance* and *accident* and more general his theory of 'categories' is a direct reflection of ancient Greek. It may even be that his theory and his logic would have been different had his native language been different. I have pointed out that a similar claim can be made with respect to the work of Polish logicians. In addition, I am claiming that some linguistic particularities of their native language prevented Polish logicians from obtaining solutions to technical problems which in other linguistic contexts are easier and more natural.*

*CNRS, Paris*

## NOTES

*The following persons helped me, voluntarily or not, with the present version of this paper: Ross Charnock, Henry Hiż, Ed Keenan and Brenda Laca. Most of them are not responsible for the majority of the remaining errors.

1 This is only a label which is not meant to imply anything about the scope of the work done by Polish logicians. It will soon become clear in what kind of work of the logicians of the Lvov-Warsaw School I am interested. For a good overview of the work done by the School see Woleński (1985).

2 Strictly speaking Polish lacks an infinite number of definite articles: for any $n$ Polish lacks the expression corresponding to **the** $n$.

3 Strictly speaking, in many cases such constructions are possible but their acceptability seems dependent on the sub-categorisation of common nouns into denoting artifacts, animate, natural kind terms, etc. My point is that such constructions are excluded with some common nouns.

4 The nouns *wódka* (vodka) and *winko* (wine) are obvious counterexamples to this claim. I think, however that in the first case we can say that there is not really a productive application of the diminutive suffix. For the case of *winko* one should probably say that there are two types of mass terms: those which refer to dissective stuff and those which refer to non-dissective stuff. My observation concerns the latter.

## REFERENCES


Ajdukiewicz, K.: 1928, *Główne zasady metodologii nauk i logiki formalnej* (Fundamental Principles of the Methodology of Sciences and of Formal Logic), Warsaw.

Ajdukiewicz, K.: 1935, 'Die syntaktische Konnexität', *Studia Philosophica* **1**, pp. 1–27; trans. as 'Syntactic connection', in S. McCall (ed.): *Polish Logic 1920–1939*, Clarendon Press, Oxford 1967, pp. 207–231.

Ajdukiewicz, K.: 1967a, 'Proposition as the Connotation of Sentence', *Studia Logica* **20**, pp. 87–98.

Ajdukiewicz, K.: 1967b, 'Intensional Expressions', *Studia Logica* **20**, pp. 63–86.

Bach, E. *et al.* (eds.): 1995, *Quantification in Natural Languages*, Kluwer Academic Publishers, Dordrecht.

Bar-Hillel, Y.: 1953, 'A quasi-arithmetical notation for syntactic description', *Language* **29**, pp. 47–58; reprinted in Y. Bar-Hillel, *Language and Information*, Addison-Wesley, Reading, MA, 1964, pp. 61–74.

Barwise, J., Cooper, R.: 1981, 'Generalised Quantifiers and Natural Language', *Linguistics and Philosophy* **4/2**, pp. 159–220.

van Benthem, J.: 1991, *Language in Action*. Studies in Logic, vol. 130, North Holland Publishing Company, Amsterdam.

Buszkowski, W.: 1995, 'Rachunek Ajdukiewicza a notacja polska' (The Ajdukiewicz Calculus and Polish Notation), in Pogonowski, J. (ed.), *Eufonia i Logos*, Wydawnictwo Naukowe UAM, Poznań, pp. 337–350.

Chomsky, N.: 1986, *Knowledge of Language: Its Nature, Origin and Use*, Praeger, New York.

Cooper, R.: 1988, 'Facts in situation theory: representation, psychology, or reality', in Kempson, R. (ed.): *Mental representations: The interface between language and reality*, Cambridge University Press, Cambridge, pp. 49–61.

Cresswell, M.: 1975, 'Hyperintensional logic', *Studia Logica* **34**, pp. 25–38.

Cresswell, M.: 1984, *Structured Meanings*, MIT Press, Cambridge, MA.

Czelakowski, J., Malinowski, G.: 1984, 'Key Notions of Tarski's Methodology of Deductive Systems', *Studia Logica* **44–4**, pp. 321–351.

Czeżowski, T.: 1949, *Logika. Podręcznik dla studiujących nauki filozoficzne* (Logic. Handbook for Students of Philosophy), PWN, Warsaw.

Geach, P.T.: 1971, 'A program for syntax', *Synthese* **22**, pp.3–17.

Gold, M.E.: 1967, 'Language identification in the limit', *Information and Control* **10**, pp. 447–474.

Kanazawa, M.: 1996, 'Identification in the Limit of Categorial Grammars', *Logic, Language and Information* **5**, pp. 115–155.

Keenan, E.L.: 1987, 'Lexical Freedom and Large Categories', in Groenendijk, J.: *et al.* (eds.) *Studies in Discourse Representation Theory and the Theory of Generalized Quantifiers*, Foris Publications, Dordrecht-Holland/Cinnaminson-USA, pp. 27–52.

Keenan, E.L.: 1993, 'Natural language, sortal reducibility and generalized quantifiers', *Journal of Symbolic Logic* **58**, pp. 314–325.

Keenan, E.L., Faltz, L.: 1985, *Boolean Semantics for Natural Language*, Reidel Publishing Company, Dordrecht.

Keenan, E.L., Moss, L.: 1985, 'Generalized Quantifiers and the Expressive Power of Natural Language', in J. van Benthem *et al.* (eds.): *Generalized Quantifiers in Natural Languages*, Foris Publications, Dordrecht-Holland/Cinnaminson-USA, pp. 71–124.

Keenan, E.L., Stavi, J.: 1986, 'A semantic characterization of natural language determiners', *Linguistics and philosophy* **9**, pp. 253–326.

Kotarbiński, T.: 1961, *Elementy teorii poznania, logiki formalnej i metodologii nauk*, PWN,

Warsaw. Eng. trans. as *Gnosiology. The Scientific Approach to the Theory of Knowledge*, Pergamon Press, Oxford, Zakład Narodowy im. Ossolińskich, Wroclaw, 1966.

Lambek, J.: 1958, 'The mathematics of sentence structure', *American Mathematical Monthly* **65**, pp. 154–169.

Larson, R., Ludlow, P.: 1995, 'Interpreted Logical Forms', *Synthese* **95**, pp. 305–355.

Lemmon, E.J.: 1965, 'Some results on finite axiomatizability in modal logic', *Notre Dame Journal of Formal Logic* **6**, pp. 301–308.

Lewis, D.: 1972, 'General Semantics', in Davidson, D. *et al.* (eds.): *Semantics of Natural Language*, Reidel Publishing Company, Dordrecht, pp. 169–218.

Lindström, P.: 1966, 'First-Order Predicate Logic with Generalized Quantifiers', *Theoria* **32**, pp. 186–195.

Lyons, J.: 1968, *Introduction to theoretical linguistics*, Cambridge University Press, Cambridge.

McGee Wood, M.: 1993, *Categorial Grammars*, Routledge, London and New York.

Oehrle, R.T. *et al.* (eds.): 1988, *Categorial Grammars and Natural Language Structures*, Reidel Publishing Company, Dordrecht.

Mostowski, A.: 1957, 'On a generalization of quantifiers', *Fundamenta Mathematicae* **44**, pp. 12–36.

Partee, B.: 1987, 'Noun Phrase Interpretation and Type-Shifting Principles', in Groenendijk, J. *et al.* (eds.): *Studies in Discourse Representation Theory and the Theory of Generalized Quantifiers*, Foris Publications, Dordrecht-Holland/Cinnaminson-USA, pp. 115–144.

Pelc, J.: 1971, 'Funkcjonalne podejście do semiotyki logicznej języka naturalnego', in Pelc, J. (ed.): *Semiotyka Polska: 1894–1969*, PWN, Warsaw, 1979, pp. 396–430. Eng. trans. as 'A Functional Approach to the Logical Semiotics of Natural Language' in Pelc (1979), pp. 342–375.

Pelc, J.: 1979, *Semiotics in Poland 1894–1969*, D. Reidel Publishing Company, Dordrecht, PWN, Warsaw.

Pelletier, F.J. (ed.): 1979, *Mass Terms: Some Philosophical Problems*, Reidel Publishing Company, Dordrecht.

Steedman, M.: 1990, 'Gapping as constituent coordination', *Linguistics and Philosophy* **13**, pp. 207–263.

Suszko, R.: 1958, 'Syntactic structure and semantical reference I', *Studia Logica* **3**, pp. 213–244.

Suszko, R.: 1960, 'Syntactic structure and semantical reference II', *Studia Logica* **9**, pp. 63–93.

Suszko, R.: 1964, 'O kategoriach syntaktycznych i denotacjach wyrażeń w językach sformalizowanych' (On Categories of Expressions and Denotations in Formalised Languages), in *Rozprawy logiczne. Księga pamiątkowa ku czci profesora Kazimierza Ajdukiewicza* (Logical Dissertations. Essays in Honour of Professor Kazimierz Ajdukiewicz), PWN, Warsaw, pp. 193–204.

Tokarz, M.: 1994, 'Non-axiomatizability of Grice's Implicature', *Studia Logica* **53–2**, pp. 343–349.

Wexler, K., Culicover, P.: 1980, *Formal Properties of Language Acquisition*, MIT Press, Cambridge, MA.

Williams, E.: 1983, 'Semantic vs. syntactic categories', *Linguistics and Philosophy* **6**, pp. 423–446.

Woleński, J.: 1985, *Filozoficzna szkoła lwowsko-warszawska*, PWN, Warsaw. Eng. trans. as *Logic and Philosophy in the Lvov-Warsaw School*, Kluwer Academic Publishers, Dordrecht, 1989.

Zuber, R.: 1991, 'Règles d'extensionalité en grammaire catégorielle', *Comptes Rendus de l'Acad. des Sc* T. 313, S. I, pp. 547–550.

Zuber, R.: 1992, 'A note on unary rules and a complete syntactic characterization of proposi-

tional calculi', *Bulletin of the section of Logic* **21**, pp. 163–167.
Zuber, R.: 1994, 'Some metalogical constraints on flexible categories', in Martin-Vide, C. (ed.): *Current Issues in Mathematical Linguistics*, Elsevier Science, pp. 309–318.
Zuber, R.: 1995, 'O językach i ich wyrazów szykach' (On Languages and Their Word Order), in Pogonowski, J. (ed.): *Eufonia i Logos*, Wydawnictwo Naukowe UAM, Poznań, pp. 681–690.

# PART IV

# LOGIC AND THE FOUNDATIONS OF MATHEMATICS

WOJCIECH BUSZKOWSKI

# THE AJDUKIEWICZ CALCULUS, POLISH NOTATION AND HILBERT-STYLE PROOFS

Current research programs in the logic of language and computational linguistics employ certain methods characteristic of logical proof theory, a branch of logic devoted to fine structures of proofs. For instance, formal semantics of natural language applies the Curry-Howard correspondence between natural deduction proofs and typed lambda-terms to describe the process of computing semantic representations of expressions. Cut-free systems in the Gentzen form are used in standard axiomatizations of logical calculi of categorial grammar, closely related to substructural logics. Cut elimination and interpolation theorems play a significant part in some proofs of equivalence theorems for different kinds of formal grammar and in some proofs of completeness theorems for substructural logics. Proofs by resolution and unification, known from logic programming, are also a basic technique of unification systems in computational linguistics, as e.g. Definite Clause Grammars and Categorial Unification Grammars. Unification can also be used in learning algorithms for categorial grammars; see Buszkowski and Penn (1990), Kanazawa (1994), Marciniec (1994) and (1997). A thorough account of interrelations between mathematical linguistics and proof theory can be found in Buszkowski (1997) (also see van Benthem, 1991).

The present paper is devoted to some less popular areas on the borderline of linguistics and proof theory. We consider Hilbert-style proofs based on *modus ponens* as the only inference rule. The classical system of type reduction, introduced by Ajdukiewicz (1935), yields a generalized form of such proofs. The main objective of Ajdukiewicz was to extend Polish notation, invented by J. Łukasiewicz for propositional logics, to formal languages of arbitrary order. (For a thorough discussion of the role of Kazimierz Ajdukiewicz in the development of formal logic, see Batóg (1994), while his philosophical achievements are best characterized in Woleński (1985).) We show that the categorial and structural unambiguity of Polish notation strictly corresponds to the frontier property of (generalized) Hilbert-style proofs: each proof tree is uniquely determined by its frontier, i.e. the plane string of types which appear on the leaves of the tree. We point out a gap in Ajdukiewicz's justification of the unambiguity of Polish notation and sketch a correct proof of the frontier property. Further, we also discuss other links between Hilbert-style systems and formal grammar: unification based learning algorithms and a grammatical consequence operation.



*K. Kijania-Placek and J. Woleńsky (eds.),*
*The Lvov–Warsaw School and Contemporary Philosophy,* 241–252.
© 1998 *Kluwer Academic Publishers.*

## 1. MP-PROOFS AND POLISH NOTATION

Polish notation, due to J. Łukasiewicz, is the prefix notation (functors are positioned before arguments) without structure markers (parentheses, dots etc.).

- Infix: $(p \Rightarrow q) \Leftrightarrow (\neg q \Rightarrow \neg p)$
- Polish: $ECpqCNqNp$

In propositional languages, i.e. languages of order 0, Polish expressions are uniquely parsed if *arities* of functors are fixed. Ajdukiewicz (1935) observes that, for languages of higher order, one also needs types of all symbols.

- Infix: $(f \circ g)(x + y)$
- Polish: $\circ fg + xy$
- types: $x, y : n$; $f, g : n \to n$; $+ : nn \to n$; $\circ : AA \to A$, where $A = n \to n$

The unique structure is $[[\circ fg][+xy]]$. With arities only, one could produce $[\circ f[g[+xy]]]$ which means $\circ$ would take arguments $f$ and $[g[+xy]]$.

Types are formed out of atomic types, e.g. $n$ (individual), $s$ (sentence), by the following formation rule: if $A_1, \ldots, A_n, B$ are types, then $(A_1 \ldots A_n \to B)$ is also a type. This is the type of functional expressions (functors) which take arguments of types $A_1, \ldots, A_n$ and give rise to a complex expression of type $B$.

### 1.1. *The Ajdukiewicz algorithm*

We describe the Ajdukiewicz algorithm for checking whether a string of symbols (supplied with types) represents a grammatically correct expression. $V$ denotes a finite lexicon (alphabet). By $v : A$ we symbolize the assignment of type $A$ to lexem (symbol) $v$. We also write $v_1 \ldots v_n : A$, if type $A$ is assigned to string $v_1 \ldots v_n$.

The (higher-order) language is defined by assigning one type to each symbol $v \in V$ (the lexical assignment). The goal of the Ajdukiewicz algorithm is as follows: for the input string $v_1 \ldots v_n$, check whether there is a type $B$ such that $v_1 \ldots v_n : B$ and, if so, find this type $B$ and the grammatical structure of the resulting expression.

Ajdukiewicz assumes the *left-first* routine. Let $v_i : A_i$ be lexical assumptions. Consider the string $A_1, \ldots, A_n$. Look for the left-most interval:

$$(C_1 \ldots C_k \to C), C_1, \ldots, C_k,$$

and replace this interval by $C$. Repeat this step for the resulting string, and so on. If the procedure terminates with a single type $B$, then $B$ is the outcome type, and the structure is determined by inserting brackets:

$$[(C_1 \ldots C_k \to C), C_1, \ldots, C_k].$$

Otherwise, the input string is incorrect.

Since there is precisely one left-first reduction of the input string, then the outcome type and the outcome structure (if they exist) are unique, which is required for the unambiguity of Polish notation.

### 1.2. *Failure*

The above algorithm rejects some well-formed expressions! Consider the lexical assignment:

$$I : (n \to n)nn \to s;\ \ f, g : n \to n;\ \ x, y : n.$$

$Ifxy$ means $f(x) = y$. $Ifxgy$ is well-formed; the structure is $[Ifx[gy]]$, the meaning is: $f(x) = g(y)$. Now, consider the corresponding string of types:

$$((n \to n)nn \to s),\ n \to n,\ n,\ n \to n,\ n.$$

According to the left-first routine, we must reduce the second and the third type to $n$, then the last two types to $n$, which yields the irreducible string:

$$((n \to n)nn \to s),\ n,\ n.$$

Thus, $Ifxgy$ is to be *rejected*! Clearly, the appropriate reduction follows the intended structure of this expression: first, reduce the last two types to $n$, and then, reduce the whole to $s$.
MORAL: To guarantee the adequacy of the Ajdukiewicz algorithm, the left-first routine must be abandoned. One needs a nondeterministic procedure: reductions are to be executed at any possible place. But, for the nondeterministic algorithm, different runs might lead to different outcome types and outcome structures, hence the plausibility of Polish notation is not obvious.

### 1.3. *MP-deductions*

Polish notation is plausible for higher-order languages. To prove this claim we must show that all successful runs of the nondeterministic Ajdukiewicz algorithm for the given input string always yield the same outcome type (*categorial unambiguity*) and the same outcome structure (*structural unambiguity*).

Ajdukiewicz reductions are essentially *proofs* based on the rule *modus ponens* (MP), and precisely, multiple-argument versions of MP:

$$\text{(MP-}k\text{)}\ \frac{C_1 \ldots C_k \to C;\ C_1; \ldots ; C_k}{C}.$$

The *frontier* of the proof tree is the string of assumptions appearing at leaves of the tree. The plausibility of Polish notation is equivalent to the following *frontier property* of (multiple-argument) MP-trees:

(FP) each MP-tree is uniquely determined by its frontier.

We sketch the proof of (FP). For types $A, B$, we define the relation $A < B$:

($<$ 1) $A < (A_1 \ldots A_n \rightarrow A)$,

($<$ 2) if $A < B$ and $B < C$, then $A < C$.

If $A < B$, then $A$ is a proper subtype of $B$. The following properties of $<$ are useful:

($<$ 3) $A \not< A$

($<$ 4) if $A < C$ and $B < C$, then $A < B$ or $A = B$ or $B < A$,

($<$ 5) $(A_1 \ldots A_m \rightarrow A) \not< (B_1 \ldots B_n \rightarrow A)$, for $m \geq 1, n \geq 0$.

$\Gamma, \Delta$ denote finite strings of types. $\Gamma \vdash A$ means: there is an MP-tree with the frontier $\Gamma$ and the outcome $A$. $\vdash$ is a sequential consequence relation; that means, it satisfies the following conditions:

(C1) $A \vdash A$ (identity),

(C2) if $\Gamma, A, \Gamma' \vdash B$ and $\Delta \vdash A$, then $\Gamma, \Delta, \Gamma' \vdash B$ (transitivity).

The following lemmas are crucial (proofs are given in (1995)):

(L1) if $A_1 \ldots A_n \vdash A$, $(n > 1)$, then $A < A_1$,

(L2) if $A < B$, then $B, \Delta \vdash A$, for some nonempty $\Delta$,

(L3) if $\Gamma \vdash A$, then there is no nonempty string $\Delta$ such that $\Gamma, \Delta \vdash A$.

We prove the uniqueness of outcome types:

$$\text{if } \Gamma \vdash A \text{ and } \Gamma \vdash B, \text{ then } A = B.$$

Assume $\Gamma \vdash A, \Gamma \vdash B$ and $A \neq B$. Let $C$ be the first type in $\Gamma$. Since $A \neq B$, then $\Gamma \neq C$, hence $A < C$ and $B < C$, by (L1). Then, $A < B$ or $B < A$, by ($<$ 2). If $A < B$, then there is a nonempty string $\Delta$ such that $B, \Delta \vdash A$, by (L2), and consequently, $\Gamma, \Delta \vdash A$, by (C2), which contradicts (L3). The case $B < A$ is analogous.

We prove (FP). Let $\Gamma$ be the common frontier of MP-trees $t, t'$. We show $t = t'$. We proceed by induction on the length of $\Gamma$. By the above, both $t$ and $t'$ have a common outcome type $A$. If $\Gamma$ is of length 1, then $\Gamma = A$ and $t = t'$. Consider $\Gamma$ of length $> 1$. Then, $t$ and $t'$ are nontrivial. Let the down-most applications of MP in $t$ and $t'$ be:

$$\frac{B_1 \ldots B_m \rightarrow A;\ B_1; \ldots ;\ B_m}{A},$$

$$\frac{C_1 \ldots C_n \rightarrow A;\ C_1; \ldots ;\ C_n}{A}.$$

Using (L1) and properties of $<$, we show $m = n$ and $B_i = C_i$, for all $i = 1, \ldots, n$, which means that both MP-patterns are identical. Thus,

$$t = [t_0, t_1, \ldots, t_n],\ t' = [t'_0, t'_1, \ldots, t'_n],$$

where $t_i, t'_i$ are subtrees of $t, t'$, respectively, with the outcome type $B_i$ ($B_0$ stands for the first premise of the rule pattern). By $\Gamma_i, \Gamma'_i$ we denote the frontiers of $t_i, t'_i$, respectively. We show $\Gamma_i = \Gamma'_i$, for all $i = 0, \ldots, n$. Otherwise, consider the least $i$ such that $\Gamma_i \neq \Gamma'_i$. Clearly, one of these strings must be a proper initial interval of the other, but they have the same outcome type, which contradicts (L3). Now, we apply induction to get $t_i = t'_i$, for all $i = 0, \ldots, n$, which yields $t = t'$. The proof of (FP) is finished.

The relation $\vdash$ can also be axiomatized in a Gentzen-style form. Axioms are $A \vdash A$ and MP-rules are replaced by the $\rightarrow$- introduction rules:

$$\frac{A, \Delta \vdash C;\ \Gamma_1 \vdash B_1; \ldots ;\ \Gamma_n \vdash B_n}{(B_1 \ldots B_n \rightarrow A), \Gamma_1, \ldots, \Gamma_n, \Delta \vdash C}.$$

Using (L3), one shows that premises of this rule are uniquely determined by the conclusion, and consequently, each derivable sequent $\Gamma \vdash A$ possesses a unique derivation in the latter system. The only derivation of $\Gamma \vdash A$ induces a structure on $\Gamma$. By (FP), this structure must be equal to the structure determined by the only MP-tree with frontier $\Gamma$.

### 1.4. *The left-first routine modified*

For one-argument types, the left-first routine is adequate.

(LFR) if $\Gamma \vdash A$, then $\Gamma$ reduces to $A$ according to the left-first routine,

for all $\Gamma$ and $A$ containing one-argument types only (so, MP-trees are standard MP-trees in the logical sense). (LFR) is proven by induction on the length of $\Gamma$. The inductive step goes as follows. Let

$$\frac{B \rightarrow A;\ B}{A}$$

be the down-most application of MP in the proof of $\Gamma \vdash A$. Then, $\Gamma = \Gamma', \Gamma''$ with $\Gamma' \vdash B \rightarrow A$, $\Gamma'' \vdash B$. By induction, there is a left-first reduction of $\Gamma'$ to $B \rightarrow A$, which can be presented as a left-first reduction of $\Gamma$ to the string $B \rightarrow A, \Gamma''$. By induction again, there is a left-first reduction of $\Gamma''$ to $B$. Let

$$\Gamma'' = \Gamma_0;\ \Gamma_1; \ldots ;\ \Gamma_n = B$$

be the sequence of intermediate strings of the latter reduction. We show that the sequence

$$B \rightarrow A, \Gamma_0;\ B \rightarrow A, \Gamma_1; \ldots ;\ B \rightarrow A, \Gamma_n;\ A$$

is a left-first reduction of $B \rightarrow A, \Gamma''$ to $A$. That is obvious for $n = 0$. For $n > 0$, we use (L1) to see that $B$ differs from the first type in $\Gamma_i$, for each

$i = 0, \ldots, n-1$, and consequently, one cannot reduce the two initial types of the above strings before the last step. Joining the two left-first reductions defined above, we get the required reduction of $\Gamma$ to $A$. For multiple-argument types, (LFR) does not hold. Nevertheless, it can be preserved for some modification of the Ajdukiewicz procedure. We use his idea of barriers: some types are written $\mid A_1 \ldots A_n \rightarrow B$, which means that these types can be employed as functors but not arguments in reductions. Ajdukiewicz introduces barriers for types of variable binding operators. For instance, $\forall x$ is assigned type $\mid s \rightarrow s$, which blocks argument roles of this operator. As it happens, barriers can help us to adjust the left-first routine to arbitrary types. We replace (MP-$k$) by:

$$\frac{C_1 \ldots C_k \rightarrow C;\; C_1}{\mid C_2 \ldots C_k \rightarrow C},$$

and similarly with $\mid C_1 \ldots C_k \rightarrow C$ as the first premise, for $k > 1$. We also admit:

$$\frac{C_1 \rightarrow C;\; C_1}{C},$$

and similarly with $\mid C_1 \rightarrow C$ as the first premise. Informally, the new system admits *partial* reductions, and the barrier $\mid$ is to block argument roles of types resulting from such partial reductions, until one has executed all partial reductions for arguments $C_1, \ldots, C_k$. Using the methods discussed above, one proves (LFR) for this modified reduction system.

Since the left-first routine can be simulated by a deterministic push-down acceptor, then each language of the above form (with finitely many symbols) is a deterministic context-free language, and consequently, it can be parsed in time $n^2$.

### 1.5. *The world of Polish notation*

We mention some typical formal languages to which Polish notation can be applied.

1. Functor-argument prefix languages of arbitrary order, e.g. typed combinatory terms.
2. Also with variable binding operators, as e.g. quantifiers, integrals, etc., having a fixed type. (L3) is crucial, and barriers can block argument roles of these operators, as has been proposed by Ajdukiewicz (1935).
3. Polymorphic operators, as e.g. lambdas, need a declaration of the argument type. If $t : B$ and $x : A$, then $\lambda^B x.t : A \rightarrow B$.

### 1.6. *Other rules*

We briefly discuss the status of (FP) for systems with other rules. First, consider the rule of Conditional Syllogism:

$$\text{(SYL)}\ \frac{A \to B;\ B \to C}{A \to C}$$

SYL is associative. Structural unambiguity fails, but categorial unambiguity holds.

MP+SYL. Both unambiguity properties fail; the following are two proof trees with the same frontier which have different structures and different outcomes.

$$[(A \to A) \to ((A \to A) \to B),\ [A \to A,\ A \to A]] : (A \to A) \to B$$

$$[[(A \to A) \to ((A \to A) \to B),\ A \to A],\ A \to A] : B$$

The same situation appears for MP+$A/\Box A$ (Necessitation), MP+$A/\forall x A$ (Generalization), and so on.

After Lambek (1958), categorial grammars use types with two conditionals $\to$ and $\leftarrow$, and the corresponding MP-rules are:

$$\text{(MP}\!\to\!\text{)}\ \frac{A;\ A \to B}{B}$$

$$\text{(MP}\!\leftarrow\!\text{)}\ \frac{B \leftarrow A;\ A}{B}$$

Systems based on these rules are categorially and structurally ambiguous, as exemplified by:

$$[[(B \leftarrow (A \to A)) \leftarrow A,\ A],\ A \to A] : B,$$

$$[(B \leftarrow (A \to A)) \leftarrow A,\ [A,\ A \to A]] : B \leftarrow (A \to A).$$

Still $\Gamma \vdash A$ and $\Gamma \vdash B$ entail $A = B$, for atomic $A, B$, since each atomic type must have an even number of occurrences in any derivable sequent $\Gamma \vdash A$.

## 2. STRATIFICATION AND LEARNING

In logic, axioms are not particular formulas but schemes. The logic of positive implication is axiomatized by two axiom schemes:

(K) $A \to (B \to A)$,

(S) $(A \to (B \to C)) \to ((A \to B) \to (A \to C))$,

where $A, B, C$ are arbitrary formulas.

MP-trees are not determined by strings of (names of) schemes, attached to leaves of the tree. For instance, SKK admits two bracketings: [[SK]K] and [S[KK]]. The former determines a most general proof of $A \rightarrow A$, and the latter determines a proof of the scheme:

$$(C \rightarrow A) \rightarrow (C \rightarrow (B \rightarrow A)).$$

In general, if schemes are used instead of particular formulas, MP-trees are determined by bracketed frontier strings; the bracketing encodes the tree structure. We find the most general proof (if it exists) determined by the bracketed string of schemes by proceeding along the tree (from the frontier to the root) and applying at each internal node the *condensed* MP-rule:

$$\text{(CMP)}\ \frac{A \rightarrow B;\ A'}{\sigma(B)},$$

where $\sigma$ is a most general unifier of $A$ and $A'$. Recall that a unifier of types $A, A'$ is a substitution $\sigma$ (of types for variables) such that $\sigma(A) = \sigma(A')$, and a most general unifier is a unifier $\sigma$ such that, for every unifier $\sigma'$ of these types, there is a substitution $\alpha$, fulfilling $\sigma' = \alpha \circ \sigma$.

The above procedure can be credited to Curry (see Curry and Feys, 1958) who considers the *stratification* problem for combinators: find a most general type scheme for a type-free combinator.

In Buszkowski (1987) and Buszkowski and Penn (1990) a similar procedure was applied to the task of determining (learning) a most general categorial grammar compatible with a given finite set of linguistic postulates. Postulates are of the form $X : A$, where $X$ is a bracketed string of words, and $A$ is a type. For natural language, postulates can be a finite set of sample sentences (or: sentences and noun phrases etc.), and the goal is to find a most general categorial grammar fulfilling these postulates, which means assigning type $s$ to each sample sentence, type $n$ to each sample noun phrase, and so on.

In systems with (MP$\rightarrow$) and (MP$\leftarrow$), bracketings must indicate the direction of rules. We write $[XY]_1$ for (MP$\leftarrow$) (functor on the left) and $[XY]_2$ for (MP$\rightarrow$) (functor on the right). These bracketings are called *functor-argument structures*. The sample expressions will be represented as such structures, which is needed for the determinacy of our algorithm. If they are represented as strings of symbols, one must test all possible bracketings, which is time costly.

## 2.1. *Learning a categorial grammar*

We shall illustrate the learning procedure by a simple example. The input of our algorithm is the set:

$$[\text{John } [\text{likes Jane}]_1]_2 : s$$

$$[\text{Jane [likes [poor John]}_1]_1]_2 : s$$

One assigns a different variable to each occurrence of an argument substructure:

$$\text{John}: x, \text{Jane}: y,$$

$$\text{Jane}: z, [\text{poor John}]_1 : u, \text{John}: v.$$

Then, one derives types of functor substructures by the rules:

$$\frac{[XY]_1 : B;\ Y : A}{X : B \leftarrow A}, \quad \frac{[XY]_2 : B;\ X : A}{Y : A \rightarrow B},$$

which are reversals of MP.

$$[\text{likes Jane}]_1 : x \rightarrow s, [\text{likes [poor John]}_1]_1 : z \rightarrow s,$$

$$\text{likes}: (x \rightarrow s) \leftarrow y, \text{likes}: (z \rightarrow s) \leftarrow u, \text{poor}: u \leftarrow v.$$

One collects types of each word:

- John: $x, v$,
- Jane: $y, z$,
- poor: $u \leftarrow v$,
- likes: $(x \rightarrow s) \leftarrow y$, $(z \rightarrow s) \leftarrow u$.

The grammar so obtained is called *the general form* determined by the input. It generates a language whose elements are precisely the sample sentences.

At the end, one finds a most general substitution $\sigma$ which unifies all types of each atom (not types of different atoms). The outcome is:

- John, Jane: $x$,
- poor: $x \leftarrow x$,
- likes: $(x \rightarrow s) \leftarrow x$.

One can interpret $x$ as the type of noun phrase. The resulting grammar generates an infinite language with sentences *poor John likes Jane, poor poor Jane likes poor John* etc. It is the ultimate outcome of our procedure.

A categorial grammar is *deterministic* (or: rigid) if it assigns exactly one type to each word. A main result of Buszkowski (1987) is:

(T1) There is a deterministic categorial grammar satisfying the input assumptions if, and only if, the above procedure yields an outcome. Further, each deterministic categorial grammar satisfying the input assumptions is a substitution instance of the outcome.

Consequently, the above algorithm yields the most general deterministic categorial grammar fulfilling the assumptions ('most general' means that every deterministic categorial grammar satisfying initial postulates is a substitution instance of the outcome grammar).

In Buszkowski and Penn (1990) this method was generalized for nondeterministic grammars. Let $\mathcal{T} = \{T_1, \ldots, T_n\}$ be a finite family of finite sets of types (think of $T_i$ as the collection of all types assigned to atom $v_i$ by the general form). An *optimal unifier* for $\mathcal{T}$ is a substitution $\sigma$ fulfilling the conditions:

(OU.1) if $A, B \in T_i$ and $\sigma(A) \neq \sigma(B)$, then types $\sigma(A), \sigma(B)$ are not unifiable,

(OU.2) $\sigma$ is an mgu of the family $\mathcal{T}/\mathrm{Ker}(\sigma)$ (each $T_i$ is replaced be the family of equivalence classes of types in $T_i$ modulo the relation $A \sim B \Leftrightarrow \sigma(A) = \sigma(B)$).

A categorial grammar is said to be *minimal* in a class of categorial grammars if no grammar in this class assigns the same or a smaller number of types to each word and a smaller number of types to at least one word. The main result in Buszkowski and Penn (1990) is:

(T2) For any given input, one effectively finds the finite family of optimal unifiers of the family of sets of types provided by the general form. Each minimal categorial grammar fulfilling the assumptions is a substitution instance of some outcome of the algorithm.

Marciniec (1994) considers negative assumptions $X$ : non-$A$ (the expression $X$ may not be assigned type $A$) which improve the adequacy of the procedure. A general account of negation in categorial grammars is given in Buszkowski (1995a). Kanazawa (1994) develops an advanced learning theory for categorial grammars based on these methods.

## 2.2. *Grammatical Consequence*

$FS(V)$ denotes the set of all functor-argument structures over the lexicon $V$. By $\mathcal{D}$ we denote the family of all deterministic languages $L \subset FS(V)$, i.e. functor-argument languages generated by deterministic categorial grammars. We also assume that the total language $FS(V)$ is in $\mathcal{D}$, although it is not deterministic. Using the methods discussed above, one can prove:

(T3) $\mathcal{D}$ is a closure family, meaning, for any subfamily $\mathcal{D}' \subset \mathcal{D}$, we have $\bigcap \mathcal{D}' \in \mathcal{D}$.

Equivalently, for each set $L \subset FS(V)$, there is the smallest set $L' \in \mathcal{D}$ such that $L \subset L'$. The proof, given by Marciniec in (1997) uses unification of finite families of infinite sets of types; as above, one finds the general form determined by $L$ (now, it can be an infinite grammar which assigns infinitely many types to one word), and the outcome grammar, resulting from the unification of the general form, generates $L'$. We denote $L' = C(L)$. The operator $C$ fulfils the equivalence:

$A \in C(L)$ if, and only if, $A$ belongs to all deterministic languages $L'$ such that $L \subset L'$.

EXAMPLE: $L$ consists of two structures:

$$[\text{John } [\text{likes Jane}]_1]_2, [\text{Jane } [\text{likes } [\text{poor John}]_1]_1]_2.$$

Then, $[[\text{poor John}]_1 [\text{likes Jane}]_1]_2$ is in $C(L)$.

A LOGICAL EXAMPLE: $L$ consists of one structure: $[[Cp][[C[Np]]p]]$, which is the correct logical structure of $CpCNpp$ (a propositional formula in Polish notation); we omit functor subscripts, always equal to 1. Then, $C(L)$ is the set of all well-formed (structured) propositional formulas.

It can be shown that $C$ satisfies the Tarski conditions (see Marciniec, 1997):

(1) $L \subset C(L)$,
(2) if $L \subset L'$, then $C(L) \subset C(L')$,
(3) $C(C(L)) = C(L)$,
(4) $C(L) = \bigcup\{C(L') : L' \subset L \;\&\; L' \text{ is finite}\}$,

hence it is a compact consequence operator. Thus, in the world of deterministic languages, one can define (structured) languages by means of a consequence operator, with no direct address to formal grammars.

Deterministic languages are the class of languages especially interesting for mathematical logic, since they contain almost all formal languages of logical and mathematical systems. In a sense, one may identify *formal* languages with *deterministic* languages (see Buszkowski (1989) for discussion). Most problems dealt with in this paper are concerned with this class. The author believes that a general theory of deterministic languages following the lines sketched here is a reasonable research program for both mathematical linguistics and the foundations of logic.

*Faculty of Mathematics and Computer Science,*
*Adam Mickiewicz University, Poznań*

## REFERENCES

Ajdukiewicz, K.: 1935, 'Die syntaktische Konnexität', *Studia Philosophica* **1**, pp. 1–27.

Batóg, T.: 1994, 'Ajdukiewicz and the development of formal logic', *Poznań Studies in the Philosophy of Sciences and the Humanities* **40**, pp. 53–67.

van Benthem, J.: 1991, *Language in Action. Categories, Lambdas and Dynamic Logic*, Elsevier, North Holland, Amsterdam.

Buszkowski, W.: 1987, 'Solvable Problems for Classical Categorial Grammars', *Bull. Polish Academy of Sciences: Mathematics* **35**, pp. 373–382.

Buszkowski, W.: 1989, 'Principles of Categorial Grammar in the Light of Current Formalisms', in K. Szaniawski, (ed.), *The Lvov-Warsaw School and the Vienna Circle*, Nijhoff, Dordrecht, pp. 113–137.

Buszkowski, W.: 1995, 'Grammatical Structures and Logical Deductions', *Logic and Logical Philosophy* **3**, pp. 47–86.

Buszkowski, W.: 1995a, 'Categorial Grammars with Negative Information', in Wansing, H. (ed.), *Negation. A Notion in Focus*, de Gruyter, Berlin, pp. 107–126.

Buszkowski, W.: 1997, 'Mathematical Linguistics and Proof Theory', in J. van Benthem and A. der Meulen (eds.), *Handbook of Logic and Language*, Elsevier, North Holland, pp. 683–736.

Buszkowski, W., Penn, G.: 1990, 'Categorial Grammars Determined from Linguistic Data by Unification', *Studia Logica* **49**, pp. 431–454.

Curry, H. B., Feys, R.: 1958, *Combinatory Logic*, vol. 1, North Holland, Amsterdam.

Kanazawa, M.: 1994, *Learnable Classes of Categorial Grammars*, Ph.D. Thesis, Stanford University.

Lambek, J.: 1958, 'The mathematics of sentence structure', *The American Mathematical Monthly* **65**, pp. 154–170.

Marciniec, J.: 1994, 'Learning categorial grammars by unification with negative constraints', *Journal of Applied Non-Classical Logics* **4**, pp. 181–200.

Marciniec, J.: 1997, 'Infinite set unification with application to categorial grammar', *Studia Logica* **58**, pp. 339–355.

Woleński, J.: 1985, *Filozoficzna szkoła lwowsko-warszawska*, PWN, Warsaw. Eng. trans. as Woleński (1989).

Woleński, J.: 1989, *Logic and Philosophy in the Lvov-Warsaw School*, Kluwer Academic Publishers, Dordrecht.

ANDRZEJ INDRZEJCZAK

# JAŚKOWSKI AND GENTZEN APPROACHES TO NATURAL DEDUCTION AND RELATED SYSTEMS

It is well known that Jaśkowski and Gentzen independently published their results concerning Natural Deduction (ND for short) in 1934. But details of their work are not well known, as one can see when faced with many publications concerning ND. In what follows we briefly describe two systems of ND due to Jaśkowski (1929), (1934) and Gentzen (1934), then the Sequent Calculus (SC) from Gentzen (1934) and finally one more ND system from Gentzen (1936). We focus on similarities and differences in the approaches of both authors. Moreover we try to show their influence (not always clearly recognised) on further works on ND and related systems. This picture is necessarily somewhat sketchy and far from being complete so we would like to apologise in advance to the many logicians who are not even mentioned in this paper, but should be.

Łukasiewicz posed the problem in his seminar in 1926 of how to describe, in a formally proper way, methods applied in practice by mathematicians. Hilbert's proof theory, although theoretically satisfying, was practically inadequate and artificial. The *Deduction Theorem* proved by Herbrand (1930) and, at the same time, the axiomatics of the *Consequence Theory* of Tarski (1930), can be seen as the first bridge between theory and practice. In response to Łukasiewicz's problem Jaśkowski presented his first results on ND in 1927, at the First Polish Mathematical Congress in Lvov; these were mentioned in (1929). A final solution to the problem was found in (1934). It is worth saying that in practice ND-procedures were applied in the twenties by Leśniewski, Tarski, Salamucha, as is evident from their papers. Gentzen published his famous paper (1934) in the same year but the first results are present in his (1932). Gentzen was influenced by Hertz (1929), where a tree-format notation for proofs, as well as sequents were applied.

Before we proceed with the presentation of their systems, some terminology and conventions should be established. One can very often find in the literature names like 'Gentzen format', 'Fitch style' and so on; these are applied informally, with reference to several features that distinguish one system from another. In order to make comparisons in a unified way we provisionally introduce some notions. We are interested here not in *logics*, which are standardly treated as *Consequence Operations* defined on some fixed language, but in *Deductive Systems* (DS), here considered as syntactic forms of presentation of any logic. One should distinguish between the *content* of a DS and its *repre-*



*K. Kijania-Placek and J. Woleński (eds.),*
*The Lvov–Warsaw School and Contemporary Philosophy*, 253–264.
© 1998 *Kluwer Academic Publishers.*

*sentation*. The content consists of two nonempty sets $Seq$ and $R$. $Seq$ contains items of the form $\Gamma \vdash \Delta$ called sequents, where $\Gamma$, $\Delta$ are finite sets of wffs, antecedent and a succedent of a sequent. $R$ is a set of rules allowing the inference of some sequents from a finite set of given sequents. This definition is not very standard, because the basic items are not wffs but sequents. But it is very handy to make such a generalisation if we want to capture all the varieties of different DSs. Sequents may be reduced in many ways, with antecedent (succedent) limited to only one wff or even empty; in fact most of the systems are defined only on sequents that are so reduced.

The level of representation refers to practice, not to a theoretical basis. It is a set of instructions of how to build a proof and apply rules, sometimes with additional technical devices. These two levels are independent, e.g. two different but equivalent axiomatisations differ with respect to content, but usually not with respect to representation. On the other hand textbooks present many different representations of essentially the same (with respect to content) ND systems.

Because we are concerned mainly with ND, some additional remarks on possible forms of representation are in order. Basically, an ND system is considered here as one in which special rules allow for the introduction and elimination of both assumptions and logical constants. One can decompose or build more complex wffs, in contrast to SC, where only rules for composing (introducing constants) are present. Note that ND is often defined more broadly and SC is also treated as a kind of ND system. The rules of an ND system may be defined on wffs, but also directly on sequents. This has practical consequences, hence we distinguish *formula-systems* and *sequent-systems*; the latter notion is more general than the SC, although SC is a sequent-system by definition. Finally, at least in the case of ND formula-systems, it matters if proofs are defined as sequences or as trees, hence we distinguish *L-systems* (for linear) and *T-systems*. In an $L$-system one must have some bookkeeping device for showing that when some assumption is discharged (eliminated, not in force) then the part of proof in its scope (depending on it) is no longer available. If it is not forbidden in some way, one can still use some wff not in force, because in $L$-proofs we can use the same wff many times. As a result we can derive anything, a phenomenon well known to all teachers of elementary logic courses. In $T$-proofs this is impossible; premises of any rule must always be displayed directly over the conclusion, so we cannot use any occurrence of a wff more than once in a proof, hence we cannot use any wff which is not admissible at the moment (I mean discharged assumptions and dependent formulae). Of course we can enter such wffs again but it will be a new undischarged assumption.

The basic ND system of Jaśkowski is a formula- and $L$-system in an implication-negation language with an additional prefix $S$ (for supposition). It contains four rules:

A) introduction of any wff prefixed with $S$ into the proof as an assumption

$\rightarrow$E) $\varphi, \varphi \rightarrow \psi \vdash \psi$

$\rightarrow$I) if $\Gamma, \varphi \vdash \psi$, then $\Gamma \vdash \varphi \rightarrow \psi$

$\neg$E) if $\Gamma, \neg\varphi \vdash \bot$, then $\Gamma \vdash \varphi$

where $\varphi, \psi$ denote any wff (also prefixed with S), and $\bot$ denotes contradiction. Because of the linearity of proofs Jaśkowski needed some bookkeeping device to avoid the problems mentioned above. The first solution to the problem consisted in making boxes for each assumption and dependent part of a proof. An application of $\rightarrow$I) or $\neg$E) was connected with closing a box, and the inferred wff was put outside. Jaśkowski did not use the prefix $S$ in this variant to mark assumptions; it was always the first wff in the box. He also used the additional rule $R$) (repetition) to shift a wff from outer open box inside the inner; transition in the other direction was of course forbidden.

In his official version (1934) Jaśkowski applied an apparatus of numeric prefixes, instead of boxes. These are finite sequences of natural numbers written before each wff in the proof. Each time we enter an assumption we extend the prefix with an additional number; each application of $\rightarrow$I) or $\neg$E) is connected with subtraction of the last number in the prefix. Jaśkowski thought of prefixes as indicators of domains in which such prefixed wffs are valid, thus formulae with empty prefixes were ordinary theses and those prefixed are theses relative to some domain in which some suppositions are postulated as valid. Prefixes are then records of dependency of a wff in the context of a proof. One can find an exact technical description in Orłowska (1975) or Woleński (1985); here we illustrate both versions with an example of a proof.

| | |
|---|---|
| 1. | $Sp$ |
| 1.1. | $S\neg\neg\neg p$ |
| 1.1.1. | $S\neg\neg p$ |
| 1.1. | $\neg p$ |
| 1. | $\neg\neg p$ |
| | $p \rightarrow \neg\neg p$ |

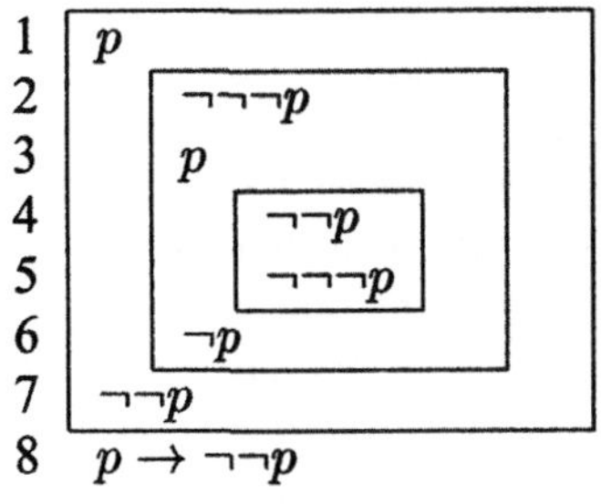


Jaśkowski modified his propositional system to capture Positive logic (by omitting $\neg$E)) and the Intuitionistic logic of Kolmogorov by using a weakened rule $\neg$I) if $\Gamma, \varphi \vdash \bot$, then $\Gamma \vdash \neg\varphi$. He also considered adding quantifiers to propositional logic and sketched a system for First Order logic. We omit the details connected with the former but take a look at his functional calculus. Jaśkowski only used the universal quantifier, predicate symbols and individual variables as terms. He added an extra rule for adding *term-supposition*:

T) introduction of a new term $\alpha$ prefixed with $T$ (for term) into the proof as term-supposition (this variable is called *valid* in this domain)

∀E) $\forall\beta\varphi(\beta), T\alpha \vdash \varphi(\beta/\alpha)$, where $\varphi(\beta)$ is any formula containing free occurrences of variable $\beta$ and $(\beta/\alpha)$ denotes uniform substitution of variables with usual restrictions

∀I) if $\Gamma, T\alpha \vdash \varphi$, then $\Gamma \vdash \forall\alpha\varphi$ ($\alpha$ not free in $\Gamma$, by restriction in T))

It should be noted that this is the first system of Inclusive logic, which allows empty domains (see Bencivenga (1986) for more details).

Jaśkowski also established some standard forms of proof extensively used by many logicians in *Soundness* proofs of ND-systems. Briefly, for each prefixed formula we build its *development*, which is a descending implication with suppositions for each number in the prefix as antecedents and the formula itself as a succedent. Then we can check that each rule expressed in terms of these developments is admissible in this logic.

One thing is common in both of Jaśkowski systems. The essential idea lies in dividing a proof into ordered and partly separated subproofs. It appears as the most popular solution in hundreds of textbooks where ND-techniques are applied. His first version, although abandoned by the author himself, is much more popular nowadays. It has many variants but there is always some graphic device added to the linear sequence of formulae in a proof. The original format of boxes was used by Kalish and Montague (1957) and (1964), but with some adjustments which make this system one of the most flexible in practice. A simplified account, where each assumption is entered with the vertical bar, is due to Fitch (1952). Borkowski and Słupecki (1958) followed the second route but also with some simplifications; prefixes are mixed together with the numbers of lines. Generally, we can call all of these systems, following Jaśkowski, the *ordered assumption approach*.

Because the parts of a proof are separated in this approach it proved especially useful with respect to many nonclassical logics formalised via ND systems. In these formalisations, some of the subproofs are marked as strict (modal) and transition of formulae from preceding subproofs is in some way restricted. Fitch (1952) extended his system to Modal logics (see Fitting (1983) for an extensive treatment), Anderson and Belnap (1975) to Relevant Logics. In a similar way it may be extended to Temporal (Indrzejczak, 1994) or Conditional Logics (Thomason, 1970).

Gentzen (1934) presented two quite different systems: NK-Natürliche Kalkül, that is an ND system proper, and LK-Logistiche Kalkül, that is a sequent calculus, both for Classical and Intuitionistic logic. NK consists of the following rules:

A), →E), →I) as in Jaśkowski (1934)

∧I) $\varphi, \psi \vdash \varphi \wedge \psi$

$\wedge$E) $\varphi \wedge \psi \vdash \varphi \quad \varphi \wedge \psi \vdash \psi$
$\vee$I) $\varphi \vdash \varphi \vee \psi$
$\vee$E) if $\Gamma, \varphi \vdash \zeta$ and $\Delta, \psi \vdash \zeta$, then $\Gamma, \Delta, \varphi \vee \psi \vdash \zeta$
$\neg$I) if $\Gamma, \varphi \vdash \bot$, then $\Gamma \vdash \neg\varphi$
$\neg$E) $\varphi, \neg\varphi \vdash \bot$ and $\bot \vdash \varphi$
$\forall$I) if $\Gamma \vdash \varphi(\sigma)$, then $\Gamma \vdash \forall\alpha\varphi(\sigma/\alpha)$, provided the *eigenvariable* (parameter) $\sigma$ is not present in $\Gamma$ (he distinguished between variables (always bound) and parameters (free occurrences)).
$\forall$E) $\forall\alpha\varphi(\alpha) \vdash \varphi(\sigma)$
$\exists$I) $\varphi(\sigma) \vdash \exists\alpha\varphi(\alpha)$
$\exists$E) if $\Gamma, \varphi(\sigma) \vdash \psi$, then $\Gamma, \exists\alpha\varphi(\sigma/\alpha) \vdash \psi$, provided $\sigma$ not present in $\Gamma$ and $\psi$
LEM) the law of excluded middle is added in order to obtain Classical logic.

In the case of the implication-negation fragment it is essentially the same as in Jaśkowski (1934) (he considers identical rules for conjunction). There is an important difference in representation, however. NK is also a formula-system but with $T$-proofs, so Gentzen did not need to bother about technical devices for blocking nonvalid deductions. Such a format is no doubt very useful in theoretical investigations. It nicely shows the structure of a finished proof, but it is not suitable for actual derivation. Gentzen himself was well aware of this fact, when he wrote that "we are deviating somewhat from the analogy with actual reasoning. This is so, since in actual reasoning we necessarily have (1) a linear sequence of propositions due to the linear ordering of our utterances, and (2) we are accustomed to applying repeatedly a result once it has been obtained, whereas the tree form permits only of a single use of a derived formula" (Gentzen, 1934, citation from Szabo, 1969). No wonder his original format is applied mainly in theoretical papers; a good example is Prawitz (1965).

Gentzen's LK-system or sequent calculus (SC) was designed to describe some properties of the logics in question and prove the so called Hauptsatz, or *Cut-elimination Theorem*. Its operational character sometimes is so overestimated that many logicians tend to treat SC as a kind of metalogical system. This view is mistaken for at least two reasons: First, it is not consistent with Gentzen's original idea – in what sense can we prove the equivalence of LK with the ordinary axiomatic system if the former is on the metalevel?; Secondly, in LK we can prove only theorems and argument forms of some logic: we are not able to prove or even express any facts about logic itself, in striking contrast to Tarski's Consequence Theory and its expressive power. Being only a logical system, LK is still extremely useful, especially in these logics where the *Cut-elimination Theorem* is provable. This is mainly because the basic items of LK are sequents, not formulae, and we have a direct transition from the theoretical level of content to the practical level of representation. In what follows $\Gamma \Rightarrow \Delta$ denotes sequents as items of representation, in contrast to $\Gamma \vdash \Delta$ denoting elements of content. It should be noted that $\Gamma, \Delta$ in $\Gamma \Rightarrow \Delta$ are not necessarily sets, but may denote different things like multisets, where the number of occurrences of a formula is important, or sequences. In the context of $\Gamma \vdash \Delta$ they always represent finite sets. In fact, in Gentzen's original formulation sequents are ordered pairs of finite sequences and he needs special structural rules, beside usual logical ones, for connectives. Here is original formulation:

AX) $\quad \varphi \Rightarrow \varphi$

W $\Rightarrow$) (weakening) $$\frac{\Gamma \Rightarrow \Delta}{\varphi\,\Gamma \Rightarrow \Delta}$$ $\Rightarrow$ W) $$\frac{\Gamma \Rightarrow \Delta}{\Gamma \Rightarrow \Delta\,\varphi}$$

C $\Rightarrow$) (contraction) $$\frac{\varphi\,\varphi\,\Gamma \Rightarrow \Delta}{\varphi\,\Gamma \Rightarrow \Delta}$$ $\Rightarrow$ C) $$\frac{\Gamma \Rightarrow \Delta\,\varphi\,\varphi}{\Gamma \Rightarrow \Delta\,\varphi}$$

P $\Rightarrow$) (permutation) $$\frac{\Pi\,\varphi\,\psi\,\Gamma \Rightarrow \Delta}{\Pi\,\psi\,\varphi\,\Gamma \Rightarrow \Delta}$$ $\Rightarrow$ P) $$\frac{\Gamma \Rightarrow \Delta\,\psi\,\varphi\,\Pi}{\Gamma \Rightarrow \Delta\,\varphi\,\psi\,\Pi}$$

Cut) $$\frac{\Gamma \Rightarrow \Delta\,\varphi \qquad \varphi\,\Pi \Rightarrow \Sigma}{\Gamma\,\Pi \Rightarrow \Delta\,\Sigma}$$

$\neg \Rightarrow$) $$\frac{\Gamma \Rightarrow \Delta\,\varphi}{\neg\varphi\,\Gamma \Rightarrow \Delta}$$ $\Rightarrow \neg$) $$\frac{\varphi\,\Gamma \Rightarrow \Delta}{\Gamma \Rightarrow \Delta\,\neg\varphi}$$

$\wedge \Rightarrow$) $$\frac{\varphi(\text{or }\psi)\Gamma \Rightarrow \Delta}{\varphi \wedge \psi\,\Gamma \Rightarrow \Delta}$$ $\Rightarrow \wedge$) $$\frac{\Gamma \Rightarrow \Delta\,\varphi \qquad \Gamma \Rightarrow \Delta\,\psi}{\Gamma \Rightarrow \Delta\,\varphi \wedge \psi}$$

$\vee \Rightarrow$) $$\frac{\varphi\,\Gamma \Rightarrow \Delta \qquad \psi\,\Gamma \Rightarrow \Delta}{\varphi \vee \psi\,\Gamma \Rightarrow \Delta}$$ $\Rightarrow \vee$) $$\frac{\Gamma \Rightarrow \Delta\,\varphi(\text{or }\psi)}{\Gamma \Rightarrow \Delta\,\varphi \vee \psi}$$

$\rightarrow\Rightarrow$) $$\frac{\Gamma \Rightarrow \Delta\,\varphi \qquad \psi\,\Pi \Rightarrow \Sigma}{\varphi \rightarrow \psi\,\Gamma\,\Pi \Rightarrow \Delta\,\Sigma}$$ $\Rightarrow\rightarrow$) $$\frac{\varphi\,\Gamma \Rightarrow \Delta\,\psi}{\Gamma \Rightarrow \Delta\,\varphi \rightarrow \psi}$$

$\forall \Rightarrow$) $$\frac{\varphi(\sigma)\,\Gamma \Rightarrow \Delta}{\forall\alpha\varphi(\sigma/\alpha)\,\Gamma \Rightarrow \Delta}$$ $\Rightarrow \forall$) $$\frac{\Gamma \Rightarrow \Delta\,\varphi(\sigma)}{\Gamma \Rightarrow \Delta\,\forall\alpha\varphi(\sigma/\alpha)}$$

$\exists \Rightarrow$) $$\frac{\varphi(\sigma)\,\Gamma \Rightarrow \Delta}{\exists\alpha\varphi(\sigma/\alpha)\,\Gamma \Rightarrow \Delta}$$ $\Rightarrow \exists$) $$\frac{\Gamma \Rightarrow \Delta\,\varphi(\sigma)}{\Gamma \Rightarrow \Delta\,\exists\alpha\varphi(\sigma/\alpha)}$$

(of course $\sigma$ not present in conclusion-sequent of $\exists \Rightarrow$) and $\Rightarrow \forall$))

Usually rules for $\vee$ and $\wedge$ are presented in the form due to Ketonen (1944):

$$\wedge\Rightarrow)\ \frac{\varphi\,\psi\,\Gamma\Rightarrow\Delta}{\varphi\wedge\psi\,\Gamma\Rightarrow\Delta} \qquad \Rightarrow\wedge)\ \frac{\Gamma\Rightarrow\Delta\,\varphi \qquad \Pi\Rightarrow\Sigma\,\psi}{\Gamma\,\Pi\Rightarrow\Delta\,\Sigma\,\varphi\wedge\psi}$$

$$\vee\Rightarrow)\ \frac{\varphi\,\Gamma\Rightarrow\Delta \qquad \psi\,\Pi\Rightarrow\Sigma}{\varphi\vee\psi\,\Gamma\,\Pi\Rightarrow\Delta\,\Sigma} \qquad \Rightarrow\vee)\ \frac{\Gamma\Rightarrow\Delta\,\varphi\,\psi}{\Gamma\Rightarrow\Delta\,\varphi\vee\psi}$$

I remind you that the rule Cut), although essential, is the only one in Gentzen's original system which does not satisfy the so called *subformula-property*, that is, anything which is present in the premises must be present, as a subformula, in the conclusion. Having Cut), Gentzen was able to prove that his system is equivalent to the axiomatic one, but its elimination gave him the possibility of obtaining many important results, including the decidability of propositional and the semidecidability of First-Order logic, a consistency proof for arithmetic without the principle of induction, and others.

The sequent calculus has had a great impact on the development of proof methods. One of the reasons seems to be the possibility of many interesting interpretations of the notion of a sequent. Gentzen's original interpretation was rather trivial: $\Gamma \Rightarrow \Delta$ is understood as a $\wedge\Gamma \rightarrow \vee\Delta$. This device plus Cut) lets us get simple proofs of equivalence of axiomatic systems. The most popular solution is to read sequents in terms of falsifiability. Because Cut) is eliminable and all other rules preserve the subformula property, we may build the proof upside-down, starting with some sequent $\Gamma \Rightarrow \Delta$ and applying all the rules in reverse order. This is in accordance with the reading: if the conclusion-sequent is unsatisfiable then the (first or second) premise-sequent is also unsatisfiable. An axiomatic-sequent here is a kind of elementary inconsistency. If each branch ends with such inconsistency then the starting sequent must be valid, if not, we have at least one falsifying branch, and the analysed sequent is nonvalid.

This way of doing things admits many simplifications, explored by Gentzen's followers. All structural rules are dispensable if we read $\Gamma, \Delta$ as sets and generalise an axiom as any sequent $\Gamma \Rightarrow \Delta$, where $\Gamma \cap \Delta \neq \emptyset$ . We can get rid of the very notion of sequent and operate on sets of formulae, where all members of $\Delta$ are interpreted as negations (Hintikka, 1955). Beth (1955) introduced two-sided tables, hence there is no need to rewrite all parametric formulae (those not directly involved in the application of a rule) all the time; the technique later extended by Kripke (1959) (and especially (1963)) to modal logics by considering families of tables instead of single ones. Finally, Smullyan (1968) introduced one-sided trees of formulas, signed with T(true) or F(false), which seems to be the most radical simplification.

There is a variant of this technique based on the fact that all rules of SC are doubly sound (the validity of the conclusion-sequent guarantees the validity of premise(s)) at least in Classical logic, and in the variant with uniform parametric wffs in all sequents involved. It was used by Rasiowa, Sikorski (1963) and extended by Orłowska (1992), (1995) to many nonclassical logics. It is dual to the falsifiability interpretation, and equivalent to looking for the *conjunctive normal form*, whereas the latter is equivalent to looking for the *disjunctive normal form* (of course in Classical logic).

One should mention that all these essentially cut-free procedures (Cut) not primitive) forced logicians to find new strategies for obtaining a completeness proof. Strategies based on Lindenbaum's Lemma need to apply modus ponens or other special forms of Cut). The proofs of Hintikka or Smullyan make use of a weaker construction, namely *downward saturated sets* (or Hintikka sets). No form of Cut) is needed to prove that if a formula belongs to such a set it is satisfied in a suitable model. All these methods form a strong trend in *Automated theorem proving* investigations. It is interesting to note that the main stream, connected with *Resolution systems* is, on the contrary, based on the use of some special form of Cut) (resolution itself). See Avron (1993) for comments on the relations between these two main approaches in Automated theorem proving.

On the other hand the Gentzen sequential approach was followed by many logicians looking for a general proof-theoretical apparatus. Fundamental works of Schütte (1960) and Takeuti (1975) may serve as examples. The introduction of Szabo (1969) contains an excellent summary of the earlier stage of investigations, especially on consistency problems. Instead of reducing sequents to (pairs of) sets of formulae, and dispensing with structural rules altogether, it proved useful to pay attention to them. A hint is that in the sequent calculus the difference between Classical and Intuitionistic logic is structural: intuitionistic sequents are limited to one formula in succedent. Various experiments with structural rules can lead to sequential formalisations for different types of Relevant logics (see Anderson and Belnap, 1975), or Many-valued logics (limited contraction in Prijatelj, 1993). Strongly weakened versions of SC appeared adequate with respect to interesting algebraic models (see Došen, 1988, 1989, or Buszkowski, 1986, 1996). These fruitful investigations have lately been grouped together under the heading *Substructural logics*.

Investigations of this kind led Girard (1987) to *Linear logic*. It is based on the simple observation that in the absence of Weakening and Contraction, the original Gentzen rules and the more popular Ketonen variants formalise two different conjunctions and disjunctions, *additive* (extensional) and *multiplicative* (intensional). A surprising number of interesting applications of this system made it very popular and influential, especially among computer scientists.

A different approach, but also in this structural spirit, is due to Belnap (1982). In his *Display logic* a basic family of *structural constants* is introduced that make explicit what is implicit in the Gentzen system. Thus, instead of a comma in Gentzen sequents we have a binary operation, in effect sequents are always made from one formula only (or rather *structure*, because of these structural connectives) in the antecedent and succedent. Different Relevant and Modal logics are then obtainable by careful choice of suitable structural rules (dealing with the structural constants) with an invariant set of rules for logical constants. Belnap was strongly influenced by the step-by-step analysis of SC of Curry (1963).

Different attempts at extending the sequential approach to various nonclassical logics have led to interesting generalisations of the very idea of a sequent. Rousseau (1967) introduced sequents with more than two elements (antecedent, succedent respectively) to handle Many-valued logics. Došen (1985) applied nested (or higher order) sequents in Modal logics, where members of the antecedent/succedent are not only wffs but also other sequents. Using more than one type of a sequent proved very fruitful in Modal logics, see Curry (1963), Zeman (1973), Indrzejczak (1996a), (1996b).

Sequent calculi are also applied not only to logics but sometimes to quite different things. A good example is Lambek (1958), who devised a cut-free system for transforming categories of expressions in *categorial grammar*. His calculus seems to be one of the earliest substructural constructions.

Finally we must mention Gentzen (1936). This contains a system which is a kind of compromise between his LK and NK system. It consists of sequents with succedents limited to one formula. The only operations on antecedents are permutation, addition and elimination of wffs but in the succedents we have defined rules both for the introduction and elimination of logical constants. Here are the rules (but only for logic, without his number theoretic rules):

A) $\varphi \vdash \varphi$
P) if $\Gamma, \varphi, \psi, \Delta \vdash \zeta$ , then $\Gamma, \psi, \varphi, \Delta \vdash \zeta$
C) if $\varphi, \varphi, \Gamma \vdash \psi$ , then $\varphi, \Gamma \vdash \psi$
W) if $\Gamma \vdash \psi$ , then $\varphi, \Gamma \vdash \psi$
$\wedge$I) if $\Gamma \vdash \varphi$ and $\Delta \vdash \psi$ , then $\Gamma, \Delta \vdash \varphi \wedge \psi$
$\wedge$E) if $\Gamma \vdash \varphi \wedge \psi$ , then $\Gamma \vdash \varphi$ or $\Gamma \vdash \psi$
$\vee$I) if $\Gamma \vdash \varphi$ , then $\Gamma \vdash \varphi \vee \psi$ or $\Gamma \vdash \psi \vee \varphi$
$\vee$E) if $\Gamma \vdash \varphi \vee \psi$ and $\varphi, \Delta \vdash \zeta$ and $\psi, \Pi \vdash \zeta$ , then $\Gamma, \Delta, \Pi \vdash \zeta$
$\rightarrow$I) if $\varphi, \Gamma \vdash \psi$ , then $\Gamma \vdash \varphi \rightarrow \psi$
$\rightarrow$E) if $\Gamma \vdash \varphi$ and $\Delta \vdash \varphi \rightarrow \psi$ , then $\Gamma, \Delta \vdash \psi$
$\neg$I) if $\varphi, \Gamma \vdash \psi$ and $\varphi, \Delta \vdash \neg\psi$ , then $\Gamma, \Delta \vdash \neg\varphi$
$\neg$E) if $\Gamma \vdash \neg\neg\varphi$ , then $\Gamma \vdash \varphi$
$\forall$I) if $\Gamma \vdash \varphi(\sigma)$ , then $\Gamma \vdash \forall\alpha\varphi(\sigma/\alpha)$, provided $\sigma$ is not in $\Gamma$
$\forall$E) if $\Gamma \vdash \forall\alpha\varphi(\alpha)$ , then $\Gamma \vdash \varphi(t)$, where $t$ is any term
$\exists$I) if $\Gamma \vdash \varphi(t)$ , then $\Gamma \vdash \exists\alpha\varphi(\alpha)$, where $t$ is any term

$\exists$E) if $\Gamma \vdash \exists\alpha\varphi(\sigma/\alpha)$ and $\varphi(\sigma), \Delta \vdash \psi$ , then $\Gamma, \Delta \vdash \psi$, provided $\sigma$ is not in $\Gamma, \Delta$ and $\psi$

This system turned out to be very influential in the field of applied ND systems, because it may be used in linear format (in Gentzen (1936) it was still in tree format) but there is no need for any bookkeeping devices to indicate dependence of wffs on assumptions. In some sense this approach gives more flexibility in carrying derivations than Jaśkowski's approach, at least in the case of Classical logic. ND systems of this kind are to be found in Suppes (1957) and Hermes (1963); in the former we apparently operate on formulae not on sequents, because instead of antecedents we simply write down the numbers of the lines where suitable assumptions appeared for the first time. This device was introduced by Feys and Ladriere (1955) in their French translation of Gentzen's paper (1934). Contrary to Jaśkowski's approach the order of assumptions plays no role and proof is not divided; we can call this the *recorded assumptions approach*.

Following the tracks of both Jaśkowski and Gentzen on the development of proof methods, we cannot forget about the possibility of mixing different techniques. Two examples make this clear. Anderson and Belnap's (1975) ND system is basically of Jaśkowski's type (Fitch's bars for assumptions) but formulae are indexed in a way which make them essentially Suppes sequents. The prefixed analytic tableaux of Fitting (1983) mix Smullyan's tree system with Jaśkowski's device of prefixing each formula; here it is a record of a suitable world in a Kripke model (and all preceding ones, via accessibility relation). The very idea of using indexed wffs in a Gentzen system appeared in rudimentary form in Kanger's (1957) system for S5, in a general form it is present in Gabbay's (1991) *Labelled deductive systems* (see also Wansing, 1995).

There is still more to say on the subject, e.g. *Refutation systems* and its connections with Sequent calculi (see Bryll (1996) for comments) or relations with *Typed lambda calculus* via the Curry-Howard isomorphism, but it is impossible to continue in one short paper.

We can conclude with this remark: both authors laid the foundations for further investigations of ND but in a slightly different fashion. Jaśkowski seemed to be more concerned with practical aspects of deduction and his general approach as well as his technical solutions are in common use. On the other hand Gentzen was more theoretically oriented; his investigations led him to profound results in general proof theory. Some special approach to applied ND is rather a by-product of his paper (1936).

*Department of Logic, University of Łódź*

## REFERENCES

Anderson, A.R., Belnap, N.D.: 1975, *Entailment: the Logic of Relevance and Necessity* **1**, Princeton University Press, Princeton.

Avron, A.: 1993, 'Gentzen-type Systems, Resolution and Tableaux', *Journal of Automated Reasoning* **10**, pp. 265–281.

Belnap, N.D.: 1982, 'Display Logic', *Journal of Philosophical Logic* **11**, pp. 375–417.

Bencivenga, E.: 1986, 'Free Logics', in D. Gabbay, F. Guenthner (eds.): *Handbook of Philosophical Logic* **3**, D. Reidel Publishing Company, Dordrecht, pp. 373–426.

Beth, E.W.: 1955, *Semantic Entailment and Formal Derivability*, Mededelingen der Koninklijke Nederlandse Akademie van Wetenschappen, new series 18, North-Holland Publishing Company, Amsterdam, pp. 309–342.

Borkowski, L., Słupecki, J.: 1958, 'A logical system based on rules and its application in teaching mathematical logic', *Studia Logica* **7**, pp. 71–113.

Bryll, G.: 1996, *Metody odrzucania wyrażeń* (Methods of Rejecting Expressions), Akademicka Oficyna Wydawnicza PLJ, Warsaw.

Buszkowski, W.: 1986, 'Completeness Results for Lambek Syntactic Calculus', *Zeitschrift für Mathematische Logik und Grundlagen der Mathematik* **32**, pp. 13–28.

Buszkowski, W.: 1996, 'Representation Theorems for Implication Structures', *Bulletin of the Section of Logic* **25/4**, pp. 33–39.

Curry, H.B.: 1963, *Foundations of Mathematical Logic*, McGraw-Hill, New York.

Došen, K.: 1985, 'Sequent-systems for modal logic', *Journal of Symbolic Logic* **50**, pp. 149–159.

Došen, K.: 1988, 'Sequent systems and groupoid models I', *Studia Logica* **47**, pp. 353–386.

Došen, K.: 1989, 'Sequent systems and groupoid models II', *Studia Logica* **48**, pp. 41–66.

Feys, R., Ladriere, J.: 1955, Supplementary notes in *Recherches sur la deduction logique*, French translation of Gentzen 1934, Presses Univ. de France, Paris.

Fitch, F.B.: 1952, *Symbolic Logic: An Introduction*, Ronald Press, New York.

Fitting, M.: 1983, *Proof Methods for Modal and Intuitionistic Logics*, D. Reidel Publishing Company, Dordrecht.

Gabbay, D.: 1991, *Labelled Deductive Systems*, CIS-Munich, Munich.

Gentzen, G.: 1932, 'Über die Existenz unabhängiger Axiomensysteme zu unendlichen Satzsystemen', *Mathematische Annalen* **107**, pp. 329–350.

Gentzen, G.: 1934, 'Untersuchungen über das logische Schliessen', *Mathematische Zeitschrift* **39**, pp. 176–210, 405–431.

Gentzen, G.: 1936, 'Die Widerspruchsfreiheit der reinen Zahlentheorie', *Mathematische Annalen* **112**, pp. 493–565.

Girard, J.Y.: 1987, 'Linear Logic', *Theoretical Computer Science* **50**, pp. 1–102.

Herbrand, J.: 1930, *Recherches sur la theorie de la demonstration*, Travaux de la Societe des Sciences et des Lettres de Varsovie, Classe III, Sciences Mathematiques et Physiques, Warsaw.

Hermes, H.: 1963, *Einführung in die mathematische Logik*, Teubner, Stuttgart.

Hertz, P.: 1929, 'Über Axiomensysteme für beliebige Satzsysteme', *Mathematische Annalen* **101**, pp. 457–514.

Hintikka, K.J.J.: 1955, 'Form and Content in Quantification Theory', *Acta Philosophica Fennica* **8**, pp. 11–55.

Indrzejczak, A.: 1994, 'Natural Deduction System for Tense Logics', *Bulletin of the Section of Logic* **23/4**, pp. 173–179.

Indrzejczak, A.: 1996a, 'Cut-free sequent calculus for S5', *Bulletin of the Section of Logic* **25/2**, pp. 95–102.

Indrzejczak, A.: 1996b, 'Generalised Sequent Calculus for Regular Modal Logics', to appear.
Jaśkowski, S.: 1929, 'Teoria dedukcji oparta na dyrektywach założeniowych' (Theory of Deduction Based on Suppositional Directives), in *Księga Pamiątkowa I Polskiego Zjazdu Matematycznego* (Essays from the I Meeting of Polish Mathematicians), the Jagiellonian University, Krakow.
Jaśkowski, S.: 1934, 'On the rules of suppositions in formal logic', *Studia Logica* **1**, pp. 5–32.
Kalish, D., Montague, R.: 1957, 'Remarks on Descriptions and Natural Deduction', *Archiv. für Mathematische Logik und Grundlagen Forschung* **3**, pp. 50–64, 65–73.
Kalish, D., Montague, R.: 1964, *Logic: Techniques of Formal Reasoning*, Harcourt, Brace & World, New York.
Kanger, S.: 1957, *Provability in Logic*, Almquist & Wiksell, Stockholm.
Ketonen, D.: 1944, 'Untersuchungen zum Prädikatenkalkül', *Annales Academiae Scientiarum Fennicae*, ser. A 1. Mathematica-physica 23, Helsinki.
Kripke, S.: 1959, 'A Completeness Theorem in Modal Logic', *Journal of Symbolic Logic* **24**, pp. 1–14.
Kripke, S.: 1963, 'Semantical analysis of Modal Logic I', *Zeitschrift für Mathematische Logik und Grundlagen der Mathematik* **9**, pp. 67–96.
Lambek, J.: 1958, 'The mathematics of sentence structure', *American Mathematical Monthly* **65**, pp. 154–170.
Orłowska, E.: 1975, 'On the Jaśkowski's method of suppositions', *Studia Logica* **39**, pp. 187–200.
Orłowska, E.: 1992, 'Relational Proof System for Relevant Logics', *Journal of Symbolic Logic* **57**, pp. 1425–1440.
Orłowska, E.: 1995, 'Relational Proof Systems for Modal Logics', in H. Wansing (ed.): *Proof Theory of Modal Logics*, Kluwer Academic Publishers, Dordrecht pp. 55–77.
Prawitz, D.: 1965, *Natural Deduction*, Almqvist & Wiksell, Stockholm.
Prijatelj, A.: 1993, *Bounded Contraction and Many-Valued Semantics*, ILLC Prepublication Series for Mathematical Logic and Foundations ML–93–04, University of Amsterdam.
Rasiowa, H., Sikorski, R.: 1963, *The Mathematics of Metamathematics*, PWN, Warsaw.
Rousseau, G.: 1967, 'Sequents in Many-Valued Logic', *Fundamenta Mathematicae* **LX**, pp. 22–29.
Schütte, K.: 1960, *Beweistheorie*, Springer, Berlin.
Smullyan, R.M.: 1968, *First-Order Logic*, Springer, Berlin.
Suppes, P.: 1957, *Introduction to Logic*, Van Nostrand, Princeton.
Szabo, M.E. (ed.): 1969, *The Collected Papers of Gerhard Gentzen*, North-Holland Publishing Company, Amsterdam.
Takeuti, G.: 1975, *Proof Theory*, North-Holland Publishing Company, Amsterdam.
Tarski, A.: 1930, 'Fundamentale Begriffe der Methodologie der deduktiven Wissenschaften', *Monatschefte für Mathematik und Physik* **37**, pp. 361–404.
Thomason, R.H.: 1970, 'A Fitch-style formulation of Conditional Logic', *Logique et Analyse* **13**, pp. 397–412.
Wansing, H.: 1995, 'Strong Cut-Elimination for Constant Domain First-order S5', *Journal of the IGPL* **3/5**, pp. 797–810.
Woleński, J.: 1985, *Filozoficzna szkoła lwowsko-warszawska*, PWN, Warsaw. Eng. trans. as Woleński, J.: 1989, *Logic and Philosophy in the Lvov-Warsaw School*, D. Reidel Publishing Company, Dordrecht.
Zeman, J.J.: 1973, *Modal Logic*, Clarendon, Oxford.

ROMAN MURAWSKI

# THE CONTRIBUTION OF POLISH LOGICIANS TO RECURSION THEORY*

## 1. INTRODUCTION

The first need for a systematic study of functions whose values can be calculated by a finite process (usually called computable) can be found in the Hilbert school. It was connected with the decision problem for first-order logic (and in general, for first-order theories) considered by Hilbert and his students in connection with the Hilbert program. The aim of this program was to justify classical mathematics by finitistic means.

One usually shows that a given decision problem has a positive solution simply by providing an appropriate algorithm. No precise definition of a finitistic method is needed here. On the other hand, if we want to prove that a given specific problem does not admit an algorithmic solution then a precise general definition of such a method is necessary.

Historically the first precise definition of a class of computable functions was given by Kurt Gödel in his famous paper 'Über formal unentscheidbare Sätze der 'Principia Mathematica' und verwandter Systeme. I' from 1931 (cf. Gödel, 1931) in which he proved the essential incompleteness of the arithmetic of natural numbers. The class of functions defined by Gödel is today called the class of primitive recursive functions. It consists of those functions on the set of natural numbers **N** which can be obtained from a specified set of initial functions by the operations of substitution and recursion.

Since then several other definitions have been proposed. It is significant that all of them proved to be equivalent. Let us mention here definitions using the notion of representability (Church, Gödel), arithmetical definitions (Kleene), definitions using canonical systems (Post), abstract machines (Turing) and algorithms (Markov). These definitions led to the notion of a general recursive function and to the founding of a new domain of mathematical logic — recursion theory.

Investigations continued in two directions. On the one hand subclasses of the class of recursive functions were studied. Those subclasses were obtained by putting various restrictions on operations on functions admitted in the process of constructing new functions from given ones. This led, for example, to the notion of elementary functions (L. Kalmár) or to the study of various forms of multiple recursion (R. Péter). On the other hand there were attempts to generalize the notions of computability and recursiveness to objects of higher types, in



*K. Kijania-Placek and J. Woleński (eds.),*
*The Lvov–Warsaw School and Contemporary Philosophy*, 265–282.
© 1998 *Kluwer Academic Publishers.*

particular to functionals, i.e., to operations defined on $n$-tuples of natural functions (functions from $\mathbf{N}^n$ to $\mathbf{N}$) and $k$-tuples of elements of $\mathbf{N}$ taking values from $\mathbf{N}$.

Precise definitions of various types of computable functions provided a basis for further logical classifications of mathematical and logical notions. Using recursive functions and quantifiers one can classify non-recursive arithmetical notions. This is the idea which lies behind the Kleene and Mostowski hierarchies and their extensions.

Another approach to the classification of notions (which became an independent domain of study) was provided by the theory of degrees of unsolvability. The subject of this theory is degrees defined as equivalence classes of various reducibility relations.

Observe that the study of computable functions and relations is strictly connected with constructivistic trends in mathematics. The tendency to restrict mathematics to constructive objects and methods was given a precise tool by recursion theory — the vague notion of constructivity could be precisely defined. It gave new impetus to the study of constructive foundations of mathematics, in particular of analysis, leading to various systems of computable analysis.

One should also add that computable functions (and consequently recursion theory) are important for the philosophy of mathematics too, viz., they provide a basis for a nominalistic philosophy of mathematics which rejects such abstract objects as sets and functions, and accepts only those objects which can be named. Recursive functions and relations (and various variants of them) can be considered as such objects and therefore can apparently be accepted by the nominalists. On the other hand one should note that there are obviously nominalistic theories of mathematics which admit a broader class of objects than the class of computable functions. We refer here to descriptive set theory, i.e., the theory of Borel sets and of analytical sets initiated by the French semi-intuitionists (Borel, Baire, Lebesgue) and developed in the thirties by set-theoretically minded mathematicians grouped around Lusin in Moscow and around Sierpiński in Warsaw. Semi-intuitionists started from the notion of a real number and insisted that only nameable sets and functions defined on the set of all reals are admissible. It has turned out that descriptive set theory has many analogies with the theory of computable (recursive) functions (cf. e.g. Addison, 1954, 1958).

The present paper is devoted to the contribution of Polish logicians to the research described above. Section 2 concerns studies of subclasses of the class of recursive functions — first of all Grzegorczyk's hierarchy of primitive recursive functions. Section 3 is devoted to a hierarchy of notions introduced independently by Kleene and Mostowski and called today the Kleene-Mostowski hierarchy or the arithmetical hierarchy. Examples of the classification of particular mathematical notions in this hierarchy are also given. In Section 4 studies of the generalizations of the notion of recursiveness to the case of functionals are discussed. Section 5 concerns Banach-Mazur's and Grzegorczyk's work on the constructive foundations of mathematics, in particular of analysis. The final section, 6, is devoted to model theoretical applications of recursion theory, in particular to the discussion of results concerning the class (in the Kleene-Mostowski hierarchy) of possible models of theories.

## 2. GRZEGORCZYK'S HIERARCHY

The hierarchy of the primitive recursive functions, called today the Grzegorczyk hierarchy, was introduced in Andrzej Grzegorczyk's (1953). Grzegorczyk defined an increasing sequence $\mathcal{E}^0, \mathcal{E}^1, \mathcal{E}^2, \ldots$ of classes of recursive functions and proved that they have several properties. Let us recall some basic notions and theorems.

Consider the following sequence of functions:

$$f_0(x,y) = y+1,$$

$$f_1(x,y) = x+y,$$

$$f_2(x,y) = (x+1)(y+1),$$

and for $n \geq 2$:

$$f_{n+1}(0,y) = f_n(y+1, y+1),$$

$$f_{n+1}(x+1,y) = f_{n+1}(x, f_{n+1}(x,y)).$$

One can show that the functions $f_n$ for $n > 0$ are strictly increasing with respect to both arguments. Using those functions Grzegorczyk defined the classes $\mathcal{E}^n$ ($n \in \mathbf{N}$) of functions in the following way: $\mathcal{E}^n$ is the smallest class of functions including the functions $x+1, U_1(x,y) = x, U_2(x,y) = y, f_n(x,y)$ as initial functions and closed under the operations of substitution and limited recursion.

It was proved that the class $\mathcal{E}^3$ coincides with the class of elementary (computable) functions introduced by L. Kalmár in (1943) (cf. also Péter, 1951, p. 60) and characterized as the smallest class of functions including as initial functions $x+1, x+y, x \dot{-} y$ and closed under the operations of substitution, limited summation and limited multiplication (it can be shown that this class is also closed under the operation of limited minimum). The main properties of the functions belonging to the class $\mathcal{E}^0$ are the following: (1) for each function $f \in \mathcal{E}^0$ there exists a number $k_0$ such that for every $n$, $f(n) < n + k_0$ and (2) each recursively enumerable set is enumerated by a certain function of the class $\mathcal{E}^0$.

The classes $\mathcal{E}^n$ ($n \in \mathbf{N}$) form a hierarchy of functions, i.e., in particular for every $n$: (1) $\mathcal{E}^n \subseteq \mathcal{E}^{n+1}$ and (2) the function $f_{n+1}(x, x)$ increases faster than any function of the class $\mathcal{E}^n$ and consequently $\mathcal{E}^n \neq \mathcal{E}^{n+1}$. One can also prove that for $n > 2$ the class $\mathcal{E}^{n+1}$ contains a universal function for the class $\mathcal{E}^n_1$, i.e., for the class of unary functions of the class $\mathcal{E}^n$.

Grzegorczyk's hierarchy is a hierarchy of primitive recursive functions. In fact it can be shown that the class of primitive recursive functions $\mathcal{P}$ is equal to the sum of all the classes $\mathcal{E}^n$ ($n \in \mathbf{N}$). This leads to the corollary that in the definition of the class $\mathcal{P}$ the operation of recursion cannot be eliminated or exchanged for the operation of limited recursion.

Grzegorczyk's hierarchy was extended to transfinite ordinals by M.H. Löb and S.S. Wainer (cf. Wainer and Lob, 1970a, 1970b; Wainer, 1970). They introduced a general procedure for generating hierarchies that can be applied to a wide variety of classes of number-theoretic functions (even to classes containing non-recursive functions) and proved that the hierarchies may be extended through the ordinals of Cantor's second number class without collapsing. In this way a proper extension of Grzegorczyk hierarchy can be obtained — in fact the extension and the original Grzegorczyk hierarchy coincide at the level $\omega$. The most interesting part of the Löb-Wainer hierarchies is below the ordinal $\varepsilon_0$.

When discussing the contribution of Polish logicians to general recursion theory one should also mention the paper by Antoni Janiczak (1955) (edited by Grzegorczyk from the notes left by the author who died prematurely in 1951). Some interesting properties of the class of partially recursive functions are studied there. The class **PR** of partially recursive functions is defined as follows: $f \in \mathbf{PR}$ iff there exist two recursive functions $g$ and $h$ such that

$$\forall z\, [z \in D^*(f) \equiv \exists x\, (h(z, x) = 0)],$$

$$\forall z\, [z \in D^*(f) \longrightarrow f(z) = g((\min x)[h(z, x) = 0])],$$

where $D^*(f)$ is the set of arguments of $f$. Janiczak proves that: (a) there exists a function $f \in \mathbf{PR}$ which assumes only two values 0 and 1 and which cannot be extended to any recursive function, (b) there exists a function $f \in \mathbf{PR}$ which cannot be majorized by any recursive function $g$, (c) if the set $D^*(f)$ is the complement of a recursively enumerable set and $f$ can be extended to a function $g \in \mathbf{PR}$ then $f$ can be extended to a recursive function $h$, hence, in particular, if $D^*(f)$ is a recursive set and $f \in \mathbf{PR}$ then $f$ can be extended to a recursive function.

## 3. KLEENE-MOSTOWSKI HIERARCHY

In Section 1 we mentioned the group of set-theoretically minded mathematicians around W. Sierpiński in Warsaw. Their work contributed to the development of descriptive set theory. Considerable space was devoted in their studies to the estimation of the Borel class or projection class of given sets.

Those researches were an inspiration for Andrzej Mostowski. Using an analogy between the operation of projection (applied in the descriptive set theory) and the strictly logical operation of the existential quantifier he constructed a hierarchy of arithmetical notions (today called the Kleene-Mostowski hierarchy or the arithmetical hierarchy).

Mostowski's work was done during the war. His results could be published only in 1947, and they appeared in the paper 'On definable sets of positive integers' in *Fundamenta Mathematicae*. Earlier, in 1943, a paper 'Recursive predicates and quantifiers' by S.C. Kleene appeared which contained results similar to those of Mostowski. This paper was of course unknown in Poland at the time — it became available only when Mostowski's paper was in press (cf. Mostowski, 1947, p. 112). Mostowski admited also in (1947) that A. Tarski informed him that he had found results very similar to his in 1942.

Though the papers by Kleene and by Mostowski contained similar results, obtained independently (hence the name the Kleene-Mostowski hierarchy), it should be stressed here that both authors had quite different motivations. Kleene was motivated by some recursion-theoretic considerations rooted in Gödel's incompleteness theorem. But he also saw some analogies between the logical operations of existential and universal quantifiers and the geometrical operations of projection and intersection, resp., (cf. Kleene, 1943, p. 50). Mostowski on the other hand was motivated by the results of descriptive set theory. The analogy between the properties of classes of the hierarchy he defined and the properties of classes of projective sets is mentioned several times in his paper (1947).

Mostowski developed his theory starting not from the notion of general recursivity (as was the case for Kleene) but, for the convenience of the reader, from a notion of a decidable propositional function. He introduced the classes $P_n^{(k)}$ and $Q_n^{(k)}$ (for $k, n \in \mathrm{N}$) forming a hierarchy. The index $k$ denotes here the arity of the classified notions (relations) and the index $n$ the level of the hierarchy. Hence $P_n^{(k)}$ and $Q_n^{(k)}$ are classes of $k$-ary relations of the $n^{\text{th}}$ level (in today's notation the upper index is omitted and the classes $P_n^{(k)}$ and $Q_n^{(k)}$ are denoted as $\Sigma_n$ and $\Pi_n$, resp., or as $\Sigma_n^0$ and $\Pi_n^0$, if one wants to distinguish this hierarchy from the analytical hierarchy — see below).

The initial classes $P_0^{(k)}$ and $Q_0^{(k)}$ are equal and are defined as the class of $k$-ary decidable propositional functions. Next, classes are introduced inductively. A relation $R$ is said to belong to the class $P_{n+1}^{(k)}$ iff there exists a relation $R_1$ of the class $Q_n^{(k+1)}$ such that for any $k$-tuple $a_1, \ldots, a_k$:

$$R(a_1, \ldots, a_k) \equiv \exists x\, R_1(a_1, \ldots, a_k, x).$$

A relation $R$ is of the class $Q^{(k)}_{n+1}$ iff its complement is of the class $P^{(k)}_{n+1}$.

Classes defined in this way possess several interesting properties. Mostowski proved in (1947) theorems concerning the sum, product and Cartesian product of relations of a given class. It was also shown that for $n \geq 0$ the common part of the classes $P^{(k)}_n$ and $Q^{(k)}_n$ is a field of sets and that

$$P^{(k)}_n \subseteq P^{(k)}_{n+1} \cap Q^{(k)}_{n+1},$$

$$Q^{(k)}_n \subseteq P^{(k)}_{n+1} \cap Q^{(k)}_{n+1}.$$

Using Cantor's diagonal theorem one can prove that

$$P^{(k)}_n \neq P^{(k)}_{n+1} \quad \text{and} \quad Q^{(k)}_n \neq Q^{(k)}_{n+1}$$

as well as

$$P^{(k)}_n \text{ non} \subseteq Q^{(k)}_n \quad \text{and} \quad Q^{(k)}_n \text{ non} \subseteq P^{(k)}_n,$$

hence the classes $P^{(k)}_n$ and $Q^{(k)}_n$ for $n \in \mathbf{N}$ form a hierarchy.

Mostowski's paper (1947) also contains various applications. First of all it is shown that certain generalizations of Gödel's incompleteness theorem hold for systems based on infinitary rules. In particular it is proved (Theorem 4.21) that if the system $S$ is $\omega$-consistent and fulfils a certain condition $C_s$ then there exists a sentence $\vartheta$ which is undecidable in $S$. The condition $C_s$ is the conjunction of the following conditions: the (arithmetized counterpart of the) function of substitution is of the class $P^{(k+2)}_s$, the (arithmetized counterpart of the) function of negation is of the class $P^{(2)}_s$, the (arithmetized counterpart of the) relation of provability is of the class $P^{(2)}_s$ and the (arithmetized counterpart of the) relation of being a theorem is of the class $P^{(1)}_s$. It is obvious that systems with finitary inference rules (and based on a recursive set of axioms) fulfil the condition $C_0$. The conditions $C_s$ ($s > 0$) can be fulfilled by systems with infinitary rules (an example of such a system is Rosser's system introduced in (1937)).

It is also proved in Mostowski (1947) that systems $S$ which are $\omega$-consistent and fulfil the condition $C_s$ are not closed under the rule of infinite induction (i.e., $\omega$-rule). Furthermore, it is shown that if $S$ fulfils the condition $C_s$ ($s \geq 0$) then the set of sentences which can be obtained from the axioms of $S$ by $n$ iterations of the rule of infinite induction is not closed under this rule and is incomplete if it is $\omega$-consistent.

The hierarchy of relations $P^{(k)}_n$ and $Q^{(k)}_n$ ($n \in \mathbf{N}$) is extended in Mostowski (1947) to the case of functions. A function $f$ is said to be of the given class iff its graph belongs to this class. Mostowski also proves Post's theorem and says that "it is an exact analogue of the well-known Souslin theorem concerning sets which are analytical together with their complements" (p. 106). In the footnote

(p. 82) it is explained that after having finished the first draft of this paper the author became acquainted with Post's paper (Post, 1944) where the same result is published.

The results of (1947) were extended by Mostowski in the sequel paper of 1948. Assuming that primitive recursive $k$-ary relations belong to the class $P_0^{(k)}$ and that the relation $qB_\varphi n$ (where $q$ is the Gödel number of a formal proof of $\varphi(n)$) is primitive recursive, it is shown that $P_0^{(k)}$ is the class of general recursive $n$-ary relations. The main result concerns the class $R_1^{(k)}$ defined as the smallest finitely additive field of sets containing the classes $P_1^{(k)}$ and $Q_1^{(k)}$. It is proved that $R_1^{(k)} \neq P_2^{(k)} \cap Q_2^{(k)}$. Again the analogy with projective sets is stressed. The author says: "These theorems are quite analogous to the following well known results concerning projective sets: if a set $B$ is an **A**-set as well as a **CA**-set, it must be Borelian; but a set which is a **PCA**-set as well as a **CPCA**-set does not necessarily belong to the smallest (denumerably additive) field of sets over **A** + **CA**" (p. 115).

Mostowski applied the hierarchy introduced in the 1947 paper to the classification of various mathematical and logical notions. One should mention here his papers (1949), (1951), and (1955).

In Mostowski (1951) the hierarchy is extended up to the first nonconstructive ordinal $\omega_1$ (denoted today by $\omega_1^{CK}$) and is used in the classification of certain theories defined in a semantical way. A logical system $S$ is said to be of the class $P_\alpha$ or $Q_\alpha$ iff the set of (numbers which correspond to) true sentences of $S$ belongs to the class $P_\alpha^{(1)}$ or $Q_\alpha^{(1)}$, resp. It is known that the propositional calculus is a system of the class $P_0$ and the first-order predicate calculus is a system of the class $P_1$. Mostowski also shows in (1951) that the system $S_e$ of elementary arithmetic introduced by Carnap in (1934) is not recursively enumerable and that the class of this system is exactly $Q_1$. The theory $S_e$ is defined as a first-order system with infinitely many predicates and function symbols (for every primitive recursive relation and function there is an appropriate symbol), with one connective (binegation) and bounded quantifiers only. It is also shown that the usual first-order system $S_a$ of general arithmetic (arbitrary quantifications are allowed) is exactly of the class $P_\omega$ and that the class of the system of arithmetic of real numbers $S_r$ (in the language of $S_a$ enriched by set variables and the membership relation $\in$) cannot be determined unless we extend the definition of classes $P_\alpha^{(k)}$ and $Q_\alpha^{(k)}$ beyond $\omega_1$.

It is worth mentioning here that in (1951) Mostowski explicitly formulated his philosophical views. He says (on p. 247):

We shall not consider logical systems as void schemata deprived of any interpretation. On the contrary we shall assume the objective existence of a kind of "mathematical reality" (e.g., of the set of all integers or of the set of all real numbers). By objective existence we mean existence independently of all linguistic constructions. The role of logical and mathematical systems is to describe this reality.

Mostowski's most well known and important result on the classification of notions is contained in his (1955). He considers there fractions of the form $10^{-x}\alpha(x,y)$ where $\alpha$ is a primitive recursive function and investigates the set of those integers $y$ for which $\lim_x 10^{-x}\alpha(x,y)$ exists and belongs to a preassigned class of real numbers. A typical result is stated, for example, in Theorem 7. It says that the set $Z_\alpha^{(5)}$ of those $y$s for which $\lim_x 10^{-x}\alpha(x,y)$ exists and is integral is the most general set of the class $Q_3^{(1)}$, i.e., if $\alpha$ ranges over the set of primitive recursive functions then the set $Z_\alpha^{(5)}$ ranges over the whole class $Q_3^{(1)}$. Results obtained in the paper allow the construction of examples of sets definable by means of two or three quantifiers but not definable by a smaller number of them. In particular from Theorem 7 it follows that if $U(n,x,y)$ is a (general recursive) function universal for the class of primitive recursive functions with two arguments then the set of pairs $(n,y)$ such that $\lim_x 10^{-x}U(n,x,y)$ exists and has an integral value belongs to the class $Q_3^{(2)}$ but not to the class $P_3^{(2)}$, hence the notion of the limit of a sequence is the notion of the class $P_3^{(2)}$ and cannot be simplified. This means that all three quantifiers used in its definition are indispensable and cannot be dropped or replaced by dual ones.

The arithmetical hierarchy of Kleene and Mostowski was extended in two ways. First it was expanded into the constructive transfinite — cf. Mostowski (1951) and Kleene (1955). In this way the hyperarithmetic hierarchy was obtained. Again there are several analogies between this hierarchy and descriptive set theory. They were explicitly formulated by Addison in (1958).

The second extension was made by Kleene in (1955) and led to the analytical hierarchy. It is in fact an extension of the hyperarithmetical hierarchy and is defined in a way similar to the arithmetical hierarchy. One starts here from arithmetical relations (i.e., from the class $\bigcup_{n\in\mathbf{N}}(\Sigma_n \cup \Pi_n)$) and divides relations into classes according to their definitions, taking into account the number and type of quantifiers, but now we allow formulae containing not only quantifiers over natural numbers but also quantifiers whose range consists of sets of integers. The particular classes are denoted by $\Sigma_n^1$ and $\Pi_n^1$ (in this notation the classes of the arithmetical hierarchy are denoted as $\Sigma_n^0$ and $\Pi_n^0$ instead of $\Sigma_n$ and $\Pi_n$). It can be proved that the class of hyperarithmetical relations is exactly the intersection of the classes $\Sigma_1^1$ and $\Pi_1^1$ (therefore it is usually denoted by $\Delta_1^1$).

Using the arithmetical and analytical hierarchies Mostowski distinguished various degrees of constructivism in (1959). Let $K$ be an arbitrary class of number-theoretic functions. One defines a class $\mathbf{K}$ of real numbers associated with the class $K$ in the following way: a real number $a$ belongs to the class $\mathbf{K}$ iff for every $n$:

$$\left| |a - [a]| - \frac{f(n)}{n} \right| < \frac{1}{n}$$

where $f$ is a fixed function from $K$. Choosing now an appropriate initial class $K$ one obtains classes $\mathbf{K}$ satisfying certain general theses of constructivism. In particular for $K = \Sigma_0^0 = \Pi_0^0$ we obtain the class $\mathbf{K}_0$ of recursive real numbers. It is the largest class of reals whose approximations can be calculated by means of algorithms. It should be noted here that they are accepted by most constructivists although the totality of them $\mathbf{K}_0$ is not generally accepted. The reason is that for its definition a universal quantification over the set of natural numbers is needed and this presupposes the existence of an actually infinite totality. Hence a radical constructivist will accept every particular element of the class $\mathbf{K}_0$ but not the whole set $\mathbf{K}_0$. In a similar way we define the class $\mathbf{K}_\alpha$ of reals which can be approximated by functions of the class $\Pi_\alpha^0 \cap \Sigma_\alpha^0$. It can be shown that $\mathbf{K}_\alpha$ is a real closed field and that $\mathbf{K}_\alpha \neq \mathbf{K}_\beta$ for $\alpha \neq \beta$.

Special attention should be paid to the class $\mathbf{K}_\omega$ obtained from the class of arithmetical functions $\bigcup_{n \in \mathbf{N}} \Sigma_n^0 (= \bigcup_{n \in \mathbf{N}} \Pi_n^0)$. It consists of the reals which are elementarily definable, i.e., it coincides with the universe of Hermann Weyl's constructive analysis (cf. Weyl, 1918). Elements of $\mathbf{K}_\omega$ are exactly those reals which are accepted by a constructivist who does not reject any infinite sets but who claims that all such sets should be reducible to the set of natural numbers.

The advantage of Mostowski's approach is that it enables us to give precise definitions of various notions of constructivism and in this way make possible a comparison of various senses of constructivity.

## 4. COMPUTABLE FUNCTIONALS

As mentioned in the introduction the notions of recursiveness and computability were extended in various ways. One of them consisted in a generalization of those notions to the case of objects of higher types. The first step in this direction was the 1958 paper of K. Gödel where primitive recursive functionals were considered. Another one was Kleene's paper (1959) in which computable functionals were introduced and investigated.

This trend in recursion-theoretical research was also present in the works of Polish logicians, first of all in the works of Andrzej Grzegorczyk. Studies of primitive recursive and computable functionals are connected with the study of constructive foundations of mathematics and in particular of analysis, which we will discuss in the next section (and therefore we will return, in a specific context, to problems considered here).

The notion of a primitive recursive functional was considered by Grzegorczyk in his (1964) paper, where he proposed a very elegant definition. The starting point is the definition of a very general class $\mathcal{R}$ of recursive objects of finite types. This may be considered as identical with the class of recursive functionals mentioned by Gödel in (1958) and by G. Kreisel in (1959). Grzegorczyk compares his definition of the notion of a primitive recursive functional with that of Kleene (cf. Kleene, 1959) and proves that the class defined by Kleene is a proper subclass of the class of recursive functionals assuming only natural numbers as values. It is also shown how one can apply the new notion to Gödel's interpretation of intuitionistic arithmetic (cf. Gödel, 1958). This notion makes the interpretation easier.

Grzegorczyk also wrote some papers devoted to the notion of a computable functional. In the paper (1955a) an inductive definition is given. The class of computable functionals is defined there as the smallest class containing some initial functionals and closed under the operations of substitution and of effective minimum. If one cancels in this definition the condition of effectiveness of the minimum operation then the definition of the class of elementarily definable functionals is obtained. This was considered by Grzegorczyk in his (1955b). The latter is connected with some ideas of H. Weyl expressed in (1918). Weyl proposed there a restriction of the logical methods of analysis to the elementarily definable ones. This vague notion was given a precise definition in Grzegorczyk (1955b). In it he stresses that many theorems of classical analysis can be obtained by means of elementary methods. In particular it is shown that the classical analysis of continuous functions can be reproduced in an elementary manner.

The mathematical definition of a computable functionals given in Grzegorczyk (1955a) was shown to be equivalent to a metamathematical one (cf. Grzegorczyk, 1955c). Namely, it was proved that a functional is computable iff it is representable in a certain system of arithmetic with no quantifiers but with the $\mu$-operation and with appropriate axioms defining the function-variables (this system is similar to the system S considered by A. Mostowski in his book (1952)).

## 5. CONSTRUCTIVE FOUNDATIONS OF MATHEMATICS

One of the reactions to the difficulties in the foundations of mathematics disclosed by antinomies was the proposal to exclude all general set-theoretical notions from mathematics and to limit it to the study of those objects that can be effectively defined or constructed. An extreme example of this tendency is provided by the intuitionism of L.E.J. Brouwer. Less extreme are the trends which do not challenge the classical rules of proof (as intuitionism does) and require only the restriction of the class of admissible mathematical objects to constructive ones. Since there is no unique notion of constructivity there are various (mutually conflicting) constructivistic programs. One of them is based on recursion theory and in particular on the notion of a computable (recursive) function. It was applied first to the reconstruction of the classical analysis. Its characteristic feature is the full acceptance of the notion of integer taken from classical arithmetic (this notion is not analysed any further). On the other hand several limitations are put on the notion of a real number and all other mathematical notions. The aim is to eliminate all non-constructive notions (cf. Grzegorczyk, 1959).

Various proposals in this direction have been formulated. Two main approaches can be distinguished here. The first one adopts from classical analysis the general notion of functions and among them singles out a narrower class of computable real functions. The second one, which seems to answer better the aims of computable analysis, defines the class of computable real functions by means of certain operations performed on simple primitive functions (a more general notion of a functional, i.e., of a function which assigns numerical values to every system of arguments composed of numbers and functions, is often used here).

The investigations of Stefan Banach and Stanisław Mazur belong to the first type. They were begun in 1936 (cf. Banach and Mazur, 1937). Their aim was to investigate a fragment of analysis admitting only numbers whose expansions into decimal fractions are represented by primitive recursive functions. The studies were continued by Mazur — the most extensive presentation of the results can be found in Mazur's paper 'Computable analysis' (cf. 1963). The paper (edited by Grzegorczyk and Rasiowa) is based on notes of Mazur's lectures in the 1949/50 academic year at the Institute of Mathematics of the Polish Academy of Sciences in Warsaw. The lectures gave a systematic exposition of the results obtained by Banach and Mazur in the period 1936–39 (their earlier publication was impossible due to the Second World War) and some results obtained by Mazur after the war (in particular results concerning general recursive mathematical objects).

Banach-Mazur's computable analysis begins with the study of the notion of a recursive or computable real number. A positive real number $a$ is said to be recursive iff there exists a primitive recursive function $f$ such that for any $n = 0, 1, \ldots$:

$$\left| a - \frac{f(n)}{n+1} \right| < \frac{1}{n+1}.$$

The number $a$ is called computable iff there exists a (general) recursive function $f$ with the indicated property. It can be proved that:

(1) all algebraic numbers are recursive,

(2) the set of recursive numbers and the set of computable numbers constitute fields of numbers,

(3) these two fields are algebraically closed in the field of real numbers.

Further the notions of a recursive and computable sequence are introduced. A sequence $\{a_k\}$ is called recursive (computable) iff there exists a primitive recursive (general recursive) function $f$ such that for every $k, n \in \mathbf{N}$:

$$\left| a_k - \frac{f(k,n)}{n+1} \right| < \frac{1}{n+1}.$$

It can be shown that the set of computable sequences is closed under the four arithmetical operations: $+, -, \cdot, :$; on the other hand the set of recursive sequences is not closed under the operation of division. The sequence $\{a_k\}$ is said to be recursively (computably) convergent to $a$ iff there exists a primitive recursive (general recursive) function $h$ such that for $k > h(m)$ the following holds:

$$|a_k - a| < \frac{1}{m+1}.$$

The notion of a convergent sequence has the following properties:

(1) the four arithmetical operations performed on recursively (computably) convergent sequences give the sequences recursively (computably) convergent to the limits obtained by the four arithmetical operations,
(2) the limit of a recursively (computably) convergent sequence is recursive (computable) provided that the sequence is recursive (computable),
(3) there exists a recursive sequence which converges to 0 in a non-computable manner,
(4) for each computable sequence convergent to a computable number there exists a computable subsequence that is computably convergent,
(5) there exists a recursive monotonic convergent sequence which does not converge to a computable number — hence the Bolzano-Weierstrass theorem is not satisfied in the domain of computable sequences and numbers.

There are also other anomalies. For example the functions 'signum' and 'the integral part' lead out of the class of computable sequences.

The next main notion considered by Banach and Mazur is the notion of a recursive and computable real function. It is defined in the following way: a real function defined on a set of recursive (computable) numbers is said to be recursive (computable) iff it transforms any recursive (computable) sequence into a recursive (computable) one. We can prove that those functions possess the following properties:

(1) the four arithmetical operations performed on computable functions do not lead out of the class of computable functions,
(2) the operations of addition, multiplication and subtraction do not lead out of the class of recursive functions; however the function $1/x$ is not recursive on the set of all positive recursive numbers,

(3) the functions 'signum $x$' and 'integer-part of $x$' considered even on the set of all recursive numbers of the closed interval $[0, 1]$ are not computable,
(4) the function 'signum $x$' is computable (but not recursive) on the set of all computable (recursive) numbers different from 0,
(5) the function 'integral-part of $x$' is computable (but not recursive) on the set of all computable (recursive) non-integers,
(6) a function defined by a power-series with a recursive (computable) sequence of coefficients and with a recursive (computable) center, considered on the set of all recursive (computable) numbers from the interval of convergence of this series, is not necessarily recursive (computable) on a set of all recursive (computable) numbers from an arbitrary closed interval contained in the interval of convergence.

It is also shown that a computable function on a set of all computable numbers of an interval is continuous and has Darboux's property. On the contrary, recursive functions do not necessarily have Darboux's property.

The above mentioned property that a computable function is continuous leads to the conclusion that in computable analysis in which only computable numbers, sequences and functions are admitted one can take into consideration only those parts of classical analysis which treat exclusively of continuous functions. On the other hand it is not clear whether the entire classical theory of continuous functions can be obtained in computable analysis. The reason is that it is not known whether a computable function defined in a computable closed interval assumes the maximum of its values at a computable point. However it has been shown that numerous theorems of the theory of continuous functions are transferable to computable analysis.

Banach and Mazur's approach to constructive analysis is similar to that of E. Specker (cf. Specker, 1949). It can be proved that a function is recursive in the sense of Specker iff it is recursive and recursively uniformly continuous in the sense of Mazur.

The second approach to constructive analysis described at the beginning of this section is represented by Grzegorczyk (cf. Grzegorczyk, 1955a). It consists of using the notion of a functional. A functional $\Phi(f_1, \dots, f_k, x_1, \dots, x_n)$ is said to be computable iff it can be derived from the primitive functionals:

$$U_1(f, g, x) = f(x),$$

$$U_2(f, g, x) = g(x),$$

$$S(f, x) = x + 1,$$

$$M(f, x, y) = x - y,$$

$$P(f, x, y) = x^y$$

by a finite number of the following operations: the substitution of the functional for the numerical variable, the identification of variables and the effective minimum.

The real function $\varphi(x)$ defined on the interval $[a, b]$, where $0 < a < b$, is called computable iff there exists a computable functional $\Phi(f, n)$ such that for each $x$ $(a < x < b)$ and each function $f$ with natural values satisfying the condition

$$\left| x - \frac{f(n)}{n} \right| < \frac{1}{n}$$

for $n = 1, 2, \ldots$, the following inequality is satisfied:

$$\left| \varphi(x) - \frac{\Phi(f, n)}{n} \right| < \frac{1}{n}$$

for $n = 1, 2, \ldots$ In other words, the functional $\Phi$ transforms the function $f(n)/n$ approximating the argument $x$ into the function $\Phi(f, n)/n$ approximating in the same degree the value $\varphi(x)$.

It can be shown that a function computable in the sense of Grzegorczyk is also computable in the sense of Banach-Mazur, hence it is continuous. One can also prove that a function computable in the sense of Grzegorczyk defined in the interval $[a, b]$ with computable endpoints, assumes its maximum value at the computable point of the interval $[a, b]$.

The comparison of various notions of a computable real function can be found in Grzegorczyk (1957).

Let us add at the end of this section that one can also build a system of constructive mathematics (and in particular of constructive analysis) by replacing the class of computable functions in the definitions accepted by Banach and Mazur by other, broader, classes. For example, one can admit elementarily definable functions, i.e., functions definable by means of the quantifiers bounding the integral variables only. This approach was considered by Grzegorczyk in (1955b). It was shown there that the classical analysis of continuous functions can be reproduced in an elementary manner. However it is not clear in what exactly the definable analysis deviates from the classical one.

## 6. COMPLEXITY OF MODELS

In Section 3 we mentioned that mathematical and logical notions can be classified in the Kleene-Mostowski hierarchy. In this section an application of this hierarchy to model theory will be indicated.

From the theorems of Löwenheim-Skolem and of Gödel it follows that every consistent first-order theory $T$ has a countable model. Hence it has a model with the universe being the set $\mathbf{N}$ of natural numbers. Thus predicates and function symbols of the language $L(T)$ of $T$ can be interpreted as relations on the set **N** and as functions from $\mathbf{N}^n$ to **N**, resp. So we can classify them in the Kleene-Mostowski hierarchy. It will be said that a model

$$\mathcal{M} = <\mathbf{N}, R_1, \ldots, R_k, f_1, \ldots, f_l>$$

is of the class $\Pi^0_n$ ($\Sigma^0_n$) iff all the relations $R_i$ and functions $f_j$ are of the class $\Pi^0_n$ ($\Sigma^0_n$).

Using Lindenbaum's method of completing a theory one can show that every consistent first-order theory has a model of the type $\Sigma^0_2 \cap \Pi^0_2$. Hence there arises the problem of whether a given theory can possess a simpler model.

Mostowski, in (1953), gave an example of a finitely axiomatizable theory which has no model of the class $\Sigma^0_1$, i.e., no recursively enumerable model. This theory is a modification of the axiomatic system of set theory proposed by P. Bernays in (1937). The modification consists in allowing more primitive notions — there are in fact twenty such notions in the system. The axioms of the theory are those given by Bernays with obvious changes which are the consequence of the choice of primitive notions.

This result was strengthened by Mostowski in (1955a) where an example of a finitely axiomatizable theory (so in fact of a single formula) which has no recursively enumerable model was given. What is important here is the fact that this example is simpler than that from the paper Mostowski (1953). In fact it does not refer to axiomatic set theory and uses only tools from the theory of recursive functions. The description of it is very long and complicated — therefore it cannot be given here. Note only that, modifying it, one can obtain a finitely axiomatizable theory (a formula) which possesses no model belonging to the class $\Sigma^0_1 \cap \Pi^0_1$.

Mostowski admits in (1955a) that he attempted to find an example of a theory with no model belonging to the smallest field of sets generated by the classes $\Sigma^0_1$ and $\Pi^0_1$. This problem was formulated by Kreisel in (1953) (p. 47; cf. also Mostowski, 1955b, p. 30). It was solved by G. Hensel and H. Putnam. Putnam showed in (1965) that every axiomatizable, consistent, first-order theory based on a finite number of predicates has a model in the field $\Sigma^*_1$ generated by the recursively enumerable relations. This results was extended by Hensel and Putnam (1969) to the case of theories of the type described with identity. More precisely Hensel and Putnam showed that such a theory, if it possesses an infinite normal model, has a model in the class $\Sigma^*_1$. The exhibited model is the simplest possible in the sense that it contains Ramsey indiscernibles and only those extra elements needed for completion.

The problem of the lowest possible class of models of theories was also studied by Grzegorczyk. In (1962) he pointed out four first-order theories formalized in a language with two individual constants 0 and $S$ and one binary function $|$. They are all finitely axiomatizable and form an increasing sequence:

$F \subseteq F^{(1)} \subseteq F^{(2)} \subseteq F^{(3)}$. It is proved that: (1) the theory $F$ is essentially undecidable (cf. Grzegorczyk's 1962a) and has no primitive recursive model, (2) the theory $F^{(1)}$ has no primitive recursive model but has a general recursive model, (3) $F^{(2)}$ has no general recursive model but has an arithmetical $\omega$-standard model and (4) the theory $F^{(3)}$ has no arithmetical $\omega$-standard model.

To the list of the results described above one should add one more due to Mostowski. In the paper (1957) he considered recursive models for axiomatic systems of arithmetic and proved that under certain assumptions every such model is isomorphic to the classical model. Hence the result says that, up to isomorphism, the only recursive model of arithmetic is the standard one.

*Faculty of Mathematics and Computer Science,*
*Adam Mickiewicz University, Poznań, Poland*

* This paper was written in the framework of the research project 'Mathematical logic in Poland: origins, development, import' realized at the Jagiellonian University (Krakow)—Committee for Scientific Research (KBN) grant No. PB 1084/91. The upper bound of the period covered in the project was fixed as 1963.

REFERENCES

Addison, J.W.: 1954, *On some points of the theory of recursive functions*, unpublished doctoral dissertation, University of Wisconsin.

Addison, J.W.: 1958, 'Separation principles in the hierarchies of the classical and effective descriptive set theory', *Fundamenta Mathematicae* **46**, pp. 123–135.

Banach, S., Mazur, S.: 1937, 'Sur le fonctions calculables', *Annales de la Société Polonaise de Mathématique* **16**, p. 223.

Bernays, P.: 1937, 'A system of axiomatic set theory', Part I, *Journal of Symbolic Logic* **2**, pp. 65–77.

Carnap, R.: 1934, *Logische Syntax der Sprache*, J. Springer, Vienna.

Gödel, K.: 1931, 'Über formal unentscheidbare Sätze der 'Principia Mathematica' und verwandter Systeme. I', *Monatshefte für Mathematik und Physik* **38**, pp. 173–198.

Gödel, K.: 1958, 'Über eine noch nicht benützte Erweiterung des finiten Standpunktes', *Dialectica* **12**, pp. 280–287.

Grzegorczyk, A.: 1953, 'Some classes of recursive functions', *Rozprawy Matematyczne* **4**, pp. 1–46.

Grzegorczyk, A.: 1955a, 'Computable functionals', *Fundamenta Mathematicae* **42**, pp. 168–202.

Grzegorczyk, A.: 1955b, 'Elementarily definable analysis', *Fundamenta Mathematicae* **41**, pp. 311–338.

Grzegorczyk, A.: 1955c, 'On the definition of computable functionals', *Fundamenta Mathematicae* **42**, pp. 232–239.

Grzegorczyk, A.: 1957, 'On the definitions of computable real continuous functions', *Fundamenta Mathematicae* **44**, pp. 61–71.

Grzegorczyk, A.: 1959, 'Some approaches to constructive analysis', in *Constructivity in Mathematics. Proceedings of the Colloquium held at Amsterdam, 1957*, ed. by A. Heyting, North-Holland Publishing Company, Amsterdam, pp. 43–61.

Grzegorczyk, A.: 1962, 'A theory without recursive models', *Bulletin de l'Académie Polonaise de Sciences, Série des sciences math., astr. et phys.* **10**, pp. 63–69.
Grzegorczyk, A.: 1962a, 'An example of two weak essentially undecidable theories $F$ and $F^*$', *Bulletin de l'Académie Polonaise des Sciences, Série des sciences math., astr. et phys.* **10**, pp. 5–9.
Grzegorczyk, A.: 1964, 'Recursive objects in all finite types', *Fundamenta Mathematicae* **54**, pp. 73–93.
Hensel, G., Putnam, H.: 1969, 'Normal models and the field $\Sigma_1^*$', *Fundamenta Mathematicae* **64**, pp. 231–240.
Janiczak, A.: 1955, 'Some remarks on partially recursive functions', *Colloquium Mathematicum* **3**, pp. 37–38.
Kalmár, L.: 1943, 'Egyszerü példa eldönthetetlen aritmetikai problémára', *Matematikai és Fizikai Lapok* **50**, pp. 1–23.
Kleene, S.C.: 1943, 'Recursive predicates and quantifiers', *Transactions of the American Mathematical Society* **53**, pp. 41–83.
Kleene, S.C.: 1955, 'Arithmetical predicates and function quantifiers', *Transactions of the American Mathematical Society* **79**, pp. 312–340.
Kleene, S.C.: 1959, 'Recursive functionals and quantifiers of finite types. I', *Transactions of the American Mathematical Society* **91**, pp. 1–52.
Kreisel, G.: 1953, 'Note on arithmetic models for consistent formulae of the predicate calculus. II', in *Proceedings of the XIth International Congress of Philosophy*, vol. XIV, North-Holland Publishing Company, Amsterdam-Louvain, pp. 39–49.
Kreisel, G.: 1959, 'Interpretation of analysis by means of constructive functionals of finite types', in *Constructivity in Mathematics. Proceedings of the Colloquium held at Amsterdam, 1957*, ed. by A. Heyting, North-Holland Publishing Company, Amsterdam, pp. 101–128.
Löb, M.H., Wainer, S.S.: 1970a, 'Hierarchies of number-theoretic functions. I', *Archiv für Mathematische Logik und Grundlagenforschung* **13**, pp. 39–51.
Löb, M.H., Wainer, S.S.: 1970b, 'Hierarchies of number-theoretic functions. II', *Archiv für Mathematische Logik und Grundlagenforschung* **13**, pp. 97–113.
Mazur, S.: 1963, 'Computable analysis', ed. by A. Grzegorczyk and H. Rasiowa, *Rozprawy Matematyczne* **33**, pp. 1–111.
Mostowski, A.: 1947, 'On definable sets of positive integers', *Fundamenta Mathematicae* **34**, pp. 81–112.
Mostowski, A.: 1948, 'On a set of integers not definable by means of one-quantifier predicate', *Annales de la Société Polonaise de Mathématique* **21**, pp. 114–119.
Mostowski, A.: 1949, 'Sur l'interpretation géométrique et topologique des notions logiques', in *Actes du X-ème Congres International de Philosophie (Amsterdam 11–18 août 1948)*, North-Holland Publishing Company, Amsterdam, pp. 610–617.
Mostowski, A.: 1951, 'A classification of logical systems', *Studia Philosophica* **4**, pp. 237–274.
Mostowski, A.: 1952, *Sentences Undecidable in Formalized Arithmetic. An Exposition of the Theory of Kurt Gödel*, North-Holland Publishing Company, Amsterdam.
Mostowski, A.: 1953, 'On a system of axioms which has no recursively enumerable arithmetical model', *Fundamenta Mathematicae* **40**, pp. 56–61.
Mostowski, A.: 1955, 'Examples of sets definable by means of two and three quantifiers', *Fundamenta Mathematicae* **42**, pp. 259–270.
Mostowski, A.: 1955a, 'A formula with no recursively enumerable model', *Fundamenta Mathematicae* **42**, pp. 125–140.
Mostowski, A.: 1955b, 'The present state of investigations on the foundations of mathematics', in collaboration with A. Grzegorczyk, S. Jaśkowski, J. Łoś, S. Mazur, H. Rasiowa and R. Sikorski, *Rozprawy Matematyczne* **9**, pp. 1–48.

Mostowski, A.: 1957, 'On recursive models of formalised arithmetic', *Bulletin de l'Académie Polonaise de Sciences (Cl. III)* **5**, pp. 705–710.

Mostowski, A.: 1959, 'On various degrees of constructivism', in *Constructivity in Mathematics. Proceedings of the Colloquium held at Amsterdam, 1957*, ed. A. Heyting, North-Holland Publishing Company, Amsterdam, pp. 178–194.

Péter, R.: 1951, *Rekursive Funktionen*, Akadémiai Kiadó, Budapest.

Post, E.: 1944, 'Recursively enumerable sets of positive integers and their decision problems', *Bulletin of the American Mathematical Society* **50**, pp. 284–316.

Putnam, H.: 1965, 'Trial and error predicates and the solution to a problem of Mostowski', *Journal of Symbolic Logic* **30**, pp. 49–57.

Rosser, J.B.: 1937, 'Gödel's theorem for non constructive logic', *Journal of Symbolic Logic* **2**, pp. 129–137.

Specker, E.: 1949, 'Nichtkonstruktiv beweisbare Sätze der Analysis', *Journal of Symbolic Logic* **14**, pp. 145–158.

Wainer, S.S.: 1970, 'A classification of the ordinal recursive functions', *Archiv für Mathematische Logik und Grundlagenforschung* **13**, pp. 136–153.

Weyl, H.: 1918, *Das Kontinuum. Kritische Untersuchungen über die Grundlagen der Analysis*, Veit, Leipzig.

EWA ORŁOWSKA

# STUDYING INCOMPLETENESS OF INFORMATION: A CLASS OF INFORMATION LOGICS

## 1. INTRODUCTION

In this paper we explore possible foundations for reasoning about knowledge discovered from incomplete information. The basic idea is to employ Kripke's classical definitions of semantic structure and truth in a semantic structure, but to make the notion of structure more specific. Namely, we intend to present structures that are derived directly from data provided by a user and to develop logical systems for reasoning about these structures. We assume that the user's data consist of descriptions of some objects in terms of their properties, and we refer to the collection of these descriptions as explicit information. It happens that this kind of information is very often provided in application domains, and is a basis for a great variety of knowledge representation and processing methods. From explicit information we derive relationships among objects. They constitute implicit information contained in users' data. These relationships have the form of binary relations in a set of objects such that each of the relations is determined by a subset of properties of the respective objects. These relations reflect some aspects of incompleteness of explicit information. We consider relational structures, referred to as frames, that consist of a set of objects and a family of (accessibility) relations that are parameterised with subsets of a set, with the intuition that these subsets consist of properties of objects. These relations are referred to as information relations.

There are two major groups of frames: frames with indiscernibility-type relations and frames with orthogonality-type relations. Relations from the former group provide information about indistinguishability of objects in terms of their properties. These relations characterise in a nonnumerical way degrees of 'sameness' of objects. Relations from the latter group exhibit degrees of distinguishability of objects. They provide a nonnumerical characterisation of the extent to which the given objects differ.

The logics we introduce in this paper enable us: (a) to represent both explicit and implicit information given in the form of an information system and to exhibit various kinds of incompleteness of explicit information and (b) to process this information, in particular to reason about consequences of incompleteness.

The logics are an elaboration of modal logics with multiple modal operators. However, they extend and generalise the known modal systems in that,



*K. Kijania-Placek and J. Woleński (eds.),*
*The Lvov–Warsaw School and Contemporary Philosophy*, 283–300.
© 1998 *Kluwer Academic Publishers.*

among others, the accessibility relations in their models are relative to subsets of a set. The information logics and their classification presented in this paper evolved from research on formalisms for reasoning about objects characterised in terms of their properties (Balbiani, 1996; Demri, 1996a,b, 1997; Demri and Orłowska, 1996, 1996a, 1996b; Farinas del Cerro and Orłowska, 1985; Konikowska, 1987, 1994; Orłowska, 1984, 1985, 1987, 1988, 1988a, 1989; Rasiowa, 1988, 1991; Rasiowa and Skowron, 1985; Vakarelov, 1987, 1989, 1991a, 1991b).

In the last section of the paper we consider two kinds of a plain calculus of information relations (implicit information only) and we show that a system of Aristotelian syllogistic is a formal system adequate for reasoning about these relations.

## 2. INFORMATION FRAMES DERIVED FROM INFORMATION SYSTEMS

Let an information system (OB, AT, $\{VAL_a : a \in \text{AT}\}$) be given such that OB is a nonempty set of objects, AT is a nonempty set of attributes, each $VAL_a$ is a nonempty set of values of attribute $a$. Each attribute is a function $a : \text{OB} \to Sb(VAL_a)$ that assigns subsets of values to objects such that $a(x) \neq \emptyset$ for all $a$ and $x$. Any set $a(x)$ can be viewed as a property of object $x$. For example, if attribute $a$ is 'colour' and $a(x)$ = green, then $x$ possesses the property of 'being green', if $a$ is 'age' and $x$ is 25 years old, then $a(x) = 25$. However, we might know the age of $x$ only approximately, say between 20 and 28, and then $a(x) = \{20, \ldots, 28\}$. If $a(x)$ is a singleton set, then we have deterministic information about $x$, otherwise our information is nondeterministic.

Let $A$ be a subset of AT. The following classes of information relations derived from an information system reflect incompleteness of information manifested by approximate (relative to properties) indistinguishability:

Strong (weak) indiscernibility:
$(x, y) \in \text{ind}(A)$ iff $a(x) = a(y)$ for all (some) $a \in A$

Strong (weak) similarity:
$(x, y) \in \text{sim}(A)$ iff $a(x) \cap a(y) \neq \emptyset$ for all (some) $a \in A$

Strong (weak) forward inclusion:
$(x, y) \in \text{fin}$ iff $a(x) \subseteq a(y)$ for all (some) $a \in A$

Strong (weak) backward inclusion:
$(x, y) \in \text{bin}$ iff $a(y) \subseteq a(x)$ for all (some) $a \in A$

Strong (weak) negative similarity:
$(x, y) \in \text{nim}(A)$ iff $-a(x) \cap -a(y) \neq \emptyset$ for all (some) $a \in A$

Strong (weak) incomplementarity:
$(x, y) \in \text{icom}(A)$ iff $a(x) \neq -a(y)$ for all (some) $a \in A$

Intuitively, two objects are $A$-indiscernible whenever all the $A$-properties that they possess are the same, in other words up to the discriminative resources of $A$ these objects are the same. Objects are weakly $A$-indiscernible whenever some of their $A$-properties are the same. Objects are $A$-similar (weakly similar) whenever all (some) of their $A$-properties are similar in the sense that they share some values. In the case of deterministic information strong (weak) similarity coincides with strong (weak) indiscernibility, respectively. Negative similarity relates objects whenever they are similar with respect to their negative properties, e.g. 'being a non-smoker'. Forward and backward informational inclusion of objects reflects growth of information about objects.

All the relations defined above reflect an indiscernibility-type incompleteness of information that is provided in the form of an information system. Indiscernibility relations derived from an information system were introduced in Konrad *et al.* (1981). Similarity and forward inclusion were introduced in Orłowska (1985). Negative similarity was introduced in Vakarelov (1991a).

Important applications of information relations from the indiscernibility group concern the representation of approximations of subsets of objects in information systems. If $R$ is one of these relations, $A$ is a subset of AT and $X$ is a subset of OB, then the lower $R(A)$-approximation of $X$ $(L(R(A))X)$ and the upper $R(A)$-approximation of $X$ $(U(R(A))X)$ are defined as follows:

$$L(R(A))X = \{x \in \mathrm{OB}\colon \text{for all } y \in \mathrm{OB} \text{ if } (x,y) \in R(A), \text{ then } y \in X\}$$
$$U(R(A))X = \{x \in \mathrm{OB}\colon \text{there is } y \in \mathrm{OB} \text{ such that } (x,y) \in R(A) \text{ and } y \in X\}.$$

In the classical rough set theory setting, if the relation $R(A)$ is a strong indiscernibility relation, then we obtain the following hierarchy of definability of sets. A subset $X$ of OB is said to be:

| | |
|---|---|
| $A$-definable | if $L(\mathrm{ind}(A))X = X = U(\mathrm{ind}(A))X$ |
| roughly $A$-definable | if $L(\mathrm{ind}(A))X \neq \emptyset$ and $U(\mathrm{ind}(A))X \neq \mathrm{OB}$ |
| internally $A$-indefinable | if $L(\mathrm{ind}(A))X = \emptyset$ |
| externally $A$-indefinable | if $U(\mathrm{ind}(A))X = \mathrm{OB}$ |
| totally $A$-indefinable | if internally $A$-indefinable and externally $A$-indefinable. |

The other application of the above information relations is related to modeling uncertain knowledge acquired from information about objects that is given in an information system (Orłowska, 1983, 1987, 1989). Letting $X$ be a subset of OB, we define the sets of $A$-positive $(\mathrm{POS}(A)X)$, $A$-borderline $(\mathrm{BOR}(A)X)$ and $A$-negative $(\mathrm{NEG}(A)X)$ instances of $X$ as follows:

$$\mathrm{POS}(A)X = L(\mathrm{ind}(A))X$$
$$\mathrm{BOR}(A)X = U(\mathrm{ind}(A))X - L(\mathrm{ind}(A))X$$
$$\mathrm{NEG}(A)X = \mathrm{OB} - U(\mathrm{ind}(A))X.$$

Knowledge about a set $X$ of objects that can be discovered from information provided in an information system can be modelled in the following way:

$K(A)X = \mathrm{POS}(A)X \cup \mathrm{NEG}(A)X$.

Intuitively, $A$-knowledge about $X$ consists of those objects that are either $A$-positive instances of $X$ (they are members of $X$ up to properties from $A$) or $A$-negative instances of $X$ (they are not members of $X$ up to properties from $A$). We say that $A$-knowledge about $X$ is:

| | | |
|---|---|---|
| complete | if | $K(A)X = \mathrm{OB}$, otherwise incomplete |
| rough | if | $\mathrm{POS}(A)X, \mathrm{BOR}(A)X, \mathrm{NEG}(A)X \neq \emptyset$ |
| pos-empty | if | $\mathrm{POS}(A)X = \emptyset$ |
| neg-empty | if | $\mathrm{NEG}(A)X = \emptyset$ |
| empty | if | pos-empty and neg-empty. |

Clearly, the above applications can be extended to the other information relations from the indiscernibility group. Some applications of similarity relations to modeling approximations can be found in Pomykała (1987, 1988).

Incompleteness of information manifested by approximate distinguishability can be modelled with orthogonality-type information relations derived from information systems. We define the following classes of these relations:

Strong (weak) diversity:

$(x, y) \in \mathrm{div}(A)$ iff $a(x) \neq a(y)$ for all (some) $a \in A$

Strong (weak) right orthogonality:

$(x, y) \in \mathrm{rort}(A)$ iff $a(x) \subseteq -a(y)$ for all (some) $a \in A$, where - is the complement with respect to the set $VAL_a$

Strong (weak) left orthogonality:

$(x, y) \in \mathrm{lort}(A)$ iff $-a(x) \subseteq a(y)$ for all (some) $a \in A$

Strong (weak) right negative similarity:

$(x, y) \in \mathrm{rnim}(A)$ iff $a(x) \cap -a(y) \neq \emptyset$ for all (some) $a \in A$

Strong (weak) left negative similarity:

$(x, y) \in \mathrm{lnim}(A)$ iff $-a(x) \cap a(y) \neq \emptyset$ for all (some) $a \in A$

Strong (weak) complementarity:

$(x, y) \in \mathrm{com}(A)$ iff $a(x) = -a(y)$ for all (some) $a \in A$

Objects are $A$-orthogonal (weakly orthogonal) whenever all (some) of their $A$-properties or negative $A$-properties are disjoint, and they are $A$-diverse (weakly diverse) if all (some) of their $A$-properties are different. In the case of deterministic information strong (weak) orthogonality coincides with strong (weak) diversity, respectively. Objects are complementary if their sets of properties are complements of each other.

Applications of diversity relations concern algorithms for finding cores of sets of attributes. Let an information system be given, and let $A$ be a subset of AT. We say that an attribute $a \in A$ is indispensable in $A$ iff $\text{ind}(A) \neq \text{ind}(A - \{a\})$, that is, there are some objects such that $a$ is the only attribute from $A$ that can distinguish between them. A reduct of $A$ is a minimal subset $A'$ of $A$ such that every $a \in A'$ is indispensable in $A'$ and $\text{ind}(A') = \text{ind}(A)$. By the core of $A$ we mean

$$\text{CORE}(A) = \bigcap\{A' \subseteq \text{AT} : A' \text{ is a reduct of } A\}$$

For any pair $x, y$ of objects we define the discernibility set $D_{xy} = \{a \in \text{AT}\colon (x, y) \in \text{div}(\{a\})\}$. It is proved in Rauszer and Skowron (1992) that $\text{CORE}(A) = \{a \in A : \text{ there are } x, y \in \mathbf{OB} \text{ such that } D_{xy} = \{a\}\}$.

Let IREL be the class of all the information relations defined above.

PROPOSITION 1: The class IREL is closed with respect to the complement and converse of relations.

*Proof.* It is easy to see that we have:
ind = −wdiv, wind = −div, sim = −wrort, wsim = −rort, fin = −wrnim, wfin = −rnim, bin = −wlnim, wbin = −lnim, nim = −wlort, wnim = −lort, icom = −wcom, wicom = −com.
For $R$ = ind, sim, nim, icom, div, rort, lort, com, we have $R = R^{-1}$ and $wR = (wR)^{-1}$. Moreover, fin = $\text{bin}^{-1}$, and rnim = $\text{lnim}^{-1}$.

Incompleteness of information reflected by the information relations defined above is a kind of inherent incompleteness that appears in any collection of information items that provide description of objects in terms of their properties. We say that a binary relation $R$ on a set $U$ is:

| | |
|---|---|
| Reflexive | if for all $x \in U$, $(x, x) \in R$ $(I \subseteq R)$, where $I$ is the identity on $U$ |
| Weakly reflexive | if for all $x, y \in U$, if $(x, y) \in R$, then $(x, x) \in R$ $(I \cap (R; 1) \subseteq R)$, where $1 = U \times U$, and ; is the composition of relations. |
| Irreflexive | if for all $x \in U$, $(x, x) \notin R$ $(I \subseteq -R)$. |
| Symmetric | if for all $x, y \in U$, if $(x, y) \in R$, then $(y, x) \in R$ $(R \subseteq R^{-1})$. |
| Transitive | if for all $x, y, z \in U$, if $(x, y) \in R$ and $(y, z) \in R$, then $(x, z) \in R$ $(R; R \subseteq R)$. |
| Intransitive | if for all $x, y, z \in U$, if $(x, y) \in R$ and $(y, z) \in R$, then $(x, z) \notin R$ $(R; R \subseteq -R)$. |

We introduce some generalisations of reflexivity, irreflexivity and transitivity. Let $R$ be a binary relation on $U$ and $n \geq 1$. By an $R, n$-path from $x$ to $y$ we mean a sequence $x, x_1, \ldots, x_{n-1}, y$ such that $(x, x_1) \in R$, $(x_{n-1}, y) \in R$ and for all $i = 1, \ldots, n-2$ $(x_i, x_{i+1}) \in R$. Informally, an $R, n$-path consists of $n$ arrows and $n-1$ intermediate points between $x$ and $y$. We define the following properties of relation $R$:

| | |
|---|---|
| $n$-reflexive | if for every $x \in U$ there is an $R, n$-path from $x$ to $x$ ($I \subseteq R; \ldots ; R \quad n-$times). |
| $n$-irreflexive | if for every $x \in U$ there is no an $R, n$-path from $x$ to $x$ ($I \subseteq -(R; \ldots ; R) \quad n-$times). |
| $n$-transitive | if for all $x, y \in U$, if there is an $R, n$-path from $x$ to $y$, then there is the $R$, 1-path from $x$ to $y$ ($n$-times $R; \ldots ; R \subseteq R$). |

In particular, $R$ is reflexive (irreflexive) iff $R$ is 1-reflexive (1-irreflexive), $R$ is transitive iff $R$ is 2-transitive, and $R$ is 3-transitive if for all $x, y, z, t \in U$, if $(x, y) \in R$, $(y, z) \in R$ and $(z, t) \in R$, then $(x, t) \in R$. In a similar way one can generalise several other properties of binary relations.

By a frame derived from information system $S = (\text{OB}, \text{AT}, \{VAL_a : a \in \text{AT}\})$ we mean a relational system

$$K_{S,R} = (\text{OB}, \{R(A) : A \subseteq \text{AT}\})$$

where $\{R(A) : A \subseteq \text{AT}\}$ is any of the families of information relations defined above, except for forward and backward inclusions, and right and left negative similarities. For these families we define derived frames of the form $K_{S,R,Q} = (\text{OB}, \{R(A) : A \subset \text{AT}\}, \{Q(A) : A \subset \text{AT}\})$, where $R$ = fin, (wfin) and $Q$ = bin (wbin) or $R$ = rnim (wrnim) and $Q$ = lnim (wlnim), respectively. Observe that relations in these frames depend on subsets of the set AT. These subsets play the role of indices, that is we deal with families of relations indexed with subsets of a set. Usually we are interested in studying relationships between information relations that belong to different families. Hence, it is also natural to consider frames with several different families of relations.

Now we define various classes of information frames in an abstract way, without assuming that the frames are determined by an information system. Let $K = (U, \{R(P) : P \subseteq A\})$ be a relational system such that $U$ is a nonempty set, $A$ is a nonempty finite set, and each $R(P)$ is a binary relation on $U$.

$K$ is an information frame with strong relations of type $A(FS)$ if for every $P, Q \subseteq A$ the relations from $K$ satisfy the following conditions:

$$R(P \cup Q) = R(P) \cap R(Q)$$
$$R(\emptyset) = U \times U.$$

$K$ is an information frame with weak relations of type $A(FW)$ if for every $P, Q \subseteq A$ the relations from $K$ satisfy the following conditions:

$$R(P \cup Q) = R(P) \cup R(Q)$$
$$R(\emptyset) = \emptyset$$

We say that an information frame $K = (U, \{R(P) : P \subseteq A\})$ is a reflection of frame $K' = (U', \{R'(P) : P \subseteq A\})$ if $U = U'$ and for all $P \subseteq A$ it holds $R(P) = -R'(P)$, and we write $K = -K'$. We say that frame $K$ is a conversion of frame $K'$ iff $U = U'$ and for all $P \subseteq A$ it holds $R(P) = R'(P)^{-1}$, and we write $K = K'^{-1}$.

PROPOSITION 2:

(a) $K$ is a frame with strong relations iff $K$ is the frame with weak relations

(b) $K$ is a frame with strong (weak) relations iff $K^{-1}$ is the frame with strong (weak) relations

(c) $K = -K'$ iff $K' = -K$

(d) $- - K = K$

(e) $K = K'^{-1}$ iff $K' = K^{-1}$

(f) $K^{-1-1} = K$

We define more specific classes of frames postulating some conditions that the relations are supposed to satisfy.

Indiscernibility frame (IND):
an information frame with strong equivalence relations.
Similarity frame (SIM):
strong tolerance relations.
Negative similarity frame (NIM):
strong, weakly reflexive and symmetric relations.
Incomplementarity frame (ICOM):
strong symmetric relations such that for all $a \in A$, $-R(\{a\})$ is 3-transitive and $n$-irreflexive for every odd $n$.
Inclusion frame (IN):
frame of the form $(U, \{R(P) : P \subseteq A\}, \{Q(P) : P \subseteq A\})$ with two families of strong relations such that all relations $R(P)$ and $Q(P)$ are reflexive, transitive, and for all $P \subseteq A$, $R(P) = S(P)^{-1}$.

Weak indiscernibility frame (WIND):
weak tolerance relations and $-R(\{a\})$ is transitive for all $a \in A$.
Weak similarity frame (WSIM):
weak tolerance relations.
Weak negative similarity frame (WNIM):
weak, symmetric relations.
Weak incomplementarity frame (WICOM):
weak symmetric relations whose complements are 3-transitive and $n$-irreflexive for every odd $n$.

Weak inclusion frame (WIN):

frame with two families of relations $(U,\{R(P):P\subseteq A\},\{Q(P):P\subseteq A\})$ such that all relations $R(P)$ and $Q(P)$ are weak, reflexive relations, $R(\{a\})$ and $Q(\{a\})$ are transitive for all $a\in A$, and for all $P\subseteq A, R(P)=S(P)^{-1}$.

Diversity frame (DIV):

strong relations whose complements are tolerance relations and $R(\{a\})$ is transitive for all $a\in A$.

Right orthogonality fame (RORT):

strong relations whose complements are tolerance relations.

Left orthogonality frame (LORT):

strong relations whose complements are weakly reflexive and symmetric.

Complementarity frame (COM):

strong, symmetric, 3-transitive relations that are $n$-irreflexive for every odd $n$.

This characterisation of complementarity relations was proposed by S. Demri (1996), the characterisation of incomplementarity, weak complementarity and weak incomplementarity relations is its consequence (see also Demri and Orłowska, 1996b, 1996c).

Right-left negative similarity frame (RLNIM):

frame of the form $(U,\{R(P):P\subseteq A\},\{Q(P):P\subseteq A\})$ with two families of strong symmetric relations whose complements are reflexive, $-R(\{a\})$ and $-Q(\{a\})$ are transitive for all $a\in A$, and for all $P\subseteq A, -R(P)=-S(P)^{-1}$.

Weak diversity frame (WDIV):

weak relations whose complements are equivalence relations.

Weak right orthogonality frame (WRORT):

weak relations whose complements are tolerance relations.

Weak left orthogonality frame (WLORT):

weak relations whose complements are symmetric.

Weak complementarity frame (WCOM):

weak, symmetric, and for all $a\in A$, $R(\{a\})$ is 3-transitive and $n$-irreflexive for every odd $n$.

Weak rightleft negative similarity frame (WRLNIM):

frame of the form $(U,\{R(P):P\subseteq A\},\{Q(P):P\subseteq A\})$ with two families of weak symmetric relations such that complements of relations $R(P)$ and $Q(P)$ are reflexive and transitive, and for all $P\subseteq A, -R(P)=-S(P)^{-1}$.

We clearly have:

PROPOSITION 3: For any information systems $S$ the following conditions hold:

(a) $K_{S,\text{ind}} \in$ IND, $K_{S,\text{sim}} \in$ SIM, $K_{S,\text{nim}} \in$ NIM, $K_{S,\text{icom}} \in$ ICOM, $K_{S,\text{fin,bin}} \in$ IN, and the analogous conditions hold for the respective frames with weak relations

(b) $K_{S,\text{div}} \in$ DIV, $K_{S,\text{rort}} \in$ RORT, $K_{S,\text{lort}} \in$ LORT, $K_{S,\text{com}} \in$ COM, $K_{S,\text{rlnim}} \in$ RLNIM and the analogous conditions hold for the respective frames with weak relations

(c) $K \in$ IND (SIM, NIM, ICOM, IN) iff $-K \in$ WDIV(WRORT, WLORT, WCOM, WRLNIM)

(d) $K \in$ WIND(WSIM, WNIM, WICOM, WIN) iff $-K \in$ DIV (RORT, LORT, COM, RLNIM)

## 3. INFORMATION LOGICS FOR CLASSES OF FRAMES WITH PARAMETERISED RELATIONS

As usual, frames determine Kripke-style models of modal logics. Various classes of information logics have been studied in the literature. Kripke models based on frames with parameterised relations are introduced in Orłowska (1988). Logical systems with modal operators determined by parameterised strong indiscernibility and similarity relations are investigated in Konikowska (1987, 1994). A class of logics whose models are based on frames with parameterised relations is investigated in Balbiani (1997). In this paper we suggest a way of defining logical systems for all the classes of frames defined in Section 2.

Let a modal language be given such that its formulas are built from propositional variables from an infinite denumerable set VARPROP with the classical propositional connectives and modal operators $<>$, $[\,]$, $\ll\gg$, $[[\,]]$. As usual, each modal operator is determined by a binary relation, so in the modal language we admit constants that denote these relations. For the sake of simplicity we denote these constants with the same symbols as the relations in the respective models, so in the formulas we write $<R(P)>$, $[R(P)]$, $\ll R(P) \gg$, $[[R(P)]]$. Let $M = (U, \{R(P) : P \subseteq A\}, m)$ be a model based on an information frame such that meaning function $m$ assigns subsets of $U$ to propositional variables and relations $R(P)$ to the respective relational constants. In the usual way we define satisfiability of formulas by objects from $U$:

$M, x$ sat $p$ iff $x \in m(p)$ for $p \in$ VARPROP,
$M, x$ sat $\neg F$ iff not $M, x$ sat $F$,
$M, x$ sat $F \vee G$ iff $M, x$ sat $F$ or $M, x$ sat $G$,
$M, x$ sat $F \wedge G$ iff $M, x$ sat $F$ and $M, x$ sat $G$,
$M, x$ sat $F \rightarrow G$ iff $M, x$ sat $\neg F \vee G$,
$M, x$ sat $F \leftrightarrow G$ iff $M, x$ sat $(F \rightarrow G) \wedge (G \rightarrow F)$,

$M, x$ sat $[R(P)]F$ iff for all $y \in U$, $(x, y) \in R(P)$ implies $M, y$ sat $F$,
$M, x$ sat $<R(P)>$ $F$ iff there is $y \in U$ such that $(x, y) \in R(P)$ and $M, y$ sat $F$,
$M, x$ sat $[[R(P)]]F$ iff for all $y \in U$, $M, y$ sat $F$ implies $(x, y) \in R(P)$,
$M, x$ sat $\ll R(P) \gg$ $F$ iff there is $y \in U$ such that not $M, y$ sat $F$ and $(x, y) \notin R(P)$.

A formula $F$ is true in model $M = (U, \{R(P) : P \subseteq A\}, m)(\models_M F)$ iff $M, x$ sat $F$ for all $x \in U$. A formula $F$ is true in a frame $K = (U, \{R(P) : P \subseteq A\})$ $(\models_K F)$ iff $\models_M F$ for every model $M$ based on $K$. A formula $F$ is true in a class $C$ of frames iff $\models_K F$ for every frame $K \in C$. By a logic $L(C)$ of class $C$ of frames we mean the set of formulas that are true in $C$. We say that $F$ is valid in $L(C)$ $(L(C) \models F)$ iff $F$ is true in $C$. In any modal language by true and false we denote formulas of the form $F \vee \neg F$ and $F \wedge \neg F$, respectively. The operators $[[\ ]]$ and $\ll\gg$ were investigated in Gargov *et al.* (1987) and Humberstone (1983).

In what follows we omit the sign $P$ in the names of relations, the respective notions are meaningful and the properties hold for every $P$. We say that a modal operator $f$ is dual to a modal operator $g$ iff $f(R)F \leftrightarrow \neg g(R)\neg F$ is valid for any $R$. It is easy to see that we have the following pairs of dual operators:

$<>$ $[\ ]$
$\ll\gg$ $[[\ ]]$

Hence, in the presence of negation in the language we can use either of $[\ ]$ and $<>$ and/or either of $[[\ ]]$ and $\ll\gg$.

Operators $f$ and $g$ are said to be conjugated iff $f(R)F \leftrightarrow g(-R)\neg F$ is valid for any $R$. The following are the conjugated pairs of operators:

$<>$ $\ll\gg$
$[\ ]$ $[[\ ]]$

A modal operator $f(R)$ is said to be normal if $f(R)(F \vee G) \leftrightarrow f(R)F \vee f(R)G$ and $f(R)(\text{false}) \leftrightarrow$ false are valid, and conormal if $f(R)(F \wedge G) \leftrightarrow f(R)F \vee f(R)G$ and $f(R)(\text{true}) \leftrightarrow$ false are valid.

PROPOSITION 4: If $R$ is a relation from any frame defined in section 2, then operators $<R>$ are normal and operators $\ll R \gg$ are conormal.

If we built a logic for class of frames with two families of relations, then we admit in the underlying language pairs of modal operators determined by relations from each of the families.

Modal formulas characterise properties of relations. Below we give a list of formulas that correspond to properties of relations from the information frames considered in this paper. The list should be read as follows: for each formula-property pair, the formula is true in a frame iff the underlying relation from the frame possesses the property.

A family $\{R(P) : P \subseteq A\}$ of relations is a family with strong relations iff for all $P, Q \subseteq A$ the following formulas are true in frame $(U, \{R(P) : P \subseteq A\})$:

(s1) $\ll R(P \cup Q) \gg F \leftrightarrow \ll R(P) \gg F \vee \ll R(Q) \gg F$

(s2) $\neg \ll R(\emptyset) \gg$ false

A family $\{R(P) : P \subseteq A\}$ of relations is a family with weak relations iff for all $P, Q \subseteq A$ the following formulas are true in frame $(U, \{R(P) : P \subseteq A\})$ :

| | | |
|---|---|---|
| (w1) | $<R(P \cup Q)> F \leftrightarrow <R(P)> F \vee <R(Q)> F$ | |
| (w2) | $\neg <R(\emptyset)>$ true | |
| (ref) | $F \rightarrow <R> F$ | $R$ reflexive |
| (wref) | $<R>$ true $\rightarrow (F \rightarrow <R> F)$ | $R$ weakly reflexive |
| ($n$-ref) | $F \rightarrow <R> \ldots <R> F$, $n$ operators $<R>$ | $R$ $n$-reflexive |
| (irref) | $F \rightarrow \ll R \gg \neg F$ | $R$ irreflexive |
| ($n$-irref) | $F \rightarrow \ll R; \ldots ; R \gg \neg F$, | $R$ $n$-irreflexive |
| (sym) | $<R> \neg <R> F \rightarrow \neg F$ | $R$ symmetric |
| (tran) | $<R><R> F \rightarrow <R> F$ | $R$ transitive |
| ($n$-tran) | $<R> \ldots <R> F \rightarrow <R> F$, $n$ operators $<R>$ | $R$ $n$-transitive |
| (intran) | $<R><R> F \rightarrow \ll R \gg \neg F$ | $R$ intransitive |
| (inv) | $F \rightarrow \neg <R> \neg <Q> F$ | $R^{-1} \subseteq Q$ |

We clearly have: $R$ is reflexive iff $-R$ is irreflexive, and $R$ is symmetric iff $-R$ is symmetric.

An axiomatisation of the logics of the classes of information frames introduced in the paper should contain the following formulas. For each operation $<R(P)>$, $\ll R(P) \gg$ and $[R(P)]$, $[[R(P)]]$ we admit the axioms:

(n) $[R(P)]F \wedge [R(P)](F \rightarrow G) \rightarrow [R(P)]G$

(cn) $[[R(P)]]F \wedge [[R(P)]](\neg F \wedge G) \rightarrow [[R(P)]]G$

Moreover, for each of the logics we provide some specific axioms. The axioms denoted by (-conditions) are obtained from the respective axioms denoted by (conditions) through replacement of $R$ by $-R$.

| | |
|---|---|
| *L*(IND): | (s1), (s2), (ref), (sym), (tran). |
| *L*(SIM): | (s1), (s2), (ref), (sym). |
| *L*(NIM): | (s1), (s2), (wref), (sym). |
| *L*(ICOM): | (s1), (s2), (sym), and for $R(\{a\})$ for all $a \in A$ (-3-tran) and (-$n$-irref) for all odd $n$. |
| *L*(IN): | (s1), (s2), (ref), (tran), (inv) for all $R(P)$ and $Q(P)$. |

| | |
|---|---|
| $L$(WIND): | (w1), (w2), (ref), (sym), (tran) for $R(\{a\})$ for all $a \in A$. |
| $L$(WSIM): | (w1), (w2), (ref), (sym). |
| $L$(WNIM): | (w1), (w2), (sym). |
| $L$(WICOM): | (w1), (w2), (sym), (-3-tran), (-$n$-irref) for all odd $n$. |
| $L$(WIN): | (w1), (w2), (ref) for $R(P)$ and $Q(P)$, (tran), (inv) for $R(\{a\})$ and $Q(\{a\})$ for all $a \in A$. |

| | |
|---|---|
| $L$(DIV): | (s1), (s2), (-ref), (-sym), (-tran) for $R(\{a\})$ for all $a \in A$. |
| $L$(RORT): | (s1), (s2), (-ref), (-sym). |
| $L$(LORT): | (s1), (s2), (-wref), (-sym). |
| $L$(COM): | (s1), (s2), (sym), (3-tran), ($n$-irref) for all odd $n$. |
| $L$(RLNIM): | (s1), (s2), (-ref), (sym), (-tran), (inv) for $R(P)$ and $Q(P)$. |

| | |
|---|---|
| $L$(WDIV): | (w1), (w2), (-ref), (-sym), (-tran). |
| $L$(WRORT): | (w1), (w2), (-ref), (-sym). |
| $L$(WLORT): | (w1), (w2), (-sym). |
| $L$(WCOM): | (w1), (w2), (sym), and for $R(\{a\})$ for all $a \in A$ (3-tran) and ($n$-irref) for all odd $n$. |
| $L$(WRLNIM): | (w1), (w2), (-ref), (sym), (inv) for $R(P)$ and $Q(P)$, (-tran) for $R(\{a\})$ and $Q(\{a\})$ for all $a \in A$. |

If in the axioms of a logic both $<>$ (or [ ]) and $\ll\gg$ (or [[ ]]) are needed, then the axiom set should contain the axioms of the Boolean modal logic (Gargov *et al.*, 1987). In all the logics defined above the rules of inference are modus ponens, the necessitation rule for operators [ ] and/or rules of the form $\vdash \neg F$ implies $\vdash$ [[ ]]$F$.

As usual, a completeness proof proceeds by the construction of a canonical model. Let $L$ be an information logic with modal operators $\{f(R(P)) : P \subseteq A\}$ for some finite set $A$. By a canonical frame for $L$ we mean the system $(U, \{R_f(P) : P \subseteq A\})$, where $U$ is the set of all the maximal theories in the language of $L$ and binary relations $R_f(P)$ in set $U$ are defined as follows:

If $f$ is a normal operator $<>$, then $(u, v) \in R_f(P)$ iff for all formulas $F$, if $F \in v$, then $<R(P)> F \in u$.

If $f$ is a conormal operator $\ll\gg$, then $(u, v) \in R_f(P)$ iff for all formulas $F$, if $F \in v$, then $\ll R(P) \gg \neg F \in u$.

The presented logics have two kinds of models: general models based on abstract frames, and standard models based on frames derived from information systems. The applicability of a logic to practical reasoning is strongly connected with what is called informational representability of frames. Let $C$ be a class of frames. Intuitively, a frame $K$ from $C$ is informationally representable if there is an information system $S$ and a frame $K_S$ derived form this system such that $K_S$ is in $C$ and, moreover, $K$ is similar (e.g. isomorphic, modally equivalent etc.) to $K_S$. In Demri and Orłowska (1996b) the notion of informational representability of frames is introduced in a formal way, a method of proving representability is given and representability results are proved for various classes of frames. Inspiration for these results has been provided by a representability theorem for the logic NIL and its proof given in Vakarelov (1987). If models of an information logic are based on frames that are informationally representable, then the deduction system of the logic enables us to process information given in an information system. This suggests that such a logical formalism is adequate for practical reasoning about concrete data. In the extended version of Demri and Orłowska (1996b) (in preparation) we present proofs of representability for all the classes of frames defined in the present paper.

## 4. CALCULUS OF INFORMATION RELATIONS AND ARISTOTELIAN SYLLOGISTIC

It has been observed by Sotirov and Vakarelov (1996) that Aristotelian syllogistic might have a semantics based on information systems. We will show a relationship between a system of syllogistic presented in Shepherdson (1956) and a calculus of information relations introduced in this paper. Let $S = (\text{OB}, \text{AT}, \{VAL_a : a \in \text{AT}\})$ be an information system and let $a$ be an attribute from AT. By the $S, a$-calculus of information relations we mean the algebra $C(S, a)$ of binary relations generated by relations $R(\{a\})$ derived from $S$, for $R$ = ind, sim, fin, bin, nim, icom, with operations $-, \cup, \cap, ^{-1}$ of complement, union, intersection, converse, respectively. The relations from the orthogonality group are obtained from the generators applying the complement operator, that is div = -ind, rort = -sim, lort = -nim, rnim = -bin, icom = -com (from now on, for the sake of simplicity, we omit the sign $a$ of attribute in the names of relations). By a logic $CI$ of calculi $C(S, a)$ we mean an equational logic whose formulas are equalities between terms denoting relations from a $C(S, a)$. These terms will be referred to as $CI$-terms. $CI$ is a (individual) variable-free logic.

We now define a logic $LI$ of information relations whose formulas represent assertions of the form: a pair of objects stands in a relation. Formulas of the logic are built with individual variables, symbols denoting atomic relations from a $C(S, a)$, and classical propositional connectives. Atomic formulas are of the form $xRy$, where $x, y$ are variables and $R$ is a symbol denoting an atomic relation. Compound formulas are built from atomic formulas with propositional operations. Semantic structures for $LI$ are determined by information systems. By a model for $LI$ we mean a system $M = (S, a, m)$, where $S = (\text{OB}, \text{AT}, \{VAL_a : a \in \text{AT}\})$ is an information system, $a$ is an attribute from AT, and $m$ is a meaning function that assigns the respective information relations derived from $S$ and determined by the attribute $a$ to symbols of atomic relations. By a valuation over $M$ we mean a mapping $v$ that assigns objects from OB to individual variables. Satisfiability of formulas in a model is defined as follows. For atomic formulas we put:

$M, x$ sat $xRy$ iff $(v(x), v(y) \in m(R))$, where $R=$ind, sim, fin, bin, nim, icom.

For compound formulas we define:

$M, x$ sat $\neg F$ iff not $M, v$ sat $F$,
$M, v$ sat $F \vee G$ iff $M, v$ sat $F$ or $M, v$ sat $G$,
$M, v$ sat $F \wedge G$ iff $M, v$ sat $F$ and $M, v$ sat $G$,
$M, v$ sat $F \rightarrow G$ iff $M, v$ sat $\neg F \vee G$.

A formula $F$ is true in a model $M$ if for every valuation $v$ over $M$ we have $M, v$ sat $F$. $F$ is $LI$-valid if it is true in all models.

The alphabet of the language of system $A_1$ of syllogistic contains variables from a nonempty denumerable set VAR, an operation symbol $'$, the relation symbol $A$, and symbols $\neg, \vee, \wedge, \rightarrow$ of propositional connectives. The terms of the language are defined as follows: Each variable is a term, and if $x$ is a variable, then $x'$ is a term. These terms will be referred to as $A_1$-terms. The atomic formulas of the language are of the form $Axy$, where $x, y$ are $A_1$-terms. The compound formulas are built in a standard way from the atomic formulas with propositional connectives. The axiom system for $A_1$ has been given in Wedberg (1948) where $x, y$ are $A_1$-terms:

A1. $Axx''$
A2. $Ax''x$
A3. $Axy \wedge Ayz \rightarrow Axz$
A4. $Axy \rightarrow Ay'x'$
A5. $Axx' \rightarrow Axy$.

The rule of inference is modus ponens.

By an $A_1$-structure we mean a system $K = (X,', A)$ where $X$ is a set, $'$ is a unary operation on $X$ and $A$ is a binary relation on $X$ such that A1, ... ,A5 are satisfied for any $x, y, z$ from $X$. By a standard $A_1$-structure we mean an $A_1$-structure such that $X$ is a $'$-closed family of subsets of a set $V$, $'$ is complement with respect to $V$, and $A$ is inclusion of sets. It is proved in Shepherdson (1956) that every $A_1$-structure is quasi-isomorphic with a standard $A_1$-structure in the following sense.

PROPOSITION 5: (Shepherdson): Let $(X,', A)$ be an $A_1$-structure. Then there is a standard $A_1$-structure $(Y, -, \subseteq)$ and a mapping $f : X \rightarrow Y$ such that $f(x') = -f(x)$ and $Axy$ iff $f(x) \subseteq f(y)$.

By a model for $A_1$ based on an $A_1$-structure $K$ we mean a system $M = (K, m)$ such that $m$ is the meaning function that provides the natural interpretation of symbols $'$ and $A$ as the respective objects from $K$, that is $m(') =\, '$ and $m(A) = A$. Clearly, on the right hand side of these equalities we have objects from $K$. Let $v$ : VAR$\rightarrow X$ be a valuation over $M$ that assigns elements of set $X$ to variables. We extend $v$ to all the $A_1$-terms: $v(x') = v(x)^{m(')}$. We define satisfiability of formulas by a valuation in a model:

$M, x$ sat $Axy$ if $(v(x), v(y)) \in m(A), x, y$ are $A_1$-terms.

For the compound formulas their satisfiability is defined according to the standard definition of the classical propositional connectives.

A formula $F$ is true in a model $M$ if for every valuation $v$ over $M$ we have $M, v$ sat $F$. $F$ is true in an $A_1$-structure $K$ if it is true in every model based on $K$. $F$ is $A_1$-valid if it is true in every $A_1$-structure. It follows that a formula $F(x_1, \ldots, x_n)$ containing only the variables $x_1, \ldots, x_n$ is true in a structure $(X, ', A)$ if for any assignment of elements from $X$ to variables $x_1, \ldots, x_n$ and under the interpretation of propositional connectives as the classical propositional operations, the resulting sentence is true in the classical propositional calculus. Formulas $F$ and $G$ are $A_1$-equivalent whenever $F$ is $A_1$-valid iff $G$ is $A_1$-valid. The following completeness theorem holds for system $A_1$:

PROPOSITION 6: (Shepherdson): The following conditions are equivalent:

(a) A formula $F$ is provable in system $A_1$

(b) $F$ is $A_1$-valid

(c) $F$ is true in all standard $A_1$-structures.

We will show that the logic $CI$ is interpretable in system $A_1$. Let $t$ be a translation mapping from $CI$-terms into formulas of $A_1$-defined as follows. Let $Ixy$ denote the formula $\neg Axy'$, and let $x, y$ be variables such that $x \neq y$.

$t$(atomic):

| | | |
|---|---|---|
| $t$(ind) | = | $Axy \wedge Ayx$ |
| $t$(sim) | = | $Ixy$ |
| $t$(fin) | = | $Axy$ |
| $t$(bin) | = | $Ayx$ |
| $t$(nim) | = | $Ix'y'$ |
| $t$(icom) | = | $Ixy \vee Iy'x'$ |

$t$(reverse atomic):

| | | |
|---|---|---|
| $t(\text{ind}^{-1})$ | = | $t$(ind) |
| $t(\text{sim}^{-1})$ | = | $Iyx$ |
| $t(\text{fin}^{-1})$ | = | $t$(bin) |
| $t(\text{bin}^{-1})$ | = | $t$(fin) |
| $t(\text{nim}^{-1})$ | = | $Iy'x'$ |
| $t(\text{icom}^{-1})$ | = | $Iyx \vee Ix'y'$. |

For the compound $CI$-terms built with $-, \cup, \cap$, we define the translation in a homomorphic way using $\neg, \vee, \wedge$, respectively:

$t$(Boolean):

$$t(-R) = \neg t(R), t(R \cup S) = t(R) \vee t(S), t(R \cap S) = t(R) \wedge t(S).$$

If a $CI$-term is built with the operation $^{-1}$, then, first, we push this operator to the atomic $CI$-terms, according to the well known relational laws, and we obtain:

$$t((-R)^{-1}) = t(-(R^{-1})), t((R \cup S)^{-1}) = t(R^{-1} \cup S^{-1}),$$

$$t((R \cap S)^{-1}) = t(R^{-1} \cap S^{-1}), t((R^{-1})^{-1}) = t(R).$$

Next, we get the required translation by using the translation laws $t$(Boolean), $t$(atomic) and $t$(reverse atomic). For example,
$t((\mathrm{ind} \cap \mathrm{sim})^{-1}) = t(\mathrm{ind}^{-1} \cap \mathrm{sim}^{-1}) = t(\mathrm{ind}^{-1}) \wedge t(\mathrm{sim}^{-1}) = Axy \wedge Ayx \wedge Ixy$

PROPOSITION 7: For any $CI$-term $R$ the following conditions are equivalent:

(a) $R = 1$ holds in every $C(S, a)$, where 1 denotes the universal relation

(b) Formula $t(R)$ is $A_1$-valid.

Now we establish the interpretability of $A_1$ in logic $LI$. Let $l$ be a translation mapping from the formulas of $A_1$ into the formulas of $LI$ defined as follows. Let $x, y$ be variables.

| $l$($A$-formulas): | | $l$($I$-formulas): | |
|---|---|---|---|
| $l(Axy)$ | $= x\mathrm{fin}y$ | $l(Ixy)$ | $= x\mathrm{sim}y$ |
| $l(Ax'y)$ | $= \neg x\mathrm{nim}y$ | $l(Ix'y)$ | $= \neg x\mathrm{bin}y$ |
| $l(Axy')$ | $= \neg x\mathrm{sim}y$ | $l(Ixy')$ | $= \neg x\mathrm{fin}y$ |
| $l(Ax'y')$ | $= x\mathrm{bin}y$ | $l(Ix'y')$ | $= x\mathrm{nim}y$ |

For formulas built with the classical connectives we define:

$l$(classical): $l(F \vee G) = l(F) \vee l(G), l(F \wedge G) = l(F) \wedge l(G),$
$l(F \to G) = l(\neg F \vee G), l(\neg F) = \neg l(F).$

Observe that every formula of $A_1$ is equivalent to a propositional combination of atomic formulas such that each of their $A_1$-terms contains at most one occurrence of sign $'$.

PROPOSITION 8: The following conditions are equivalent:

(a) The formula $F$ is $A_1$-valid

(b) $l(F)$ is $LI$-valid.

Clearly, we can also interpret $LI$ in $A_1$. The required translation is analogous to translation $t$.

We conclude that, given an information system $S = (\mathrm{OB}, \mathrm{AT}, \{VAL_a : a \in \mathrm{AT}\})$, for a fixed attribute $a$ the (variable free) calculus of the information relations derived from $S$ and determined by $a$ is embeddable into the system $A_1$. To formalise the calculus of information relations of the form $R(X)$ for subsets $X$ of AT we need an extension of $A_1$ having not a single relation $A$ (in the language and in the semantic structures) but a family $\{A(X) : X \subseteq \mathrm{AT})$ of relations. If we admit individual variables, then the respective logic of information relations is mutually interpretable with the system $A_1$.

## 5. CONCLUSION

In this paper we proposed a family of logics for reasoning about data in information systems. We defined a class of information relations that reflect relationships between data. This class of relations is closed with respect to derivability in a system of Aristotelian syllogistic. The relations form a basis for modal operators in the logics. We suggested some specific axioms for these logics which characterise properties of information relations.

*Institute of Telecommunications, Warsaw*

## REFERENCES

Balbiani, Ph.: 1997, 'Axiomatization of logics based on Kripke models with relative accessibility relations', in Orłowska, E. (ed.), *Incomplete Information: Rough Set Analysis*, Physica Verlag, Berlin, (forthcoming).

Demri, S.: 1996, private communication.

Demri, S.: 1996a, 'The validity problem for logic DALLA is decidable', *Bulletin of the PAS, Mathematics* **44**, No. 1, pp. 79–86.

Demri, S.: 1996b, 'A class of information logics with a decidable validity problem', *Proceedings of the 21st International Symposium on Mathematical Foundations of Computer Science, Krakow, Poland, September 1996*, Springer Lecture Notes in Computer Science, to appear.

Demri, S.: 1997, 'A completeness proof for a logic with an alternative necessity operator', *Studia Logica*, to appear.

Demri, S., Orłowska, E.: 1996, 'Every finitely reducible logic has the finite model property with respect to the class of <>-formulae', *ICS Research Report* **10/96**, Warsaw University of Technology, pp. 1–22.

Demri, S., Orłowska, E.: 1996a, 'Logical analysis of indiscernibility', *ICS Research Report* **11/96**, Warsaw University of Technology, pp. 1–33.

Demri, S., Orłowska, E.: 1996b, 'Informational representability of models for information logics', *ICS Research Report* **9/96**, Warsaw University of Technology, pp. 1–15.

Demri, S., Orłowska, E.: 1996c, 'Informational representability of complementary relations', manuscript.

Farinas del Cerro, L., Orłowska, E.: 1985, 'DAL-a logic for data analysis', *Theoretical Computer Science* **36**, pp. 251–264. Corrigendum: ibid. 47, 1986, 345.

Gargov, G.: 1986, 'Two completeness theorems in the logic for data analysis', *ICS PAS Report* **581**, Warsaw, pp. 1–12.

Gargov, G., Passy, S., Tinchev, T.: 1987, 'Modal environment for Boolean speculations', in Skordev, D. (ed.), *Mathematical Logic and Applications*, Plenum Press, New York, pp. 253–263.

Humberstone, I.: 1983, 'Inaccessible worlds', *Notre Dame Journal of Formal Logic* **24**, pp. 346–352.

Konikowska, B.: 1987, 'A formal language for reasoning about indiscernibility', *Bulletin of the PAS, Mathematics* **35**, pp. 239–249.

Konikowska, B.: 1994, 'A logic for reasoning about similarity', in Orłowska, E. (ed.), *Reasoning with incomplete information*, to appear.

Konrad, E., Orłowska, E., Pawlak, Z.: 1981, 'Knowledge representation systems', *ICS PAS Report* **433**, pp. 1–44.

Orłowska, E.: 1983, 'Semantics of vague concepts', in Dorn, G. and Weingartner, P. (eds.), *Foundations of Logic and Linguistics. Problems and Solutions. Selected contributions to the 7th International Congress of Logic, Methodology, and Philosophy of Science, Salzburg 1983*, Plenum Press, London & New York, pp. 465–482.

Orłowska, E.: 1984, 'Logic of indiscernibility relations', *Lecture Notes in Computer Science* **208**, Springer, Berlin & Heidelberg & New York, pp. 177–186.

Orłowska, E.: 1985, 'Logic of nondeterministic information', *Studia Logica* **44**, pp. 93–102.

Orłowska, E.: 1987, 'Logic for reasoning about knowledge', *Bulletin of the Section of Logic* **16**, No. 1, pp. 26–38. Also in *Zeitschrift fiiur Mathematische Logik und Grundlagen der Mathematik* **35**, 1989, pp. 559–572.

Orłowska, E.: 1988, 'Kripke models with relative accessibility and their application to inferences from incomplete information', in Mirkowska, G., Rasiowa, H. (eds.), *Mathematical Problems in Computation Theory*, Banach Center Publications **21**, Warsaw, pp. 329–339.

Orłowska, E.: 1988a, 'Logical aspects of learning concepts', *Journal of Approximate Reasoning* **2**, pp. 349–364.

Pomykała, J.A.: 1987, 'Approximation operations in approximation space', *Bulletin of the PAS, Mathematics* **35**, pp. 653–662.

Pomykała, J.A.: 1988, 'On definability in the nondeterministic information system', *Bulletin of the PAS, Mathematics* **36**, pp. 193–210.

Rasiowa, H.: 1988, 'Logic of approximation reasoning', *Lecture Notes in Computer Science* **239**, Springer, Berlin & Heidelberg & New York, pp. 188–210.

Rasiowa, H.: 1991, 'On approximation logics, a survey. Jahrbuch 1990', Kurt Gödel Gesellschaft, Vienna 1990, pp. 63–87.

Rasiowa, H., Skowron, A.: 1985, 'Approximation logic', in Bibel, W., Jantke, K.P. (eds.), *Mathematical Methods of Synthesis and Specification of Software Systems*, Mathematical Research **31**, Akademie Verlag, Berlin, pp. 123–139.

Rauszer, C., Skowron, A.: 1992, 'The discernibility matrices and functions in information systems', in Slowinski, R. (ed.), *Intelligent decision support. Handbook of applications and advances in the rough set theory*, Kluwer Academic Publishers, Dordrecht, pp. 331–362.

Shepherdson, J.C.: 1956, 'On the interpretation of Aristotelian syllogistic', *Journal of Symbolic Logic* **21**, pp. 137–147.

Sotirov, V., Vakarelov, D.: 1996, Private communication.

Vakarelov, D.: 1987, 'Abstract characterization of some knowledge representation systems and the logic NIL of non deterministic information', in Ph. Jorrand, V. Sgurev (eds.), *Artificial Intelligence II, Methodology, Systems, Applications*, North Holland Publishing Company, Amsterdam, pp. 255–260.

Vakarelov, D.: 1989, 'Modal logics for knowledge representation', *Lecture Notes in Computer Science* **363**, Springer, Berlin & Heidelberg & New York, pp. 257–277.

Vakarelov, D.: 1991a, 'Logical analysis of positive and negative similarity relations in property systems', *Proceedings of the First World Conference on the Fundamentals of Artificial Intelligence, Paris*, Ministere de la Recherche et de la Technologie, Paris, pp. 491–500.

Vakarelov, D.: 1991b, 'A modal logic for similarity relations in Pawlak knowledge representation systems', *Fundamenta Informaticae* **15**, pp. 61–79.

Wedberg, A.: 1948, 'The Aristotelian theory of classes', *Ajatus* **15**, pp. 299–314.

# PART V

## ONTOLOGY

## EPISTEMOLOGY, PHILOSOPHY OF SCIENCE

TIMOTHY CHILDERS AND ONDREJ MAJER

# ON ŁUKASIEWICZ'S THEORY OF PROBABILITY

## 1. INTRODUCTION

Łukasiewicz's theory of probability has been neglected. This is perhaps due to the fact that it was written at an early stage in the development of formal logic, and so Łukasiewicz's intentions are at times obscure. We argue that this neglect is unjust: by restating the theory in modern terms, and paying careful attention to Łukasiewicz's motivations, the theory is of interest, especially within its historical setting. We point out a novel notion of interpretation which can be used to explicate the theory, and we also note that the theory is a pragmatic one. We conclude by raising three puzzles, and hopefully answering them.

## 2. THE DEFINITION OF PROBABILITY: INDEFINITE PROPOSITIONS

Łukasiewicz identified probability with "the truth value of an indefinite proposition" which is "the ratio between the number of values of the variables for which the proposition yields true judgements and the total number of values of variables".[1] This definition depends crucially, obviously, on what Łukasiewicz meant by 'indefinite proposition'. Thus, we shall concentrate on this notion in the first part of this paper.

At first glance an indefinite proposition appears to be nothing more than an open sentence. Łukasiewicz's definition is that an indefinite proposition is one which contains a variable (p. 16). It might be said that such a proposition is surely indefinite since it is nothing other than an open sentence. But open sentences usually are seen as having truth values only under a given interpretation. Łukasiewicz presumably meant something more by the notion of an indefinite proposition since he continued: "I shall consider hereafter only those indefinite propositions in which the values of the variables range over a well-defined, *finite* class of individuals. For instance, it may be assumed that in the statement "$x$ is greater than 4" $x$ will stand only for integers from 1 to 6" (p. 16). (We shall follow Łukasiewicz and tradition in referring to this class as the range.) Indefinite propositions are contrasted with definite propositions, that is, propositions where a constant is substituted for the value of the variable, such as "5 is greater than 4." Such propositions, obviously, are true or false. This leads Łukasiewicz to call an indefinite proposition true if all definite propositions resulting from substitution are true, false if all substitutions yield false definite propositions, and neither true nor false otherwise.



*K. Kijania-Placek and J. Woleński (eds.),*
*The Lvov–Warsaw School and Contemporary Philosophy*, 303–312.
© 1998 *Kluwer Academic Publishers.*

An indefinite proposition seems to be nothing but a prescription for forming sets of closed sentences. Indeed, it is hard to see what they could be otherwise: it would be peculiar if an indefinite proposition were a single sentence with a kind of mixed truth value, that is, partly true and partly false. Łukasiewicz explicitly rejected such a notion, and identified an indefinite proposition as a group of sentences:

> Probability lies midway between truth and falsehood, in the same way as a proper fraction lies between 0 and 1, and gray between black and white. But it is not possible to mix truth with falsehood in order to obtain probability as it is possible to obtain a gray colour from a mixture of white and black point. No proposition can be both true and false. Something intermediate between truth and falsehood can be obtained only by forming a *group* consisting of true and false statements. This condition is satisfied precisely by indefinite propositions. (p. 37)

Once we have obtained the group of closed sentences, we can then compute the probability of the indefinite proposition by counting up the number of true sentences and dividing by the total. We will examine this more closely below, but first we should point out that there are two possible readings of Łukasiewicz's notion of indefinite propositions.

The first reading we consider can be called syntactic. It seems that this is what Łukasiewicz intended, since he never explicitly introduced semantic considerations. Computing the probability of an indefinite proposition requires that we substitute each constant from the range for the variable, and then check the truth value of each of the resulting sentences. This procedure requires that we have some list of admissible constants which will be substituted (that is, that the range is given as a list of constants), and some function which provides a truth value for each of the resulting sentences. Since we are dealing only with finite ranges, it might seem that we would not run a great risk of paradox or worse by following such an approach. This is not so: the probability function itself would have to be embedded in a richer language. We would need at the very least a cardinality predicate to measure probabilities, and to specify this syntactically we would require a second order language. This, however, seems a somewhat pointless exercise: although such an interpretation of probability as a second order concept is in keeping with some modern theories of probability as a kind of quantifier, the resulting language would be complex beyond anything imagined by Łukasiewicz.

Although, of course, Łukasiewicz did not have model theory, it seems to us simpler, and closer to his original intentions, to write as if he did, and interpret his notion of an indefinite proposition semantically (which also seems to us more in keeping with the spirit of the Lvov-Warsaw School). Interestingly, a semantic approach to Łukasiewicz's theory requires a different notion of interpretation than the usual one. The usual notion of interpretation is the assignation of a model to a theory, a sentence then being evaluated with respect to that model.[2] The sentence is compared with one model at a time: it is true or false under *an* interpretation. The sentence is of course tautologous if it is made true by all models, contradictory if made false by all, and neither otherwise.

Łukasiewicz's theory, by contrast, seems to employ the notion of a simultaneous interpretation. Let us take an indefinite proposition as a prescription for taking a set of models, the models taken depending on a prior restriction (Łukasiewicz's well-defined finite range). We can proceed as follows: We interpret extra-logical symbols used in the indefinite proposition according to the rules of interpretation for the language in which it is written (for example the relation 'greater than' or the predicate 'is an Englishman'). We then valuate the variable, assigning it all the values from the range. The result of this valuation is not unique, and hence we have an interpretation in which we consider a number of models simultaneously: roughly speaking, we are not considering truth in a model, but truth in a group of models.

The difference between Łukasiewicz's and the standard approach appears when only part of the models satisfy the open sentence corresponding to the indefinite proposition. The simultaneous valuation then makes the sentence, say, only half true, if half the models make it true. And this is a way in which Łukasiewicz can equate probability with partial truth, as opposed to the usual logical notion of partial deduction.

Another difference is that with (what we have called) Łukasiewicz's notion of interpretation an indefinite proposition can come out as logically true if it is satisfied by all values from a given limited range, while it is not a tautology in the usual sense. For example, if we restrict the range of $x$ in the indefinite proposition $x > 4$ to some (finite) set of numbers greater than four, the sentence comes out as logically true. This view may seem peculiar, but it does to some degree explain Łukasiewicz's claim to have uncovered a truly logical notion of probability.

The distinction between our two readings of Łukasiewicz may not seem so great since both will give us the same theorems of the probability calculus. But the difference is significant for probability theory: the status of the probability function is different depending on which is selected. Under the syntactic reading, the probability function is part of the language, while under the semantic it is part of the meta-level. Thus, for example, it is much easier to construct iterated probabilities under the syntactic approach, while under the semantic reading, iterating probabilities is a category mistake.

Łukasiewicz's definition of probability leads naturally to the following axioms (we rewrite them using modern notation):

1. $A = \top \Leftrightarrow p(A) = 1$
2. $A = \bot \Leftrightarrow p(A) = 0$
3. $A \vdash B \Rightarrow p(A) + p(\neg A \wedge B) = p(B)$

(Conditional probabilities are defined as usual.) The axioms would be standard, in that they would be equivalent to the so-called Kolmogorov axioms, if the 'if and only if's were replaced by 'if's. However, the difference is minor, since Łukasiewicz restricts his theory of probability to the finite case. Given this (admittedly severe) restriction, Łukasiewicz can have claimed to have discovered a means to correlate probabilistic and logical notions. For example, only contradictions will have probability 0, and probabilities add if and only if they are mutually exclusive.

## 3. SOME PUZZLES

We now turn to several puzzles connected with Łukasiewicz's conception of probability, the first of which we have raised and answered: the isomorphism between probabilistic and logical relations follows from what we call Łukasiewicz's unique notion of evaluation. The probabilistic relations appear to be much too strong. However, this is merely an artifact of the restriction of the values of the variables to finite cases. While this does allow one to switch between logical and probabilistic relations, it also severely restricts the applicability of probability theory.

The second puzzle is Łukasiewicz's notion of the Principle of Indifference. The Principle of Indifference was invoked in classical probability to get unique probabilities for statements. Łukasiewicz seems to be able to dispense with this notion altogether, and yet still get unique probability assignments. Considering this appears to be done without recourse to considerations of relative frequencies or betting quotients, this is quite a feat. In fact, however, the Principle of Indifference is implicit in Łukasiewicz's definition, as we shall show.

Consider the indefinite proposition that the value of this roll of the die will be $x$. It is hard to see what Łukasiewicz meant, but clearly the probability of getting any of the values $1, \ldots, 6$ on the $x$th roll with a fair die is $\frac{1}{6}$. (See, for example, his discussion on p. 46.) This is because each of the values of the indefinite proposition is assigned equal probability. Implicit in the definition is a symmetry principle. By counting up positive cases, and then dividing them by total cases, we are implicitly assigning each value of the variable equal probability. (Hence, even Łukasiewicz's account of frequentist probability relies on a symmetry principle.)

But surely this leads to contradictions. Consider the following famous so-called bookmark paradox. Suppose that your local library has only three colours of bookmarks: red, green and black, and that there is no more than one bookmark to a book. Then using the classic principle of indifference, the probability of drawing at random a book with a red bookmark as opposed to a non red bookmark is one-half. Similarly for the probabilities of green and black bookmarks. But then the probabilities of drawing a red, green or black bookmark add to more than one.

Keynes took pains to avoid this paradox by applying the Principle of Indifference to cases in which the sentences were non-decomposable.[3] This was similar to a restriction of von Kries, which Łukasiewicz explicitly mentions but does not include in his account of probability. Does Łukasiewicz require such a restriction? The answer is no: he applies the Principle of Indifference at another level than Keynes. Symmetry is a result of the choice of the values of the variable, and is implied by the choice of the language. No principle of indifference is required at a later stage. The bookmark paradox cannot arise, because it would require changing the range of the indefinite variable.

Avoiding the paradoxes comes at a price, however. It makes Łukasiewicz's theory of probability language dependent. Probabilities are determined by the richness of our language, and, by our current hypotheses. Łukasiewicz was surely aware of this, since he was aware of the paradoxes, as was every writer on probability at this time. Our interpretation puts Łukasiewicz much closer to the French conventionalists. In fact, this theory of probability then seems to become quite pragmatic. Whether Łukasiewicz would have accepted this conclusion is not clear. What is clear, however, is that it is unavoidable.

The final puzzle has to do with conditionalisation. Why did Łukasiewicz, Keynes and Wittgenstein all choose a measure which ruled out learning from experience?[4] That their chosen measure does so can be easily shown, that is, that the probability of any statement conditional on another statement of which it is logically independent is always equal to $\frac{1}{2}$. To be more specific, if we sample from a population $m$ times, and observe $n$ members have attribute $A$ (that is, $n$ successes), then the probability that the $n_{\mathrm{th}}+1$ sample will have attribute $A$ is $\frac{1}{2}$, for any $m$ and any $n$. Thus this measure is completely insensitive to sample size and the number of successes. (A proof of this statement for a language with a single monadic predicate and 101 individuals can be found in Carnap, 1950, p. 565. A more general proof can be found in Howson and Urbach, 1993, p. 64–65.) It might seem that they never did the calculations, and so never realised that their theory had this consequence. This would indeed be a strange state of affairs, considering that they were all competent mathematicians. It seems even stranger considering that, according to the standard picture, this was first pointed out by Carnap only as late as 1950. But, as we shall now argue, they did do the calculations.

Łukasiewicz had clearly calculated probabilities for repeated experiments. In a summary of one of his lectures from 1909,[5] he analysed a special form of Laplace's Rule of Succession, the inductive measure resulting from equiprobability of constituents (sequences of trials containing the same number of successes form a single constituent, that is, the order of successes is discounted).

It can be proved that given $m$ trials with $n$ success, the probability of success on the next trial is $\frac{n+1}{m+2}$. The specific case Łukasiewicz dealt with is where there have been $n$ successes from $n$ trials, that is, $\frac{n+1}{n+2}$. It can be shown that the probability of a further $M$ successes is

$$\frac{n+1}{n+M+1}$$

As Łukasiewicz points out, this can never rise above $\frac{1}{2}$, and so it would seem that learning from experience is ruled out. Specifically, the probability of universal statements (laws) must be 0. This might seem to be a good enough reason to reject this measure. In fact, it does not seem that the problem of assigning probability 0 to universal statements was what troubled Łukasiewicz: in his system a universal statement is a closed statement which is true or false: it is not the kind of thing to which probability can be applied.

Łukasiewicz implicitly made a distinction between an ever increasing indefinite statement, that is, one describing the situation above, where $M$ is allowed to grow arbitrarily large, and a general statement. The former could be probable and the latter was a definite sentence, and so either true or false.

We digress in order to point out that this distinction can be supported. In languages like Łukasiewicz's (and first order logic as well) there is a difference between the two types: an infinite amount of instantiations of a predicate is not equivalent to universal quantification over these instantiations, since they cannot form an infinite conjunction. However, there seems to be a tension in Łukasiewicz's account. There is a clear sense in which one approximates the other: if we consider the objects over which the probability is defined, as we let the size of this set of objects grow to that of the universe, then the probabilities of the two will converge (namely, to 0. Carnap, 1950, p. 60, mentions this.) One way to maintain the distinction would be to allow only finite, and not countable, additivity, as Łukasiewicz did.[6] The proof that the probability of an ever increasing sequence of events is equal to the probability of the limit of that sequence requires countable additivity (in fact, any such probability function satisfying countable additivity is continuous in just this sense). Showing that the probability of a statement employing the rule of succession is never greater than 0.5 requires only finite additivity, but the statement that the probability of a universal general statement is 0 requires countable additivity.[7] Issues of additivity aside, it is clear that Łukasiewicz had done the calculations for the Laplacean measure: it therefore seems quite implausible that he rejected it, and chose his measure because of efficacy with respect to inductive learning. Instead, we hypothesise that his concerns were quite different. To bring this out, we shall briefly describe how Wittgenstein constructed his theory of probability.

Wittgenstein took conditional, as opposed to absolute, probabilities as basic. For Łukasiewicz, the notion of probability was, as we have seen, not applicable to a single event: it could only come about in the context of a group of events, defined by an indefinite sentence. For Wittgenstein, probability also was not meaningful in the context of a single proposition. Instead, he speaks about "the degree of probability that the proposition $r$ gives to the proposition $s$." (5.15) Wittgenstein also made a very Łukasiewiczian pronouncement in 5.153: "In itself, a proposition is neither probable nor improbable. Either an event occurs or it does not: there is no middle way." Probability is calculated by counting up the number of rows on a truth table where a statement $r$ is true (the "truth grounds" for $r$, denoted by Wittgenstein as $T_r$) and then the rows on a truth table for $r, s, T_{rs}$, and then dividing $T_r$ by $T_{rs}$. Implicit in this definition is a Principle of Indifference: probability is equidistributed among the rows of the truth table of the basic statements of the system. (This approach, like Łukasiewicz's, also avoids the standard contradictions generated by unrestricted application of the Principle of Indifference, mentioned above. The theory is also equivalent to Łukasiewicz's if we reconstruct the unconditional probabilities, $p(A)$, by setting them equal to $p(\frac{A}{T})$ where $T$ is an arbitrary tautology). If we take the probability of independent sentences, or, as Wittgenstein put it, propositions which "have no truth-arguments in common with one another," they always give each other conditional probability 0.5. Wittgenstein in fact points this out (5.152), and was surely aware of the consequence of his definition, namely that increasing probabilities of sequences of events are ruled out a priori.

However, his view on this matter is somewhat clarified in what follows. He employs the standard example of drawing black and white balls from an urn, examining their colour and then replacing them. He notes that we may determine if the number of black balls is roughly equal to that of white balls. But this is not a matter of logic (or of "mathematics," as he puts is). But, if we determine by experiment that there are no confounding factors, that is, that the draws are in fact "independent of the circumstances of which I have no more detailed knowledge" then we can apply the definitions of probability and independence, and it will follow as a matter of logic, according to Wittgenstein, that the probability will be one half.

We cannot claim to be able to fully understand Wittgenstein's reasoning here, but it seems to be something along the following lines. Probability is a matter of logic, and is applicable only when two statements do not entail each other. Now this may seem strange (and certainly does look very strange to the modern probabilist). But a careful reading of his definition helps somewhat: if you are given only a language, and no experience, then, by the definition, you assign 0.5 probability. But if you have some experimental results which show you that the proportion of balls in the urn is not even, then the definition, the logical definition, of probability cannot be applied. The draws will not be independent of other knowledge (this requires, of course, a strict analytic/synthetic distinction).

For Wittgenstein there is a (somewhat useless) logical probability (5.156: "We use probability only in default of certainty — if our knowledge of a fact is not indeed complete, but we do know something about its form."), and there is an empirical study of frequencies (at least, this seems to be the import of 5.154). If we look at logical probability through positivist lenses, there seem to be good reasons for adopting a theory of probability independent of inductive experience. The notion of learning from experience smacks of psychologism: logic is not about learning; rather, it operates over what we have already learned. Learning is science. So there are three ways a proposition can be linked: one can entail the other, one can contradict the other, or they can be independent of one another, that is, they give each other probability 0.5. Wittgenstein said: "A tautology's truth is certain, a proposition's possible, a contradiction's impossible. (Certain, possible, impossible: here we have the first indication of the scale that we need in the theory of probability.)" (4.464)

We find in Wittgenstein, and, we hypothesise, in Łukasiewicz a distinction current in scientific philosophy at the time. Logic is apriori, and to give a logical account of probability means divorcing probability from learning. Wittgenstein, like Łukasiewicz, chose his theory of probability because it modelled certain logical relationships. This, in effect, gives a kind of three-valued logic: as we noted above, in both theories there is a one-one correlation between logical and probabilistic operations. From a modern perspective, the theory of probability seems remarkably truncated, since the correlation between logic and probability is achieved at the expense of a restriction to finite ranges in both theories. But taking into account their motivations, the choice is not one of a truncated probability, but rather a generalised logic.

The view of a distinction between logic and probability as the distinction between apriori and aposteriori knowledge was reflected as well in Richard von Mises's and Hans Reichenbach's theories of probability as relative frequency, the difference being that they chose to identify probability with science, and not with logic.

Today we see this identification of probability with the aposteriori as part of the "renunciation of determinism" (echoing de Finetti's (1931).) A general feature of the classical aprioristic theories of probability based on the Principle of Indifference is their mechanistic view of the world. Opposition to the deterministic world view, while growing at least from the time of Łukasiewicz's paper exploded in the 1930s. Scientific developments had by this time made a mechanistic view untenable: we have in mind mainly developments in statistical and quantum mechanics, although the case can be made for genetics, for example, as well. De Finetti and von Mises, both positivists, are the best known exponents of the new world view (for example de Finetti (1931) and von Mises (1957), both of whom aimed to replace deterministic with statistical laws. Von Plato (1994) provides an overview of this remarkable period).

The astonishing power of their results tends to overshadow, for probabilists, the fact that there is another path towards the renunciation of determinism, namely, the adoption of a multi-valued logic. Łukasiewicz was explicit about this motivation in his (1930 [1970]) and it is the development towards this view that we see in his earlier theory of probability. This, we feel, provides the explanation of our third puzzle. (Interestingly enough, de Finetti cites this paper approvingly in his (1933 [1993]), although for other reasons, namely, for the development of means by which probability could be investigated from a formal, logical point of view.)

## 4. CONCLUSION

In conclusion we may say that it is possible to explicate Łukasiewicz's conception of probability in modern terms, and to show that, when so translated, the theory is workable, albeit with different motivations and purposes. In fact, it is this difference which separates it from other brands of purely logical probability which, we now know, are unworkable.[8] And, no matter what the contemporary interest of this theory may be, Łukasiewicz's work on probability deserves a chapter in the history of the ongoing excavations of the foundations of probability.

*Institute of Philosophy, Academy of Sciences of the Czech Republic*

## NOTES

[1] J. Łukasiewicz (1970), p. 17. See also p. 33, where Łukasiewicz identifies probability with the truth value of an indefinite proposition. All references to Łukasiewicz are from this paper unless otherwise noted.

[2] We include the evaluation of the free variables in the interpretation. The notion of valuation is sometimes considered apart from that of interpretation: Jaroslav Peregrin in (1995) uses 'model structure' where we use 'model.'

[3] Keynes (1921).

[4] We shall not answer this question in the case of Keynes: this would take us too far afield.

[5] Our attention was drawn to this paper by Woleński (1989).

[6] Finite additivity is the requirement that $p(\bigcup_{i=1}^{n} A_i) = \sum_{i=1}^{n} p(A_i)$ while countably additivity is the stronger requirement that is $p(\bigcup_{i=1}^{\infty} A_i) = \sum_{i=1}^{\infty} p(A_i)$. We of course do not claim that Łukasiewicz was aware of these issues, which only came to the forefront much later: we only point out how to preserve the consistency of his theory. However, it is interesting that de Finetti attributes a hostility to countable additivity on the part of theorists who advocate equiprobability of basic cases: "I incidentally observe that [ . . . ] complete [countable] additivity is (implicitly or explicitly) denied by many Authors who consider successions of cases as "equally probable" " (de Finetti, 1933).

[7] Carnap, of course, is an exception, since he proposed a logical theory of inductive learning. His views on this matter are complex, and an examination of them would, once again, lead us far beyond the scope of this paper.

[8] See Howson and Urbach (1993) for some of the reasons why.

## REFERENCES

Carnap, R.: 1950, *Logical Foundations of Probability*, University of Chicago Press, Chicago.

de Finetti, B.: 1931, 'Le leggi differenziali e la rinunzia al determinismo', *Rendiconti del Seminario Matematico della Universita di Roma* **7**, pp. 63–74, repr. and trans. as 'Differential Laws and the Renunciation of Determinism' in de Finetti (1993), pp. 323–334.

de Finetti, B.: 1933, 'Sul Concetto di Probabilita', *Rivista Italiana di Statistica, Economia e Fanaza* V, pp. 723–747; repr. and trans. as 'On the Probability Concept' in de Finetti (1993), pp. 335–352.

de Finetti, B.: 1993, *Probabilita e induzione*, eds. P. Monari and D. Cocchi, Statistica, Bologna.

Howson, C., Urbach, P.: 1993, *Scientific Reasoning: the Bayesian Approach*, Open Court, Chicago.

Keynes, J.M.: 1921, *A Treatise on Probability*, repr. in 1973, *The collected writings of John Maynard Keynes* **8**, Macmillan for the Royal Economic Society, London.

Łukasiewicz, J.: 1909, 'Autoreferat', *Przegląd Filozoficzny* **12**, pp. 14–15.

Łukasiewicz, J.: 1913, 'The Logical Foundations of Probability Theory', in Łukasiewicz, J., 1970, *Selected Works*, ed. L. Borkowski, North-Holland Publishing Company, Amsterdam, PWN, Warsaw

Łukasiewicz, J.: 1930, 'Philosophische Bermerkungen zu mehrwertigen Systemen des Aussagenkalküls', *Comptes Rendes des Sciences de la Société des Sciences et de Lettres de Varsovie* **23**, pp. 51–77; repr. and trans. in Łukasiewicz, J., 1970, *Selected Works*, ed. L. Borkowski, North-Holland Publishing Company, Amsterdam, PWN, Warsaw.

Peregrin, J.: 1995, *Doing Worlds with Words: Formal Semantics without Formal Metaphysics*, Kluwer Academic Publishers, Dordrecht.

von Mises, R.: 1957, *Probability, Statistics and Truth*, 2nd rev. English ed., prepared by H. Geiringer, repr. 1981, Dover Publications, Inc., New York.

von Plato, J.: 1994, *Creating Modern Probability: Its Mathematics, Physics and Philosophy in Historical Perspective*, Oxford University Press, Oxford.

Wittgenstein, L.: 1922, *Tractatus Logico-Philosophicus*, Routledge and Kegan Paul, London.

Woleński J.: 1989, *Logic and Philosophy in the Lvov-Warsaw School*, Kluwer Academic Publishers, Dordrecht.

JAN CZERNIAWSKI

# ON WHAT THERE IS NOT — A VINDICATION OF REISM

There are many well-known objections to reism, and therefore the doctrine is widely held to be untenable. Indeed, for various reasons it is not easy to defend it as stated by Kotarbiński himself. First, so-called *semantical reism* is of doubtful use. It was invented in order to give support to *ontological reism*, but it does not adequately serve that purpose.[1] Even its main concept, that of an *onomatoid*, or apparent name, cannot be applied without knowing in advance whether the referent of the word in question exists or not. Thus, ontological reism needs some independent justification. Second, the *positive* thesis of ontological reism, stating that only things exist, implies the need to define the concept of *thing*. Otherwise, its non-tautological character may be called into question.[2] Moreover, even if some satisfactory definition of this concept were to be supplied, which is not the case as yet, the philosophical importance of the positive thesis is due to the *negative* theses which result from it and are suitable for eliminating several pseudoproblems. This strongly suggests concentrating on those negative theses rather than troubling with the positive one.

Third, Leśniewski's Ontology has not been a very fortunate choice of an underlying logic for reism. In its framework no meaningful negative existential thesis can be stated. Although one may deny the existence of the referent of a name, this very name cannot be used in any true predicative sentence which could serve for explaining its meaning. For instance, one cannot assert that *redness is a property* without presupposing the existence of the referents of the words 'redness' and 'property'.[3] This is the reason Kotarbiński's version of reism has been so vulnerable to Ajdukiewicz's criticism of the negative theses. In standard classical predicate logic the situation is even worse; one cannot even introduce an empty name. Fortunately, this is not so in *free logics* [4] — so the choice of one of them is advisable.

Since we are interested only in the negative ontological theses of reism, we do not need any definition of a thing; however, we are in need of a justification of those theses, independent of any semantic considerations. Such a justification can be reached by the method of *phenomenological analysis*, based on some ontological *intuition*, or *insight*. We can simply 'see' some ontological states of affairs. For example, we can 'see' that the *whiteness* of snow consists merely in snow's being white. Similarly, the *roundness* of an orange consists in the orange's being round. Thus, there is no such object as whiteness: rather, there is only snow and it is white; nor is there any roundness of the orange, but



*K. Kijania-Placek and J. Woleński (eds.),*
*The Lvov–Warsaw School and Contemporary Philosophy,* 313–317.
© 1998 *Kluwer Academic Publishers.*

there is an orange and it is round. In other words, no whiteness or roundness exists. It can also be seen that for similar reasons no property exists.

It might seem that the above reasoning looks like a mere repetition of some of Kotarbiński's reistic reductions.[5] What is new lies in the interpretation. Whereas Kotarbiński interprets this reasoning as supported by some linguistic intuitions, here it is interpreted as a phenomenological analysis. Thus, whereas Kotarbiński gets only *semantic* results, we reach *ontological* ones. What is more, Kotarbiński in fact needs our results in order to justify his own. We can even say that what he had done was actually an unconscious phenomenological analysis, misinterpreted as a merely semantic one. The same may be said about all of his reistic reductions. In this way most of his results may be re-established.

The question arises whether the results obtained by intuition can have any objective meaning. Contrary to a popular prejudice, the answer is positive. Were it not so, most non-trivial mathematical results would be unfounded, as Kant and Husserl pointed out. We have some insight into the contents (or *essences*) of our concepts, which is accessible to every competent and unbiased thinker. This is the reason why there are universally accepted mathematical proofs. Once understood, such a proof cannot be rejected. On the other hand, if it contains some error, understanding this error means rejecting the alleged proof with the same necessity. The same could hold for philosophical reasoning based on phenomenological analysis; unfortunately, there are many fewer competent and unbiased thinkers among philosophers than among mathematicians.

One might argue that the results of phenomenological analysis are usually ambiguous. Is it not true, e.g., that some philosophers 'see' things as consisting of properties and thus requiring the existence of properties for their own existence? This clearly contradicts the corresponding reistic intuition. But do they really 'see' things this way? No doubt there is some justification for regarding knowledge about a thing as consisting of propositions that ascribe different properties to this thing. Thus, the alleged intuition of a thing as consisting of properties may be suspected of being merely the result of an illegitimate projection of the logical analysis of the knowledge about the thing onto its subject. In this way, the credibility of the above alternative to *reistic phenomenology* may be questioned.

It is a great paradox that such excellent pieces of phenomenological analysis were done by Kotarbiński, who was very far from *phenomenology* as a philosophical movement. However, the unconscious use of this method led him to some fallacious results, which need correcting. The most important of them is his rejection of the existence of (distributive) *sets*. This result is based on the observation that the sentence '$X$ is an element of the set of $M$s' means the same as the sentence '$X$ is an $M$', where the expression 'the set of $M$s' does not appear.[6] This seems to resemble the case of the sentence 'Whiteness is a property of snow', which means the same as the sentence 'Snow is white'. Thus, if 'whiteness' is an onomatoid, then 'the set of *M*s' seems to be an apparent name, too.

The analogy between these two cases is, however, far from perfect. Whereas there are serious difficulties with identifying the referent of the word 'whiteness', the referent of the expression 'the set of *M*s' may be identified very easily: the whole built of all *M*s can play its role. Kotarbiński was aware of this possibility, but rejected it because it seems to have paradoxical consequence. The same organism may be considered either as the set of its cells, or as the set of its organs. Does this mean that the set of its organs is identical with the set of its cells?[7]

Such an inference would be based on the following formula: 'If *O* is the set of *O*'s organs and *O* is the set of *O*'s cells, then the set of *O*'s organs is identical with the set of *O*'s cells'. Its validity strongly depends on the interpretation of the copula 'is'. Certainly, were this copula interpreted as signifying the relation of identity, the formula would be valid — but why should it be so interpreted? In fact, the resulting paradox is a *reductio ad absurdum* of such an interpretation — and any other interpretation that validates the above formula. An organism is the set of its cells, but not conversely. Similarly, any whole built of non-overlapping[8] elements may be considered as the set of them — but it is something more than the mere set of its parts. If Leśniewski's Ontology implies the opposite, so much the worse for Ontology.

Another of Kotarbiński's results that seems unfounded is his rejection of the existence of space.[9] When we want to localise some thing or event, we usually point at some part of the world, which we call 'place'. This suggests that the role of the referent of the word 'space' can be played by the world, taken in its *geometrical* aspects. In other words, the world may be considered as a space, although it is not only a space, but something more. The matter is different for the problem of time, since there is no existing object that could be considered as *time*. At most, the *process* of the *history* of the world, taken in its *chronometrical* aspects, could serve this purpose — but no such process exists, as Kotarbiński correctly pointed out. For similar reasons *space-time* does not exist, either. Nevertheless, this does not mean that neither time nor space-time can be considered. The domain of *thinkable* objects is far broader than the domain of *existing* ones. While the latter contains some of the objects of *applied* mathematics, the rest of them and all the objects of *pure* mathematics belong to the former.

The last important point to be corrected concerns the existence of physical *fields*.[10] It is beyond doubt that they can be reduced neither to interactions between bodies nor to some special bodies. However, this might be harmful only for *pansomatism*,[11] as a special variety of reism. It seems reasonable to choose rather a more liberal version of reism that does not exclude some *non-corporeal* things, thus allowing for the existence of fields as such objects. There is no clear reason against such a choice.

In spite of the above objections, reism is an essentially sound ontological approach. The weak points of its original formulation are mostly due to the influence of the spirit of the Vienna Circle on its author. Kotarbiński was a great phenomenologist; it is a pity that he was not aware of this. Conscious use of the phenomenological method would probably have protected him from some erroneous conclusions. It would also have led him to many interesting new results. Moreover, it would have given him much more confidence in the truths he discovered. This paper is intended to mark a programme of further developing his ideas. The resources of reism are still far from exhausted.

*Institute of Philosophy, Jagiellonian University*

## NOTES

[1] Cf. Przełęcki (1990).
[2] Cf. Ajdukiewicz (1930).
[3] See e.g. Słupecki (1955).
[4] See Bencivenga (1986).
[5] Cf. Kotarbiński (1929), part 1, Sec. 12 and (1958).
[6] Cf. Kotarbiński (1929), part 1, Sec. 3, (1958) and (1968).
[7] Cf. Kotarbiński (1968).
[8] Cf. Simons (1987), Sec. 2.6.2.
[9] Cf. Kotarbiński (1929), part 5, Sec. 10 and (1958).
[10] Cf. Kotarbiński (1968).
[11] Cf. Kotarbiński (1935).

## REFERENCES

Ajdukiewicz, K.: 1930, Review of Kotarbiński (1929), *Przegląd Filozoficzny* **33**, pp. 140–60. Eng. trans. in Kotarbiński (1966), pp. 515–36.

Bencivenga, E.: 1986, 'Free logics', in Gabbay and Guenthner (1986), pp. 373–426.

Gabbay, D., Guenthner, F. (eds.): 1986, *Handbook of Philosophical Logic*, vol. 3, D. Reidel Publishing Company, Dordrecht.

Kotarbiński, T.: 1929, *Elementy teorii poznania, logiki formalnej i metodologii nauk* (Elements of Epistemology, Formal logic and Methodology of Sciences), Ossolineum, Lvov; revised edition, Wroclaw, 1961. Eng. trans. as Kotarbiński (1966).

Kotarbiński, T.: 1935, 'Zasadnicze myśli pansomatyzmu' (The Fundamental Ideas of Pansomatism), *Przegląd Filozoficzny* **38**, pp. 283–94. Eng. trans. in *Mind* **64**, 1955, pp. 488–500.

Kotarbiński, T.: 1958, 'Fazy rozwojowe konkretyzmu' (The Development Stages of Concretism), *Studia Filozoficzne* **4(7)**, pp. 3–13. Eng. trans. in Kotarbiński (1966), pp. 429–37.

Kotarbiński, T.: 1966, *Gnosiology. The Scientific Approach to the Theory of Knowledge*, Pergamon Press-Ossolineum, Oxford-Wroclaw. Eng. trans. of 2nd ed. of Kotarbiński (1929), with appendices.

Kotarbiński, T.: 1968, 'Reism: Issues and Prospects', *Logique et Analyse* **44**, pp. 441–58.
Przełęcki, M.: 1990, 'Semantic Reasons for Ontological Statements: the Argumentation of a Reist', in Woleński (1990), pp. 85–96.
Simons, P.M.: 1987, *Parts. An Essay in Ontology*, Oxford University Press, Oxford.
Słupecki, J.: 1955, 'S. Leśniewski's Calculus of Names', *Studia Logica* **3**, pp. 7–70.
Woleński, J. (ed.): 1990, *Kotarbiński: Logic, Semantics and Ontology*, Kluwer Academic Publishers, Dordrecht.

ANNA KANIK

# ON THE CONCEPT OF A SUBJECT OF COGNITION IN AJDUKIEWICZ'S PHILOSOPHY

I begin my discussion of a subject of cognition in the philosophy of Kazimierz Ajdukiewicz by recounting his view of science. In his paper 'Methodology and Metascience' he distinguishes three different meanings of the word 'science'. First, 'science' means all kinds of activities undertaken by scientists. Second, 'science' is a set of sentences which are the result of these activities. Science as an activity is a phenomenon which takes place in time and even in space. It has its history. But, science as a result of these activities, as a set of sentences uttered by scientists, has its history as well. Its components change over time. Here the development of science depends on the empirical data collected by scientists and the language they have invented. Thus, science as a result of research is a time construct as well.

Taking mathematics as an example, Ajdukiewicz introduces the notion of ideal science as a set of sentences of some language, even if those sentences have not been uttered by anybody. This is the third meaning of the word 'science'. Ideal science is a set of sentences which could be logically derived from some basic sentences (axioms), or a set of all those true sentences which could be expressed in the language of the given science. According to this conception, science does not have its history, it does not develop in time. Ideal science is the field of study for metascience.

Since we have three meanings of 'science', we can discuss three kinds of subjects of cognition corresponding to these meanings:

1. the empirical subject
2. the subject of science as a result
3. the subject of ideal science.

First, I shall point out the common characteristics of these three meanings. I shall then list some problems which arise when each of these conceptions is adopted. This issue is vital for at least two reasons: first, the contemporary tendency to study science in a wider context taking into account the influence of social and cultural factors, and second, the present discussion on apriorism and empirism.

The subject of cognition is a subject who actively does science. Neither is science stating the truth in itself, nor is the subject of cognition one who only knows what is truth, who only possesses the truth. Divine revelation could be sufficient to know the truth if somebody accepts its validity, but it cannot be science. A subject of cognition does science in the sense that by accepting any



*K. Kijania-Placek and J. Woleński (eds.),*
*The Lvov–Warsaw School and Contemporary Philosophy*, 319–322.
© 1998 *Kluwer Academic Publishers.*

sentence one accepts a large part of the language, because to accept a sentence is to know and assert the meaning of the sentence. Therefore, according to Ajdukiewicz's conception of meaning, the subject must know a large part of the language. A theorem is asserted if and only if a sequence of sentences of its justification is asserted as well. However, a sequence of sentences is something dynamic and requires some kind of activity on the part of the subject.

Furthermore, the subject is an individual. The concept of subject does not favour any particular group of people, such as very gifted persons or persons with rare mental powers. According to Ajdukiewicz's view on the social role of science, distinguishing any group would not be rational. Everybody should have an opportunity to critique scientific results, so that he/she could defend him/herself and society against the dangers of being governed by uncontrollable ideologies.

Now, cognition in Ajdukiewicz's philosophy is always articulated. It is always expressed in language and depends on it. In addition, the subject does not have any innate categories which generate language. If one assumes that cognition depends on certain fixed categories which are shared by all empirical subjects, then there is no need to take different subjects into account. The idealisation that identifies real scientists with one subject does not lose any important features. On the other hand, if as, Ajdukiewicz claims, cognition depends on the choice of a conceptual apparatus,[1] and this choice is not determined for the subject, i.e. the conceptual apparatus is not rigidly linked to the nature of the subject, then the identification of real subjects with one ideal subject would make it impossible to discuss such problems as, for example, the problem of choosing categories, of comparing different conceptual apparatuses or describing the role of language in our cognition and its evolution. Thus, investigations of the concept of the subject of cognition must allow us to talk about more then one subject within Ajdukiewicz's framework.

Thus, each of these kinds of subjects is an unprivileged individual who is active and who does science. There are no innate languages or categories of cognition, but the subject of cognition has a language or a set of languages at his disposal. Since we have no right to assume that there is something like a universal language, we should consider more than one language at a subject's disposal.

Let us examine problems connected with the various concepts of a subject. To consider only the empirical subjects is not sufficient because of the condition of intersubjectivity and time limitations of such subjects. The inductive and deductive methods require the use of the same language by the researchers.[2] Each premise of an induction is based on the direct experience of a concrete person. The premise is intersubjective if and only if the researcher is guided by the empirical rules of the language. Empirical subjects are not allowed to change language when they carry out research in the given science.

Hence in this case the evolution of a conceptual apparatus cannot be explained. A similar case arises with the subject of ideal science. It should be mentioned that Ajdukiewicz considers this kind of subject mainly when he engages in discussion with other philosophies, especially with idealism. The results of his investigations are magnificent and well-known. Without going into details it should be pointed out that if a subject is assigned to a language, then the object of cognition must be external to the subject with respect to Tarski's theorem on the undefinability of truth, Gödel's theorems and the law of the excluded middle.[3] This argumentation supports realism.[4] Although under this idealisation different conceptual apparatuses can be compared, one still cannot provide a satisfactory explanation of why and how the subject changes these languages. For, if we change the language, we change the subject: this is the reason why a subject who decides to change the language can be described neither as any empirical subject nor as any subject of an ideal science.

The evolution of science can be clarified when we discuss a subject of the science as a result. The subject should be treated as a member of a community. One can discuss the reasons for changing a language, one can also invent new languages and communicate them to other members of the community and afterwards one can carry out research in this language. However in this case the concept of subject needs additional explanation:

- the means of learning a language
- the possibility of introducing a new language
- the possibility of ascertaining that the same language is used
- the possibility of choosing a language

The first two points can be treated together. Ajdukiewicz assumes the behavioural theory of learning a language. What should be emphasised here is that one learns to assert sentences on the basis of other sentences in the same way that he/she learns to name objects in our everyday experience.[5] The only condition that the language must fulfil is the presence of certain regularities so that a subject could find the rules of meaning (including the deductive rules). No innate categories, in the Kantian sense, are required for the subject. However, the assumption that the subject has the innate ability to recognise these regularities has to be made.

In order to decide whether that the same language is used by two subjects one of them should pose a question and check the answer.[6] If the answer does not agree with his answer then he can be sure that their languages are different. If it does agree, he can suppose that they use the same language. Thus, this is only the necessary, but not sufficient, condition for checking whether two subjects use the same language.

Ajdukiewicz distinguishes four criteria of choosing a language in the development of science[7]: the tendency to discard a language if it turns out to be contradictory; the tendency to transform hypotheses into principles; the tendency to eliminate unsolvable problems; and the tendency toward increased empirical sensitivity of the conceptual apparatus.

Assuming a subject of cognition as a subject of science as a result, it is possible to describe the evolution of science within Ajdukiewicz's philosophy. In the philosophical tradition the dependence of cognition on time has not been taken into account which is, in my opinion, a heritage of every kind of apriorism. Each idealisation of the subject ignores this problem. Then all knowledge is available to the subject independently from the state of development of science. However, Ajdukiewicz considers science as a historical-social phenomenon as well. The question of the interdependence between a subject and the state of development of science can be explained satisfactorily only if we consider a subject to be a member of a community who can interact with its others members. The interaction is necessary with regard to the process of language learning, the intersubjective controllability, and what is most important, to the possibility of deciding that the same language is used. The last factor is a necessary condition of the methods of induction and deduction.

*Institute of Philosophy, Jagiellonian University*

## NOTES

[1] A conceptual apparatus is the totality of the expressions of a closed and connected language together with their meanings, i.e. positions in the total scope of the meaning rules (Ajdukiewicz, 1934a).

[2] See Ajdukiewicz (1962).

[3] See Ajdukiewicz (1937).

[4] For further discussion see Woleński (1993).

[5] For a detailed analysis of the issue see Ajdukiewicz (1931).

[6] Ajdukiewicz analyzes the problem in detail in (1934a).

[7] See Ajdukiewicz (1934b).

## REFERENCES

Ajdukiewicz, K.: 1931, 'On the Meaning of Expressions', in Ajdukiewicz (1978), pp. 1–34.

Ajdukiewicz, K.: 1934a, 'Language and Meaning', in Ajdukiewicz (1978), pp. 35–67.

Ajdukiewicz, K.: 1934b, 'The World-Picture and the Conceptual Apparatus', in Ajdukiewicz (1978), pp. 68–89.

Ajdukiewicz, K.: 1937, 'A Semantical Version of the Problem of Transcendental Idealism', in Ajdukiewicz (1978), pp. 140–154.

Ajdukiewicz, K.: 1962, 'Subiektywność i niepowtarzalność metody bezpośredniego doświadczenia' (Subjectivity and Uniqueness of the Method of Direct Experience', *Studia Logica* **13**, pp. 209–211.

Ajdukiewicz, K.: 1978, *The Scientific World-Perspective and Other Essays 1931–1963*, ed. by J. Giedymin, D. Reidel Publishing Company, Dordrecht.

Woleński, J.: 1993, *Metamatematyka a epistemologia* (Metamathematics and Epistemology), PWN, Warsaw.

ILKKA NIINILUOTO

# INDUCTION AND PROBABILITY IN THE LVOV-WARSAW SCHOOL

## 1. BACKGROUND

The Lvov-Warsaw School is famous for its great achievements in many fields of logic and philosophy, among them classical and non-classical logic, formal ontology, truth, philosophy of language, epistemology, philosophy of mind and praxiology. Induction and probability were not among the strongest areas where the disciples and followers of Kazimierz Twardowski worked.[1] Still, our picture of this school would be incomplete without the recognition of its studies in this domain.

Jan Woleński devotes Chapter VIII, with 20 pages, of his comprehensive *Logic and Philosophy in the Lvov-Warsaw School* (1989) to the philosophy of science, and gives an informative but brief summary of works on induction and probability.[2] When we add some links between the Polish contributions and other related investigations, already this summary suggests that there were interesting and important ideas in this field which had influence on the twentieth-century logic and philosophy of science — and which still deserve to be discussed as parts of contemporary philosophy.

Before going into detail, it is useful to have a look at some general characteristics of the Lvov-Warsaw approach to philosophy. First, the members of the school had high respect for science and scientific thinking.[3] Eloquent expressions were given by Jan Łukasiewicz. In 'Creative elements in science' (1912), he asserted that science aims at truth, with the additional value of having "the ability to arouse, or to satisfy, directly or indirectly, intellectual needs common to humanity."[4] Subjective and practical elements of science were not denied, but the emphasis was on realism, objective truth, and intellectual needs. Moreover, science is not only a reproduction of facts, but also a creative synthesis.

In his Farewell Lecture at the Warsaw University in 1918, Jan Łukasiewicz claimed that great philosophical systems "fall into nothingness when subjected to logical criticism."[5] This anti-metaphysical attitude was forcefully repeated in his lecture 'On Determinism', originally Rector's Inauguration of the academic year 1922/23 in Warsaw University: even "the degree of precision sufficient for the mathematician" does not satisfy the new generation of logicians.



*K. Kijania-Placek and J. Woleńsky (eds.),*
*The Lvov–Warsaw School and Contemporary Philosophy*, 323–335.
© 1998 *Kluwer Academic Publishers.*

Philosophical systems "fall into pieces as if they were houses of cards", since "their basic concepts are not clear, their most important theses are incomprehensible, their reasoning and proofs are inexact [ . . . ] Philosophy must be reconstructed from its very foundations; it should take its inspiration from scientific method and be based on the new logic."[6]

Secondly, there was the "Austro-Polish obsession with truth."[7] Twardowski supported the correspondence theory of truth, and wrote in 1902 an important article on absolute and relative truth. The most significant work on this topic was of course done by Alfred Tarski, who published only one brief remark on probability in 1936.[8]

Thirdly, the Polish school was mainly working on deduction, deductive inference, and deductive sciences. This focus was in harmony with its quest for exactness and interest in metamathematics. Further, the background of its members was in most cases in philosophy and mathematics, not as frequently in physics as among other similar schools in France, Austria, and Germany (e.g., Mach, Duhem, Poincaré, Schlick, Reichenbach).

Fourthly, no general synthesis was written on the philosophy of science, at least if Zygmunt Zawirski's stolen 1947 manuscript is not counted.[9] Kazimierz Ajdukiewicz's *Pragmatic Logic* (1965) contains a lot of material on scientific inference (e.g., statistical methods), but it is a late and not very original book. The later work of his students (among them, Henryk Mehlberg, Jerzy Giedymin, and Henryk Skolimowski) does not fall in the scope of the Lvov-Warsaw School any more.

Fifthly, Łukasiewicz did important early work on probability and induction, but later lost his interest in these problems, largely due to his discovery of many-valued logics in 1920. The most important studies in the field of induction and confirmation theory were made by Janina Hosiasson, born in 1899, a student of Tadeusz Kotarbiński in Warsaw in the 1920s, and wife of the logician Adolf Lindenbaum, but her life and promising career were violently interrupted by the Gestapo in Vilna in 1942.

## 2. INDUCTION IN SCIENCE

Classification of reasoning was a very important theme in the Lvov-Warsaw School.[10] Early work in 1901 was done by Twardowski, whose classification included direct vs. indirect and certain vs. probable reasoning. In 1923 Twardowski distinguished a priori, rational, deductive sciences and a posteriori, empirical, inductive sciences by their methods of justification.

Łukasiewicz in 1912 divided reasoning into deductive and reductive. According to him, deductive or direct reasoning includes inference and testing, while reductive or indirect reasoning includes explaining (incomplete induction and creating hypotheses) and argumentation. The debate was continued by Kotarbiński and Ajdukiewicz. Father Jan Salamucha in 1930 argued quite convincingly against Łukasiewicz that argumentation (from known premises) should be classified as deductive rather that reductive, and testing as reductive rather than deductive.

Ajdukiewicz in 1955 classified inferences into deductive, probabilistic (acceptance of a new statement with some degree of belief), and logically useless. His *Pragmatic Logic* (1965) argues that the "inductive sciences" employ simple induction, analogy, induction by elimination, and the testing of statistical hypotheses.[11]

Maria Kokoszyńska, who did early work on truth and meaning in the 1930s, returned in 1967 to Twardowski's theme on the difference between deductive and non-deductive sciences. According to her, deductive sciences are analytic and concern classes of models, while non-deductive sciences are synthetic and concern one distinguished model of the meaning postulates. In my view, the corresponding division should be made within mathematical theories: arithmetic has a standard model, group theory does not.[12] Kokoszyńska's application of the Tarskian model theory anticipates the later important work that Polish philosophers of science (Marian Przełęcki, Ryszard Wójcicki) did on the semantics of empirical theories.[13]

The majority opinion in Poland was that empirical sciences are inductive. Such views need not be "inductivist", as Woleński says, since often this term is understood to refer to the narrow view of inductive generalization as a method of discovery.[14] But also other positions were discussed and supported.

Inductive inference aims at giving probability or confirmation to scientific hypotheses on the basis of observational data, so that it is usually combined with stronger or weaker forms of empiricism. If inductive probabilities are assigned to scientific laws and theories, this approach is associated with some form of scientific realism. Duhem's instrumentalism and Poincaré's conventionalism (or structural realism in Giedymin's terms) regard theoretical statements or at least theoretical ontologies as lacking truth values.

The radical conventionalism of Ajdukiewicz in the 1930s anticipated Kuhn's and Feyerabend's work on incommensurability in the 1960s. Instrumentalism was supported by E. Poznański and A. Wundheiler in 1934. Also Łukasiewicz became doubtful about the possibility of probabilistic induction in his 1929 article 'On reasoning in science'.

Karl Popper's *Logik der Forschung* (1934), and its later refinement with Tarski's notion of truth, combined scientific realism with the rejection of inductive inference. In §81, added in proof, Popper remarked that the probability of hypothesis, its "degree of validity", is not relative frequency (vs. Reichenbach), but "degrees of corroboration" might have some analogies with the calculus of probability. "I am indebted to Dr. J. Hosiasson for suggesting this possibility to me." It seems obvious from this note that Popper had met Hosiasson in September 1934 during the Sixth International Congress of Philosophy in Prague.[15]

Popper's anti-inductivism did not have a strong impact on the Polish scene. A balanced criticism of Popper's views was given by Janina Kotarbińska in 1962. Like Wesley Salmon some years later, she argued that there is no real difference between deductivism and inductivism.[16]

## 3. THE NATURE OF INDUCTION AND PROBABILITY

The main approach to the theory of induction in the Lvov-Warsaw School was probabilistic. As exceptions to this rule one could mention Władysław Biegański's book on analogical reasoning in 1909, and the analysis of Mill's methods of eliminative induction by Jerzy Łoś, in his graduation thesis for Jerzy Słupecki in Lublin in 1947,[17] and Ajdukiewicz in *Pragmatic Logic* in 1965.

The study of probabilistic induction was started in 1903 by Łukasiewicz. Following the Bayesian tradition of Laplace, Jevons, and Sigwart, he considered the probability $P(h/e)$ of a hypothesis $h$ on evidence $e$ as inverse to the deductive relation from $h$ to $e$. In 'On the probability of inductive hypotheses' (1909), he generalized Laplace's rule of succession to the formula $(n+1)/(n+m+1)$, which is supposed to give the probability of a generalization, when $m$ is the size of the universe and $n$ is the number of observed favorable instances. This formula has the property that it converges to 0, when $m$ grows without limit. In 1929 Łukasiewicz was ready to reject the Bayesian approach to probabilistic induction.

The convergence property of Łukasiewicz's formula is in fact the same as in the system of inductive logic that Rudolf Carnap developed in the 1940s. Similar results, as special cases of Carnap's system, were obtained in the 1920s in Cambridge by W.E. Johnson and Richard Braithwaite, and in Vienna by Ludwig Wittgenstein and Friedrich Waismann. Popper, too, joined the claim that the inductive or logical probability of genuinely universal generalizations is zero. John Maynard Keynes in his *Theory of Probability* (1921) and Harold Jeffreys argued that scientific hypotheses should have non-zero probabilities. A satisfactory theory of inductive generalization was given by Jaakko Hintikka in 1964 as an extension of Carnap's approach. An axiomatic diagnosis of the situation was given by Hintikka and Ilkka Niiniluoto in the 1974 Conference in Jabłonna.[18]

In the Bayesian school, probability is understood as a subjective or an epistemic concept, indicating a rational degree of belief in the truth of a hypothesis on the available evidence (Laplace, Jevons, Ramsey, de Finetti) or degree of partial entailment (Keynes, Nicod, Carnap). This view was revived in Poland by Janina Hosiasson (cf. Section 5). Stefan Mazurkiewicz's axiomatic treatment of probability in 1932 belongs to this tradition of inductive logic as well. Also Ajdukiewicz treated probability as an epistemic concept linked with decision theory (cf. Section 6).

The rival approach, first formulated clearly by John Venn in 1866 and Charles Peirce in 1878, identifies probability with the long run relative frequency of an event-type in a sequence of repeated trials. In the finite frequency interpretation, this sequence is taken to be finite, while the limiting frequency interpretation equates probability with the limit of relative frequency in an infinite sequence. Well-known formulations of the latter view, besides Venn and Peirce, were given by Hans Reichenbach (from 1915) and Richard von Mises (from 1918). Łukasiewicz's work on probability in 1913 belongs to this frequentist tradition (cf. Section 4).

Another interpretation of objective probability takes probability to be a law-like disposition of a physical device (such as a dice or coin) to behave in certain ways in individual or repeated trials. Peirce in 1910 characterized such a disposition as a "would-be." This view, which Popper rediscovered as the "propensity interpretation" in the 1950s, did not receive support in the Lvov-Warsaw School.

## 4. PROBABILITY AND FREQUENCIES

Łukasiewicz took part in the seminar of Alexius Meinong in Graz in the summer of 1909 by outlining a study of probability in the exact spirit of modern logic. The result was an important monograph *Die logische Grundlagen der Wahrscheinlichkeitsrechnung* (1913).[19]

Łukasiewicz rejected the Laplacian classical definition of probability in terms of "equally possible" cases. It gives only the form of the probability function. He was also dissatisfied with 'subjective' and 'objective' interpretations of equipossibility. The former view, supported by Carl Stumpf in 1892, appeals to lack of knowledge about the alternatives, but the idea of probability as dealing with psychic phenomena and their measurement does not lead to a coherent theory. (This criticism was indeed justified before Ramsey's and de Finetti's work in the 1920s.) One version of the objective theory (sometimes regarded as a precursor of the propensity interpretation) was proposed in 1886 by Johann von Kries, who linked probability to the symmetric physical structure of the coin, but Łukasiewicz argued that his account falls back to an epistemic principle of indifference.

Łukasiewicz claimed more generally that an objective theory of probability is excluded. His deepest reason for this thesis was the conviction that the notion of objective possibility does not make sense. Here he followed the classical idea that probability is intimately connected with the notion of possibility. According to Leibniz, probability is the "degree of possibility."[20]

Łukasiewicz presented two specific arguments against objective possibility. The principle of causality, which asserts that everything occurs with necessity, leaves no room for objective possibility. The same conclusion, he contended, follows from the principle of the excluded middle: a black ball is either drawn or is not drawn, but one of these alternatives is true, and hence true at every time, and "also *before* the drawing of the ball."

The underlying basic idea was that definite propositions are either true or false, and this leaves no room for possibility. Ever since Aristotle, this has been a ground for adopting the statistical interpretation of modality for temporally indefinite sentences (e.g., 'it is raining today'): necessity means "always true", possibility "sometimes true", and impossibility "never true."[21] This view of modality was advocated by such pioneers of modern logic as Gottlob Frege and Bertrand Russell. Russell argued in 1912 that modalities cannot be assigned to propositions, which are either true or false, but only to propositional functions with a free variable (e.g., 'object $x$ is blue', 'the sun is shining at time $t$'). Łukasiewicz adopted a similar principle by arguing that probability can be assigned only to indefinite propositions which contain a free variable. He noted also the similarity of this approach to Bolzano's concept of "validity" of a sentence with changeable parts.

Let $p$ be an indefinite proposition with a free variable $x$ which ranges over a fixed finite set $V$. Then Łukasiewicz defined the truth value $w(p)$ of $p$ to be the ratio between the number of values of $x$ for which the proposition $p$ yields true judgments and the total number of values of $x$ in $V$. The relative truth value $w_q(p)$ of $p$ given $q$ is defined by $w(pq)/w(q)$, where $pq$ is the logical product (conjunction) of $p$ and $q$. An indefinite proposition is true if its truth value is 1, false if its truth value is 0, and probable if it is neither true nor false. It is then possible to prove that the function $w$ satisfies the classical properties of probability. In this way, probability as the truth value of indefinite propositions is a purely logical concept, a creative construction of the human mind, but the calculus of probability "is not a science of subjective processes and has a thoroughly objective nature."

Łukasiewicz's definition of relative truth value is essentially similar to Peirce's account of the probability of induction in terms of truth-frequencies, indicating how often an argument "carries truth with it"—even though Peirce's suggestion was not limited to finite classes. It can also be easily translated into terms more familiar in probability theory. An indefinite proposition like '$x$ is black' corresponds to an attribute (being black) of objects or events, and the range of variable $x$ is the reference class. Then his definition is equivalent to saying that probability is the relative frequency of an attribute in a reference class.[22] Hence, Łukasiewicz's account is an elegant and exact formulation of the finite frequency definition of probability. He did not raise the question of infinite reference classes, and problems with the limits of relative frequences, and did not refer to the earlier work by Venn and Peirce on these issues. And, sticking strictly to his basic principles, he did not consider (like Venn earlier and Reichenbach later) the prospects of applying probabilistic concepts to the "single case" (such as individual events or universal generalizations).

In his own later work, Łukasiewicz challenged the premises of his arguments against objective possibility. The discovery of three-valued logic in 1920 rejected the law of the excluded middle, and the 1922 article on determinism attacked necessity and predetermined truth values of future statements. In his new "indeterministic philosophy", Łukasiewicz interpreted the third truth value $\frac{1}{2}$ (besides truth 1 and falsity 0) as "possibility." This gave rise to a new modal logic, where truth, falsity, and possibility are applied to definite propositions $p$. Tarski, as a student in Warsaw in 1921, proposed the famous definition of possibility $Mp$ by the formula $\neg p \supset p$ (or $CNpp$ in the Polish notation).[23] In the 1922 system of many-valued logic, there were infinitely many degrees of possibility as in the theory of probability, but still in 1930 it was pointed out that the relation of these theories "awaits further inquiry."[24]

Hans Reichenbach started in the late 1920s to work out a probability logic on the basis of frequentist ideas. The first full exposition was published in 1932, and again in his *Wahrscheinlichkeitslehre* (1935). Reichenbach defined probability implications of the form $\forall x\,(x \in A \ni_u x \in B)$ by the condition that the relative frequency of cases of $B$ among the cases of $A$ equals $u$. If the class of $A$s is finite, this probability is the same as Łukasiewicz's 1913 definition of the relative truth value of '$x$ is $B$' given '$x$ is $A$'.[25]

Reichenbach's studies were well known to Polish logicians. Hosiasson argued that there is no clear sense in which probability logic is an extension of classical logic.[26] Mazurkiewicz and Tarski pointed out that probability is intensional (i.e., non-truth-functional). For example, the probability of a disjunction is not a function of the probabilities of the disjuncts alone. The laws of probability differ from those of many-valued logic. For example, the probability of a contradiction is 0, but for truth values $v(p \,\&\, \neg p) = \frac{1}{2}$ if $v(p) = v(\neg p) = \frac{1}{2}$.

Reichenbach's general approach in probability logic was supported by Zygmunt Zawirski and Tadeusz Czeżowski.[27] Zawirski believed that it has important applications in quantum mechanics. He argued that in many-valued logic it is necessary to distinguish two concepts of disjunction $p \vee q$, one definable by $(p \supset q) \supset q$ (Łukasiewicz), the other by $(\neg p \supset q)$. This makes the system compatible with the probability calculus, but it is not a Łukasiewiczian many-valued logic any more, since it satisfies the principles of contradiction and excluded middle. In 1951 Czeżowski studied deductive relations between probability implications.

## 5. CONFIRMATION THEORY

Kazimierz Ajdukiewicz's book on the methodology of sciences in 1928 (in Polish) contained a discussion of a simple rule of acceptance of a scientific hypothesis which is probable (or more probable than its negation) by virtue of its ability to entail some contingent consequence. But he returned to these problems only thirty years later (see Section 6).

The first and only member of the Lvov-Warsaw School who was a professional confirmation theorist by international standards was Janina Hosiasson-Lindenbaum. She had personal contacts with Cambridge, which was perhaps the most important center of the probabilistic theory of induction in the 1920s by virtue of the work of Keynes, Russell, Ramsey, and C.D. Broad.

In her first papers published in Polish, Hosiasson defended the view that inductive inference is justifiable if its conclusion is not the hypothesis itself but the assignment of a degree of confirmation to the hypothesis. This was later the basic idea of Carnap's approach to induction.

In a paper published in *Mind* in 1931, Hosiasson asked why we prefer probabilities relative to many data.[28] If probabilities are always relative to information, as Keynes asserted, why should a probability of the form $P(B/AC)$ be preferable to $P(B/A)$? Variants of this problem were later discussed by Reichenbach (the problem of the reference class) and Carnap (the requirement of total evidence). We cannot generally know that the former probability is closer to some ultimate value $U$ than the latter. However, Hosiasson argued, we can know that the probability of the claim

$$| P(B/AC) - U | < | P(B/A) - U |$$

given $AC$, is very large. In the latter part of the paper, Hosiasson referred to an unpublished paper by Ramsey, which is said to be similar to her previous thought, and outlined a "pragmatistic" analysis to show that the expected gain of probabilities on more data are larger than on less data.

In a paper presented in the Prague Congress in 1934, Hosiasson studied partial entailment relations.[29] The question is whether the probability of an inference depends on the logical weakness of the additions that are needed to make it a conclusive inference. Suppose that $s$ can not be inferred from $q$ alone, and let $z$ be an additional premise such that $s$ can be inferred from $q\& z$, $z$ is compatible with $q$, and $z$ alone does not entail $s$. Is the probability $P(s/q)$ then the larger the weaker the required additional premise $z$ is? As in fact $q \supset s$ is the weakest of such statements $z$, the question is about the possible identity or relation of $P(s/q)$ and $P(q \supset s)$. Problems of this sort have later been discussed in the debate about the probability of conditionals.

Hosiasson gave an example to show that the answer to her question is not always positive. However, the statement $z_1$ about tosses of a dice (not 1, ... , not 5) is only apparently "weaker" than the statement $z_2$ about tosses of octahedron (not 1, ..., not 5, not 8), since there is no deductive relation between them. Indeed, if information content is measured by cont($z$) = 1 - $P(z)$, as Popper and Carnap later suggested, then we have $\mathrm{cont}(z_1) = \frac{5}{6} > \mathrm{cont}(z_2) = \frac{3}{4}$, so that $z_1$ is not weaker than $z_2$ by its degree of information content. Using the cont-measure, Hosiasson's main conclusion can be reformulated as follows: assuming that $P(q) = P(q'), P(z/q) > P(z'/q')$ if and only if $\mathrm{cont}(q \supset z) < \mathrm{cont}(q' \supset z')$.

The best known article by Hosiasson-Lindenbaum is 'On Confirmation' in *The Journal of Symbolic Logic* in 1940. Her related paper on induction and analogy was published in *Mind* in 1941.[30] Following Mazurkiewicz, she expressed statements about probabilistic confirmation in metalanguage, and characterized the degree of confirmation $c(a, b)$ of a sentence $a$ with respect to sentence $b$ by axioms. These studies immediately influenced others who were starting their work on induction, like Rudolf Carnap and Georg Henrik von Wright. Carnap largely adopted Hosiasson's notation ('c' for confirmation), terminology, and axioms, but proceeded to study specific inductive probability functions and their classes.

Hosiasson formulated the problem of induction following Keynes, and she defended Keynes' main conclusion about inductive generalization against Jean Nicod. Let $g$ be a universal generalization, $b$ background knowledge, and $e_n$ the result of $n$-th observation. Assume that $g$ and $b$ entail $e_n$ for $n$= 1, 2, ... Does this continued confirmation of $g$ suffice to increase its posterior probability $P(g/b \,\&\, e_1 \,\&\, \ldots \,\&\, e_n)$ towards the maximum value 1 in the limit? The answer of Keynes and Hosiasson is yes, if $P(g/b) > 0$ (i.e., the prior probability of $g$ is non-zero), $P(e_{n+1}/e_n \,\&\, b) < 1$, and $P(e_n/\neg g \,\&\, b) \longrightarrow 0$ when $n \longrightarrow \infty$.[31]

What Hosiasson called the problem of analogy concerns a situation where a law $g$ (with background knowledge $b$) entails two independent observational statements $h_1$ and $h_2$. For example, Newton's theory entails Galileo's and Kepler's laws. This phenomenon was called the "consilience of inductions" by William Whewell in 1840. Hosiasson noted that the deductive link *via* the law $g$ does not guarantee that thereby $h_1$ and $h_2$ confirm each other, and looked for sufficient conditions for $P(h_2/h_1 \,\&\, b) > P(h_2/b)$. These observations are related to the transitivity paradoxes of confirmation, later analysed by Carl G. Hempel and Mary Hesse.[32]

Hempel, who was Reichenbach's student from Berlin, formulated in 1937 the famous "raven paradox" of confirmation. By Nicod's criterion, a generalization of the form 'All $A$s are $B$' is confirmed by its positive instances, i.e., by the observation of an individual $a$ such that $Aa \,\&\, Ba$. But this generalization is logically equivalent to 'All non-$B$s are non-$A$', and that in turn is confirmed by $\neg Ab \,\&\, \neg Bb$. So it seems that the law 'All ravens are black' can be confirmed by a white shoe. Hempel has reported that he met Hosiasson in 1937 and discussed this problem with her.[33] In 'On Confirmation' (1940), Hosiasson published the first Bayesian solution of this paradox. The essential point is that a non-black non-raven confirms the law but by a much smaller amount than a black raven, since the class of non-ravens is immensely large in comparison with the class of ravens.[34]

Hosiasson-Lindenbaum's paper on the advancement of knowledge, presented originally in Vilna in 1939, was published posthumously in *Synthese* in 1948.[35] This paper shows her interest in the growth of knowledge — a problem area that became fashionable in the 1960s through the work of Popper, Kuhn, and others. In the case of "indubitable opinions", the advancement of knowledge may take place in three ways: "enrichment" or increasing logical strength, degree of "actuality" obtained by proof, and internal "interlinking" of propositions. For probable or uncertain knowledge, advancement may also be gained by the adjustment of subjective and objective degrees of certainty, and increasing the weight of evidence. She further distinguished the aim of attaining the greatest amount of information ("the whole truth") and the greatest possible "readiness" of that knowledge — showing thus awareness of the limitations of the human capacity to store and use information.

## 6. INDUCTIVE ACCEPTANCE

Ajdukiewicz published an article on non-deductive inference in 1958. Inspiration seems to have come from Abner Shimony's 1955 paper, which continues Ramsey's and de Finetti's work on subjective probabilities as coherent betting ratios. It is also connected to the theory of inductive acceptance rules, developed by Isaac Levi and Jaakko Hintikka in the 1960s. Ajdukiewicz's article was published in English in 1978. His work was continued in Poland by Klemens Szaniawski.[36]

Ajdukiewicz proposed that the degree of certainty $P_X(A)$ with which a person $X$ accepts a sentence $A$ (for the purpose of some action) is the highest betting ratio on the truth of $A$ acceptable to $X$. If $k$ is the minimal acceptable profit of $X$ if $A$ is true, and $m$ is the maximal acceptable loss of $X$ if $A$ is false, then the value of $P_X(A)$ may be taken to be equal to $m/(k+m)$. Note that if the probability of $A$ is $p$, then the expected value of the option '$k$ if $A$ is true, $-m$ if $A$ is false' is $pk-(1-p)\,m$, and this is larger than 0 if and only if $p > m/(k+m) = P_X(A)$. If the action does not involve any risk (i.e., $k > 0, m = 0$), then the degree of certainty is minimal 0. But if the risk is infinitely high (i.e., $k = 0, m > 0$), then the required degree of certainty is maximal 1.

The "degree of infallibility" of the scheme of inference 'From $Q(x)$ infer $R(x)$' is defined by the number of values of variable $x$ that satisfy $Q(x)$ and

$R(x)$ divided by the number of values of $x$ that satisfy $Q(x)$. Ajdukiewicz thus employed a frequentist definition that is formally equivalent to Peirce's truth-frequences and Łukasiewicz's relative truth values. Similar ideas are used in the Neyman-Pearson theory of statistical inference, and in Alvin Goldman's "reliabilist" epistemology.[37]

Finally, Ajdukiewicz suggested that a method of inference is "rational from the practical point of view" if the degree of certainty of conclusions derived according to this method from true premises does not exceed its degree of infallibility. This frequentist justification of rationality, he claimed, is lacking from Carnap's theory of confirmation. It is not quite fair to claim that Carnap's system does not allow the transition from a metalinguistic statement $P(h/e) = r$ about the degree of confirmation $r$ of $h$ on $e$ to "any assertive attitude" toward $h$ after the acceptance of $e$.[38] But it is an interesting idea to try to study, in terms of truth-frequences, the reliability of rules of inference based on inductive logic.

*Department of Philosophy, University of Helsinki*

## NOTES

[1] Jan von Plato's *Creating Modern Probability* (Cambridge University Press, Cambridge, 1994), a recent account of the emergence of probability theory from a philosophical and mathematical viewpoint, does not mention any Polish authors at all. But such classics of the frequency theory as John Venn and Charles S. Peirce are not mentioned, either.

[2] J. Woleński, *Logic and Philosophy in the Lvov-Warsaw School*, Kluwer Academic Publishers, Dordrecht, 1989.

[3] Cf. Woleński, *op. cit.*, p. 273.

[4] See J. Łukasiewicz, *Selected Works*, ed. by L. Borkowski, North-Holland Publishing Company, Amsterdam, 1970, pp. 1–15.

[5] See Łukasiewicz, *op. cit.*, pp. 84–86.

[6] See Łukasiewicz, *op. cit.*, pp. 111–112. The spirit of this declaration is close to the manifesto of the Vienna Circle in 1929. In this respect, Woleński (*op. cit.*) seems slightly to overemphasize the differences between the Lvov-Warsaw School and the logical empiricism of the Vienna Circle.

[7] This phrase is used by Peter Simons and Jan Woleński, '*De Veritate*: Austro-Polish Contributions to the Theory of Truth from Brentano to Tarski', in K. Szaniawski (ed.), *The Vienna Circle and the Lvov-Warsaw School*, Kluwer Academic Publishers, Dordrecht, 1989, pp. 391.

[8] See Alfred Tarski, 'Wahrscheinlichkeit und mehrwertige Logik', *Erkenntnis* **5** (1935–36), pp. 174–175. Reprinted in A. Tarski, *Collected Papers*, vol. 2, Birkhäauser, Basel-Boston-Stuttgart, 1986, pp. 201–202.

[9] See Woleński, *op. cit.*, p. 325.

[10] See T. Kwiatkowski, 'Classifications of Reasonings in Contemporary Polish Philosophy', in F. Coniglioni, R. Poli, and J. Woleński (eds.), *Polish Scientific Philosophy: The Lvov-Warsaw School*, Poznań Studies in the Philosophy of the Sciences and the Humanities **28**, Rodopi, Amsterdam, 1993, pp. 117–167.

[11] See K. Ajdukiewicz, *Pragmatic Logic*, D. Reidel Publishing Company, Dordrecht, 1965.

[12] See M. Kokoszyńska, 'On the Difference between Deductive and Non-Deductive Sciences', in Przełęcki and Wójcicki, *op. cit.*, pp. 201–231. Cf. I. Niiniluoto, *Is Science Progressive?*, D. Reidel Publishing Company, Dordrecht, 1984, Ch. 8.

[13] See M. Przełęcki, *The Logic of Empirical Theories*, Routledge and Kegan Paul, London, 1969. Cf. Niiniluoto, *op. cit.*, Ch. 6.

[14] See Woleński, *op. cit.*, p. 294; cf. I. Niiniluoto and R. Tuomela, *Theoretical Concepts and Hypothetico-Inductive Inference*, D. Reidel Publishing Company, Dordrecht, 1973.

[15] For Popper's attempted definitions of degrees of corroboration as a function of probabilities, but not identical with probability, see K. Popper, *The Logic of Scientific Discovery*, Hutchinson, London, 1959. See also K. Popper, *Postscript to the Logic of Scientific Discovery: Realism and the Aim of Science*, Rowman and Littlefield, Totowa, NJ, 1983, p. 248. Cf. Niiniluoto and Tuomela, *op. cit.*

[16] See J. Kotarbińska, 'The Controversy: Deductivism versus Inductivism', in E. Nagel, P. Suppes, and A. Tarski (eds.), *Logic, Methodology and Philosophy of Science*, Stanford University Press, Stanford, 1962, pp. 265–274. Cf. W. Salmon, *The Foundations of Scientific Inference*, The University of Pittsburgh Press, Pittsburgh, 1966.

[17] See J. Łoś, 'The Foundations of a Methodological Analysis of Mill's Methods', in M. Przełęcki and R. Wójcicki (eds.), *Twenty-Five Years of Logical Methodology in Poland*, D. Reidel Publishing Company, Dordrecht, 1977, pp. 291–325. Later in 1963 Łoś applied model theory to give a representation theorem for the probabilities of first-order formulas (see ibid.). Woleński does not include Łoś in the Lvov-Warsaw School, but his work clearly continues its tradition.

[18] See Niiniluoto and Tuomela, *op. cit.*; J. Hintikka and I. Niiniluoto, 'An Axiomatic Foundation for the Logic of Inductive Generalization', in M. Przełęcki, K. Szaniawski, and R. Wójcicki (eds.): *Formal Methods in the Methodology of Empirical Sciences*, D. Reidel Publishing Company, Dordrecht, 1976, pp. 57–81.

[19] The original German work was published in Krakow in 1913. For the English translation, see Łukasiewicz, *op. cit.*, pp. 16–63. See also P. Simons, 'Łukasiewicz, Meinong, Many-Valued Logic', in Szaniawski, *op. cit.*, pp. 249–291.

[20] For the connection between probability and possibility, see I. Niiniluoto, 'From Possibility to Probability: British Discussions on Modality in the Nineteenth Century', in S. Knuuttila (ed.): *Modern Modalities*, D. Reidel Publishing Company, Dordrecht, 1988, pp. 275–309.

[21] See S. Knuuttila, *Modalities in Medieval Logic*, Cambridge University Press, Cambridge, 1993.

[22] cf. Salmon, *op. cit.*

[23] See Łukasiewicz, *op. cit.*, pp. 88, 167.

[24] See Łukasiewicz, op, cit., pp. 129–130, 173.

[25] Jan Woleński gives an informative account of the Polish reception of Reichenbach in 'Reichenbach's probability logic and the Lvov-Warsaw School' (forthcoming). However, he claims that Łukasiewicz's 1913 monograph has nothing to do with probability logic.

[26] See J. Hosiasson, 'La théorie des probabilités est-elle une logique généralisée?', *Actes du Congrés International de Philosophie Scientifique, Paris 1935*, Hermann, Paris, 1936, pp. 58–64.

[27] See Z. Zawirski, 'Über das Verhältnis mehrwertiger Logik zur Wahrscheinlichkeitsrechnung', *Studia Philosophica* **1** (1935), pp. 407–442; 'Über die Anwendung der mehrwertigen Logik in der empirischen Wissenschaft', *Erkenntnis* **6** (1936), pp. 430–435; T. Czeżowski, 'On Testability in Empirical Sciences', in Przełęcki and Wójcicki, *op. cit.*, pp. 93–109.

[28] J. Hosiasson, 'Why do we prefer probabilities relative to many data?', *Mind* **40** (1931), pp. 23–36. For Ramsey's unpublished work see N.-E. Sahlin, *The Philosophy of F.P. Ramsey*, Cambridge UP, 1990, p. 234.

[29] J. Hosiasson, 'Wahrscheinlichkeit und Schluss aus Teilprämissen', in *Actes du huitième Congrés International de Philosophie, Prague 1934*, Prague, 1936, pp. 192–196.

[30] J. Hosiasson-Lindenbaum, 'On Confirmation', *The Journal of Symbolic Logic* **5** (1940), pp. 133–148; 'Induction et analogie: Comparaison de leur fondement', *Mind* **50** (1941), pp. 351–365. For the impact of these articles, see R. Carnap, *The Logical Foundations of Probability*, The University of Chicago Press, Chicago, 1950, 2nd edition 1962; G.H. von Wright, *The Logical Problem of Induction*, Acta Philosophica Fennica, Helsinki, 1941, 2nd revised edition, Blackwell, Oxford, 1957.

[31] An interesting treatment of inductive generalization on the basis of Keynes was given by Zbigniew Czerwiński in his doctoral dissertation in 1957. See his article 'The Problem of Probabilistic Justification of Enumerative Induction', in Przełęcki and Wójcicki, *op. cit.*, pp. 65–80. For Czerwiński, the main hypothesis $g$ states that all $A$s are $B$, and its alternatives $g_c$ that the proportion $c$ $(0 \leq c < 1)$ of the $A$s are $B$. Then the continued observation of cases of $A$ & $B$ confirms $g$ and increases its probability to 1 in the limit. However, he does not address the special question that here the space of alternative hypotheses has the power of continuum, so that strictly speaking the prior probability distribution over it assigns probability 0 to a singular point like $g$. Cf. I. Niiniluoto, 'Inductive Logic as a Methodological Research Program', in *Logic in the 20th Century*, Scientia, Rome, 1983, pp. 77–100.

[32] The transitivity paradox was the topic of my own doctoral dissertation in 1973. See Niiniluoto and Tuomela, *op. cit.*

[33] Hempel discussed Hosiasson's solution in *Mind* in 1945. See C.G. Hempel, *Aspects of Scientific Explanation*, The Free Press, New York, 1965, pp. 20–21.

[34] For a clear Bayesian formulation, see P. Suppes, 'A Bayesian Approach to the Paradoxes of Confirmation', in J. Hintikka and P. Suppes (eds.): *Aspects of Inductive Logic*, North-Holland, Amsterdam, 1965, pp. 198–207. Suppes formulates the sampling rule: sample items with a higher prior probability of disconfirming the law.

[35] J. Hosiasson-Lindenbaum, 'Theoretical Aspects of the Advancement of Knowledge', *Synthese* **7** (1948), pp. 253–261.

[36] K. Ajdukiewicz, 'The Problem of the Rationality of Non-Deductive Types of Inference', in *The Scientific World-Perspective and Other Essays*, D. Reidel Publishing Company, Dordrecht, 1978, pp. 239–253. Cf. K. Szaniawski, 'Ajdukiewicz on Non-Deductive Inference', in V. Sinisi and J. Woleński (eds.): *The Heritage of Kazimierz Ajdukiewicz*, Poznań Studies in the Philosophy of the Sciences and the Humanities **40**, Rodopi, Amsterdam, 1995, pp. 295–302.

[37] See A.I. Goldman, *Epistemology and Cognition*, Harvard University Press, Cambridge, MA, 1986.

[38] This issue, which is still a matter of controversy among the Bayesians, is known as the principle of conditionalization. Some comments on this question were given by Z. Ziemba in 1961. See his 'Rational Belief, Probability and the Justification of Inductive Inference', in Przełęcki and Wójcicki, *op. cit.*, pp. 705–735.

TOMASZ PLACEK

# ŁUKASIEWICZ'S LOGICAL PROBABILITY AND A PUZZLE ABOUT CONDITIONALIZATION

## 1. ŁUKASIEWICZ'S 'LOGICAL PROBABILITY' VS. THE SUBJECTIVIST INTERPRETATION

Łukasiewicz's early philosophical interests centered upon probability, determinism and causation. Anticipating Russell's ideas in *Human Knowledge*, Łukasiewicz construed probability as a property assignable to *indeterminate sentences*. An indeterminate sentence is a sentential formula with a free variable together with a stipulation of the range of the variable. Any indeterminate sentence yields a number of corresponding sentences, which are either true or false, and each of which is obtained from the indeterminate sentence by replacing the variable with a name. For a given indeterminate sentence, the ratio

$$x = \frac{\#\text{true corresponding sentences}}{\#\text{all corresponding sentences}}$$

is interpreted as its probability. If $x$ equals 0, then the sentence is false; if $x$ equals 1, then the sentence is true. Thus, we have a spectrum, whose endpoints are occupied by true and false sentences, and in between we have some indeterminate sentences. Importantly, sentences without free variables, including those about future events, are not probable: they are either true or false. Łukasiewicz (1913) contrasts this picture of probabilistic sentences with subjectivist interpretations of probability, against which he stated a few objections. First, according to him it is not known how to measure beliefs or suppositions and accordingly how to calculate probability fractions. Secondly, taking probability to be a measure of certainty would mean that it changes from individual to individual and from circumstances to circumstances, which conflicts with the requirement that exactly one probability value is assigned to any sentence. Thirdly, he holds it evident that the probability calculus has nothing in common with degrees of certainty and psychic events, since otherwise "[ ... ] the entire theory of probability would have to be relegated to a psychological laboratory [ ... ]" (Łukasiewicz 1913, p. 37). What makes a present reading of Łukasiewicz annoying is the subsequent development of both the probability calculus and its subjectivist interpretation. To express this uneasiness in a



*K. Kijania-Placek and J. Woleński (eds.),*
*The Lvov–Warsaw School and Contemporary Philosophy,* 337–340.
© 1998 *Kluwer Academic Publishers.*

slogan, we have today a nice subjectivist interpretation of probability that answers Łukasiewicz's worries while on the other hand, as to his theory of logical probability, it is doubtful whether it is about *probability*. Moreover it is of no use as an account of any truly probabilistic theory (e.g., quantum mechanics) and in modeling states of opinion. For these significant developments in the probability theory I take its axiomatization by Kolmogorov and a somewhat later perception that probability can be assigned to elements of *anything* that has the structure of a Boolean algebra. Also, from the point of view of the subjectivist, an important improvement was the invention of various strategies for disclosing someone's subjective probabilities (degrees of his beliefs). Thus, the subjectivist starts with a perception that our beliefs are gradable, which he construes by assigning real numbers from the interval [0, 1] to sentences of a given language. Secondly, he realizes that as long as the logical connectives of the language behave nicely, the language can be mapped onto a Boolean algebra, and accordingly, the numbers assigned to sentences can be interpreted as probabilities. Finally, he comes up with an idea of how to measure someone's degrees of beliefs, and accordingly what probabilities should be assigned to sentences. I hold all these moves as uncontroversial; what gives a sting to subjectivist probability is the idea that our changes of degrees of beliefs should be modeled by a probabilistic rule.

## 2. THE RULE OF SIMPLE CONDITIONALIZATION

A particularly nice rule is that of simple conditionalization, which is assumed to hold in its area of application by almost every subjectivist. According to this rule, given that your initial probability for $A$ is $p(A)$ and for $B$ is $p(B)$, once you verify that $B$, your new probability for $A$ is:

$$p'(A) = p(A/B) = p(A \text{ and } B)/p(B)$$

The rule is applicable in fairly limited circumstances: it is assumed that $B$ is conclusively verified (revelation model) and that there are no significant alterations of the language (no conceptual changes). In its domain of application the rule is credible for at least two reasons. Given $B$ is verified,

1) non-$B$ and all sentences entailing non-$B$ receive zero probability;
2) for any two sentences entailing $B$, the rule leaves intact the ratio of their probabilities.

Despite these nice properties, my intention is to cast doubt on the validity of this rule, seen as a means of accommodating new data. The problem I pose is: given someone is a determinist about some subject, can he appeal in his reasoning about this subject to simple conditionalisation? If the rule turns out to be unreliable, much of the appeal of the subjectivist interpretation of probability fades away. After all, an ascription of fractions to sentences in which we believe (and calling these fractions probabilities) is equally as incontestable as it is useless. In this sense, Łukasiewicz's reservations about the subjectivist interpretation of probability can to some extent be vindicated. To criticize the rule, let us consider a case in which the rule prescribes changes of degrees of beliefs which we had rather abstained from making.

## 3. THE KRAKOW-WARSAW PARADOX

Suppose you plan to drive from Krakow to Warsaw. You put some amount of gas in the tank of your car. Knowing the car and the distance, you are not sure whether the amount is sufficient to reach the destination. To be more precise, you are certain that you will get to any place between Krakow and Grójec, but your probability of getting to any place further than Grójec decreases linearly with distance, and for Warsaw it is 0.5. Now imagine that you have arrived in a place, say Janki, pretty close to Warsaw, for which your initial probability was 0.7. The question is: should you become more convinced that you will finally reach Warsaw?

## 4. IDEALIZATIONS

To clarify the discussion let us assume a couple of idealizations. Let us agree that (1) the car is a deterministic device, whose gas consumption is a linear function of distance, (2) there are no traffic jams or any other 'irregularities' on the road and (3) the car immediately stops in a place at which it has run out of gas. In spite of these idealizations, your initial probabilities still seem to make sense: your uncertainty is due to your imprecise knowledge of the fuel consumption and the amount of gas.

## 5. CONDITIONALIST'S ANSWER

This person says that the closer to Warsaw, the greater the probability of reaching it. Recollecting that the initial probability for getting to Warsaw is 0.5, for reaching Janki (close to Warsaw) is 0.7, and assuming that in order to reach Warsaw one needs first to get to Janki (and other places not mentioned here), he conditionalizes and obtains the new probability of getting to Warsaw of 5/7 (clearly, $5/7 > 0.5$).

## 6. DETERMINIST'S ANSWER

He claims that the amount of gas determines where the car stops. He adds also that the fact that one drove some distance does not say anything about whether or not this person would drive some further distance. Accordingly, he says: no mater how close to Warsaw you are, keep cool and do not be more convinced that you will finally make it.

## 7. ARE CONDITIONALIST AND DETERMINIST RATIONAL?

By the standard of the Dutch Book, they are. Consider their new sets of probabilities which they accept in Janki. View these sets of probabilities as the sets of bets which they accept. Their sets of bets do not sabotage themselves. The result is independent of whether the sets are finite, countable, uncountable. (The bet has the form 'I will reach the place X'.)

## 8. DOES THE DIFFERENCE IN THEIR DEGREES OF BELIEFS MATTER?

The answer is 'Yes'. By introducing appropriately chosen utilities, their rational behaviors should be different. For instance, suppose that on the road from Krakow to Warsaw there is only one gas station, in Janki, which sells gas for an enormously high price. Assume further that they win if the person reaches Warsaw. (The former assumption goes back to Polish realities in the late 80s: gas shortages plus rare gas stations selling gas for Western currencies.) Given that the price of gas and the win are appropriately chosen, Conditionalist should not buy gas, while Determinist should.

## 9. CONCLUSIONS

My intuitions are deterministic. Given the above idealizations, I would stay with my old degree of belief that I will make it to Warsaw. I am also more inclined to buy gas in Janki than my friend Conditionalist. I wonder whose side you will take. Will you really conditionalize?

*Institute of Philosophy, Jagiellonian University*

## REFERENCES

Łukasiewicz J.: 1913, *Die logische Grundlagen der Wahrscheinlichkeitsrechnung*, Krakow. Eng. trans. in his *Selected Works*, ed. by L. Borkowski, North-Holland Publishing Company, Amsterdam, PWN, Warsaw, 1970, pp. 16–63.

# PART VI

# LOGIC AND PHILOSOPHY

KATARZYNA KIJANIA-PLACEK

# TRUTH AS CONSENSUS. A LOGICAL ANALYSIS*

## 1. INTRODUCTION

Most philosophers of the Lvov-Warsaw School accepted the classical concept of truth, especially after Tarski's works. The only exception are the views of Edward Poznański and Aleksander Wundheiler expressed in the paper 'The concept of truth in physics', written in 1934, where they developed a conception that combined the coherence theory with the consensus theory, arguing that the concept of absolute truth has no application in physics. By saying that truth is absolute they meant that the truth of a sentence or proposition is independent of the knowing subject, of truth of other sentences and of the state of knowledge at a given time; absolute truth is a disjunctive property, which admits no degrees. Poznański and Wundheiler claimed that instead of the absolute notion of truth, in physics and in other areas of human activity where the classical notion of truth provides no means of deciding whether a sentence is true or false, we use the so called operationalist notion of truth. On that construal, truth is closely related to the methods by which we decide whether a sentence is true or false. Their proposal consists of conjoining two criteria, the coherence criterion, which, they think, applies to a physical theory as a whole, and the common agreement criterion, which applies to elementary sentences, by which they understand sentences about the spatio-temporal ordering of phenomena and about counting.

The main advantage of this approach over the classical conception of truth is, in their opinion, the effectiveness of the criteria. I will not concentrate on their arguments. What interests me is just the content of the latter criterion of truth — consensus and its logical characteristics. In this paper I shall point to some difficulties involved in applying the criterion, which suggest weakening of the concept of common agreement. Poznański and Wundheiler propose some restrictions. I will make critical remarks on their proposals and suggest a different way of amending this criterion. It consist of allowing majority agreement instead of common agreement.

My proposal consists in treating common agreement as a special case of majority agreement and analyzing the latter in the framework of first order predicate logic with the majority quantifier $Mx$ (Most$x$) added. The semantics I propose is based on Barwise and Cooper's account 'Generalized Quantifiers and Natural Language', but the concept of generalised quantifiers can be traced



*K. Kijania-Placek and J. Woleński (eds.),*
*The Lvov–Warsaw School and Contemporary Philosophy,* 343–353.
© 1998 *Kluwer Academic Publishers.*

back to another member of the Lvov-Warsaw School — Andrzej Mostowski and his 1957 paper.

## 2. CONSENSUS IN POZNAŃSKI AND WUNDHEILER

Poznański and Wundheiler point out that we can't just demand common agreement of all people; we must place some restrictions. When we say *common* we mean a community of competent people and what *competent* means depends on the content of the sentence under consideration. They attempt to treat the requirement objectively. For example, we could stipulate that if the sentence in question belongs to physics, any physicist working in the field is *competent*. But then the authors propose that we exclude those members of the community who don't meet certain additional requirements: "common agreement [ . . . ] is common agreement of normal, disinterested and competent people".[1] By abnormal they meant mentally retarded, mentally disabled, the colour blind as far as recognition of colours is concerned, people of impaired hearing as far as ability to hear voices is in question.

There is some vagueness in almost all these terms. But an utterly vague concept is that of being disinterested. Can we ever say of any community that all its members are disinterested? The paradigmatic case of Poznański and Wundheiler is the observation sentence in physics. Those who carry out an experiment must be the candidates for the community in question. How can we test whether or not they are disinterested and what degree of partiality, if any, should be accepted? These are the questions that should be answered by proponents of the consensus criterion of truth, if it is to be made manageable.

Although Poznański and Wundheiler are aware of the fact that those additional requirements make the application of the criterion more complicated, they do not suggest how to make their crucial notions, like 'normal' or 'disinterested' precise. So despite of its declared effectiveness the common agreement criterion turns out to be hard to apply.

An alternative approach is to accept those requirements which can be stated objectively and weaken the criterion by demanding majority agreement instead of common agreement. Democratic procedures and court statements could serve as examples. In what follows I shall propose two logical accounts of the consensus criterion treated as an extreme case of majority agreement criterion.

## 3. LOGICAL FRAMEWORK FOR CONSENSUS AND MAJORITY AGREEMENT

It seems that a natural proposal for a uniform treatment of those two concepts is to treat them as quantifiers: the universal quantifier, $\forall x$,[2] and the majority quantifier M$x$ (Most$x$). Our language **L(M)** results from adding an additional symbol to the extended language of first order logic ($L_{\omega_1\omega}$)

The language **L(M)** consists of:

propositional connectives: $\vee$, $\wedge$, $\neg$
individual variables: $x, y, z, x_0, \ldots$
for each $n = 1, 2, \ldots$ some set (possibly empty) of $n$-ary relation symbols $R, S, K, \ldots$; the 1-ary relation symbols are called predicate symbols
a set of distinguished predicate symbols $A_p, A_q, A_{p_1}, \ldots$
quantifier symbols: $\mathrm{M}x, \forall x, \exists x$
parentheses: (, ) and brackets: [, ]

Formation rules:

R1. Any predicate symbol is a set term.
R2. If $\varphi$ is a formula and $x$ is a variable, then $x[\varphi]$ is a set term.
We write just $\varphi$ and $(\varphi \wedge \psi), (\varphi \vee \psi), (\neg\varphi)$ for the set terms to mean $x[\varphi]$, $x[\varphi \wedge \psi], x[\varphi \vee \psi], x[\neg\varphi]$ respectively, if no confusion is likely.
R3. If $R$ is an $n$-ary relation symbol and $t_1, \ldots, t_n$ are variables, then $R(t_1, \ldots, t_n)$ is a formula. If $\varphi$ is a set term and $t$ is a variable, then $\varphi(t)$ is a formula.
R4. If $R$ is an $n$-ary relation symbol, $t_1, \ldots, t_n$ are variables, $Qx_1, \ldots, Qx_n$ are quantifiers and $Qx_i \in \{\exists x_i, \forall x_i, \mathrm{M}x_i\}$ for $i = 1, \ldots, n$, then $Qx_1, \ldots, Qx_n R(t_1, \ldots, t_n)$ is a formula, and $Qx_2, \ldots, Qx_n R(t_1, \ldots, t_n)$ is a set term.
In particular, if $\varphi$ is a set term, then $\forall x\varphi, \exists x\varphi, \mathrm{M}x\varphi$, are formulas.
R5. The formulas are closed under the propositional connectives: $\vee$, $\wedge$, $\neg$.

Abbreviations: if $\varphi, \psi$ are formulas, then
$(\varphi \rightarrow \psi) =\mathrm{df}\ (\neg\varphi \vee \psi)$
$\mathrm{FOR}_{\mathrm{L(M)}}$ – set of formulas of **L(M)**
$\mathrm{SR}_{\mathrm{L(M)}}$ – set of relation symbols and set terms of **L(M)**

### 3.1. *A Classical Approach*

If we follow the classical approach to the semantics of generalized quantifiers, for example the Barwise and Cooper account, it would look as follows:

A model $\mathfrak{M}$ for the language **L(M)** is an ordered pair $\langle \mathrm{E}, I \rangle$ where E is some non-empty set such that $|\mathrm{E}| < \aleph_0$ and $I$ is a function which assigns interpretations to expressions of the language as follows:[3]

$I : \mathbf{L}(\mathbf{M}) \rightarrow\ \mathrm{E} \cup \bigcup_n \wp(\mathrm{E}^n) \cup \wp(\wp(\mathrm{E})) \cup \{1, 0\}$, where $\wp(Y)$ is the power set of Y

S1. If $t$ is a variable then $I(t) \in \mathrm{E}$.

S2. If $R$ is an $n$-ary relation symbol then $I(R) \subseteq \mathrm{E}^n$.
If $U$ is a set term then $I(U) \subseteq \mathrm{E}$. In particular, if $\varphi, \psi$ are formulas, then $x[\varphi \vee \psi], x[\varphi \wedge \psi], x[\neg\varphi]$ are set terms and
$I(x[\varphi \vee \psi]) = I(x[\varphi]) \cup I(x[\psi])$
$I(x[\varphi \wedge \psi]) = I(x[\varphi]) \cap I(x[\psi])$
$I(x[\neg\varphi]) = \mathrm{E} \setminus I(x[\varphi])$

S3. $I(\forall x) = \{\mathrm{E}\}$
$I(\exists x) = \{\mathrm{X} \subseteq \mathrm{E} : \mathrm{X} \neq \emptyset\}$
$I(\mathrm{M}x) = \{\mathrm{X} \subseteq \mathrm{E} : |\mathrm{X}| > |\mathrm{E} \setminus \mathrm{X}|\}$

S4. If $R$ is an $n$-ary relation symbol, then
$I(R(t_1, \dots, t_n)) = 1$ iff $\langle I(t_1), \dots, I(t_n)\rangle \in I(R)$
$I(R(t_1, \dots, t_n)) = 0$ iff $\langle I(t_1), \dots, I(t_n)\rangle \notin I(R)$
if $\varphi$ is a set term, then
$I(\varphi(t)) = 1$ iff $I(t) \in I(\varphi)$
$I(\varphi(t)) = 0$ iff $I(t) \notin I(\varphi)$

S5. If $\varphi$ is a set term and $\mathrm{Q}x$ is a quantifier, then
$I(\mathrm{Q}x\varphi) = 1$ iff $I(\varphi) \in I(\mathrm{Q}x)$
$I(\mathrm{Q}x\varphi) = 0$ iff $I(\varphi) \notin I(\mathrm{Q}x)$
in particular:
$I(\forall x\varphi) = 1$ iff $I(\varphi) = \mathrm{E}$
$I(\forall x\varphi) = 0$ iff $I(\varphi) \neq \mathrm{E}$
$I(\exists x\varphi) = 1$ iff $I(\varphi) \neq \emptyset$
$I(\exists x\varphi) = 0$ iff $I(\varphi) = \emptyset$
$I(\mathrm{M}x\varphi) = 1$ iff $| I(\varphi) | > | \mathrm{E} \setminus I(\varphi) |$
$I(\mathrm{M}x\varphi) = 0$ iff $| I(\varphi) | \leq | \mathrm{E} \setminus I(\varphi) |$

S6. If $\varphi, \psi$ are formulas, then
$I(\varphi \wedge \psi) = 1$ iff $I(\varphi) = I(\psi) = 1$
$I(\varphi \wedge \psi) = 0$ otherwise
$I(\varphi \vee \psi) = 0$ iff $I(\varphi) = I(\psi) = 0$
$I(\varphi \vee \psi) = 1$ otherwise
$I(\neg\varphi) = 1$ iff $I(\varphi) = 0$
$I(\neg\varphi) = 0$ iff $I(\varphi) = 1$

Distinguished predicate symbols $A_p, A_q, A_{p_1}, \dots$ are interpreted so that $A_p(x)$ means: an individual $x$ asserts a sentence $p$. Taking $\mathrm{M}x A_p(x)$ to mean that the majority of $x$s assert the sentence $p$, I can state the truth criterion as follows:

A sentence $p$ is true in a model $\mathfrak{M}$ if $I(\mathrm{M}x A_p(x)) = 1$;
it is false if $I(\mathrm{M}x A_p(x)) = 0$

For the formal account to be adequate, it must be in accord with our intuitions concerning assertions that, I assume, are strong enough. One of them is the conviction, at least my conviction, that it is not so that for every sentence we are ready to commit ourselves to the assertion of either that sentence or of its negation, there being cases in which we remain undecided. Is the approach stated above in accord with this intuition? Let's take a sentence $A_p(x) \to A_q(x)$, whose intended meaning is: 'an individual $x$ asserts the sentence $p \to q$', and a model in which $I(x)$ is neither in $I(A_p)$ nor in $I(A_q)$

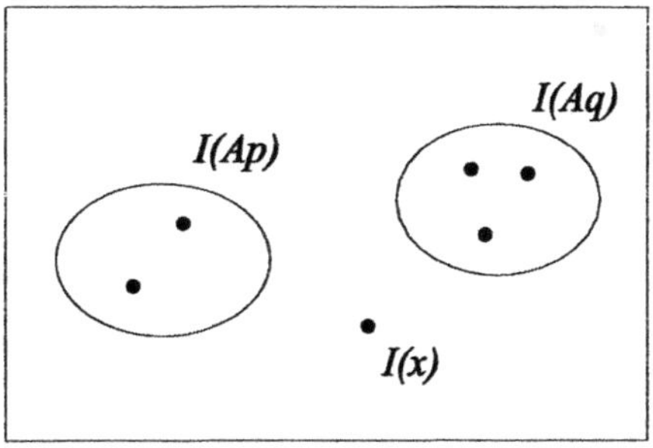


Under classical interpretation $I(A_p(x) \to A_q(x)) = 1$ because $x$ is automatically counted as rejecting $p$. In other words, the classical interpretation forces us to assume that every individual has something to say about every sentence — asserts or rejects it.

I found this consequence hardly acceptable and propose an alternative approach, which consists in treating $A_p, A_q, A_{p_1}, \ldots$ as partial predicates and providing a semantics based on partial models.

### 3.2. *Partial models*

A model $\mathfrak{M}$ for the language **L(M)** is an ordered quadruple $\langle \mathrm{E}, I, p_1, p_2 \rangle$ where E is some non-empty set such that $|\mathrm{E}| < \aleph_0$ and $I, p_1, p_2$ are functions which assign interpretations to expressions of the language as follows:[4]

$$I : \mathbf{L(M)} \to \mathrm{E} \cup \bigcup_n (\wp(\mathrm{E}^n) \times \wp(\mathrm{E}^n)) \cup (\wp(\wp(\mathrm{E})) \times \wp(\wp(\mathrm{E}))) \cup \{1, 0\}$$
$$p_1, p_2 : \mathrm{SR}_{\mathrm{L(M)}} \to \mathrm{E} \cup \bigcup_n \wp(\mathrm{E}^n) \cup \wp(\wp(\mathrm{E}))$$

S1. If $t$ is a variable then $I(t) \in \mathrm{E}$.

S2. If $R$ is an $n$-ary relation symbol then $I(R) = \langle \bar{R}, \bar{R}' \rangle$, where $\bar{R} \subseteq \mathrm{E}^n$, $\bar{R}' \subseteq \mathrm{E}^n \setminus \bar{R}$,
$p_1(R) = \bar{R}, \qquad p_2(R) = \bar{R}'$
If $U$ is a set term then $I(U) = \langle \bar{U}, \bar{U}' \rangle$, where $\bar{U} \subseteq \mathrm{E}$, $\bar{U}' \subseteq \mathrm{E} \setminus \bar{U}$,
$p_1(U) = \bar{U}, \qquad p_2(U) = \bar{U}'$.
In particular, if $\varphi, \psi$ are formulas, then $x[\varphi \vee \psi], x[\varphi \wedge \psi], x[\neg\varphi]$ are set terms and

$p_1(x[\varphi \vee \psi]) = p_1(x[\varphi]) \cup p_1(x[\psi])$
$p_2(x[\varphi \vee \psi]) = p_2(x[\varphi]) \cap p_2(x[\psi])$
$p_1(x[\varphi \wedge \psi]) = p_1(x[\varphi]) \cap p_1(x[\psi])$
$p_2(x[\varphi \wedge \psi]) = p_2(x[\varphi]) \cup p_2(x[\psi])$
$p_1(x[\neg\varphi]) = p_2(x[\varphi])$
$p_2(x[\neg\varphi]) = p_1(x[\varphi])$

S3. $I(\forall x) = \langle \{\mathrm{E}\}, \wp(\mathrm{E}) \setminus \{\emptyset\} \rangle$
$p_1(\forall x) = \{\mathrm{E}\} \qquad p_2(\forall x) = \wp(\mathrm{E}) \setminus \{\emptyset\}$
$I(\exists x) = \langle \wp(\mathrm{E}) \setminus \{\emptyset\}, \{\mathrm{E}\} \rangle$
$p_1(\exists x) = \wp(\mathrm{E}) \setminus \{\emptyset\} \qquad p_2(\exists x) = \{\mathrm{E}\}$
$I(\mathrm{M}x) = \langle \{\mathrm{X} \subseteq \mathrm{E} : |\mathrm{X}| > |\mathrm{E} \setminus \mathrm{X}|\}, \{\mathrm{X} \subseteq \mathrm{E} : |\mathrm{X}| > |\mathrm{E} \setminus \mathrm{X}|\} \rangle$
$p_1(\mathrm{M}x) = p_2(\mathrm{M}x) = \{\mathrm{X} \subseteq \mathrm{E} : |\mathrm{X}| > |\mathrm{E} \setminus \mathrm{X}|\}$

S4. If $R$ is an $n$-ary relation symbol, then
$I(R(t_1, \ldots, t_n)) = 1$ iff $\langle I(t_1), \ldots, I(t_n) \rangle \in p_1(R)$
$I(R(t_1, \ldots, t_n)) = 0$ iff $\langle I(t_1), \ldots, I(t_n) \rangle \in p_2(R)$
if $\varphi$ is a set term, then
$I(\varphi(t)) = 1$ iff $I(t) \in p_1(\varphi)$
$I(\varphi(t)) = 0$ iff $I(t) \in p_2(\varphi)$

S5. If $\varphi$ is a set term and $\mathrm{Q}x$ is a quantifier, then
$I(\mathrm{Q}x\varphi) = 1$ iff $p_1(\varphi) \in p_1(\mathrm{Q}x)$
$I(\mathrm{Q}x\varphi) = 0$ iff $p_2(\varphi) \in p_2(\mathrm{Q}x)$
in particular:
$I(\forall x\varphi) = 1$ iff $p_1(\varphi) = \mathrm{E}$
$I(\forall x\varphi) = 0$ iff $p_2(\varphi) \neq \emptyset$
$I(\exists x\varphi) = 1$ iff $p_1(\varphi) \neq \emptyset$
$I(\exists x\varphi) = 0$ iff $p_2(\varphi) = \mathrm{E}$
$I(\mathrm{M}x\varphi) = 1$ iff $| p_1(\varphi) | > | \mathrm{E} \setminus p_1(\varphi) |$
$I(\mathrm{M}x\varphi) = 0$ iff $| p_2(\varphi) | > | \mathrm{E} \setminus p_2(\varphi) |$

S6. If $\varphi, \psi$ are formulas, then
$I(\varphi \wedge \psi) = 1$ iff $I(\varphi) = I(\psi) = 1$
$I(\varphi \wedge \psi) = 0$ otherwise;
$I(\varphi \vee \psi) = 0$ iff $I(\varphi) = I(\psi) = 0$
$I(\varphi \vee \psi) = 1$ otherwise;
$I(\neg\varphi) = 1$ iff $I(\varphi) = 0$
$I(\neg\varphi) = 0$ iff $I(\varphi) = 1$

In partial semantics we can account for the fact that $x$ neither asserts nor rejects any of the sentences, $p$ and $q$ in the example stated in Section 3.1. $I(A_p(x) \to A_q(x))$ will turn out undefined in any model in which $I(x) \notin \bar{A}_p{}'$, $I(x) \notin \bar{A}_q$ and $I(x) \notin (\bar{A}_p \cap \bar{A}_q{}')$ because for it to be true $I(x)$ would have either to assert $q$ or reject $p$, while for it to be false $I(x)$ would have to assert $p$ and reject $q$ at the same time, which of course is just not the case in the models described above. Here is an example of such a model:

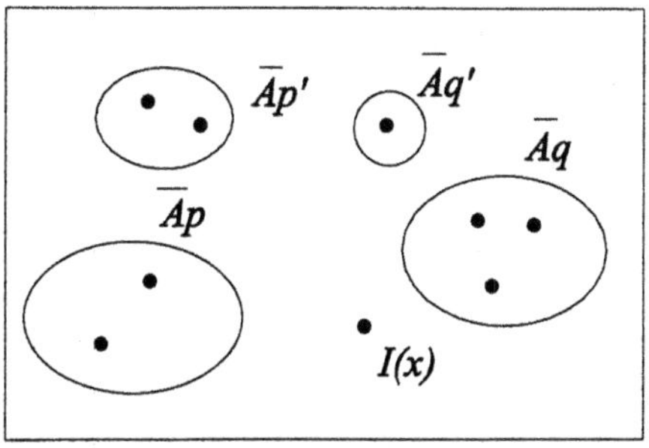


What is more, classically all sentences about which nobody has ever thought turn out to be true and, consequently, their negations false, because if $I(A_p) = \emptyset$ then $I(A_p(x)) = 0$ for all $x$, and hence for most $x$. In partial models, however, if we account for a sentence $p$ that was never thought of by anybody then $I(A_p) = \langle\emptyset, \emptyset\rangle, I(A_p(x))$ is undefined and so is $\mathrm{M}xA_p(x)$, leaving $p$ neither true nor false.

## 4. TAUTOLOGIES AND VALIDITY

The problem we face in analyzing majority agreement in terms of partial models is that not all classical tautologies are valid under this interpretation as stated. If we consider a sentence $P(x) \vee \neg P(x)$ we are likely to expect that it is assigned 1 in any model since it expresses the principle of the excluded middle, but it is not if $I(x)$ is neither in $\bar{P}$ nor in $\bar{P}'$, as the following example shows:

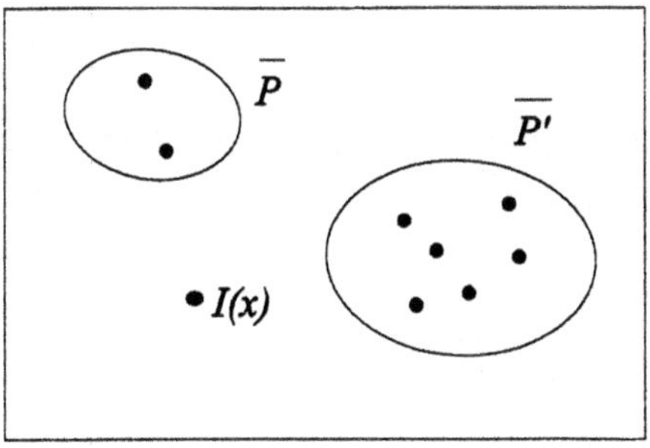


This suggests that the definition of a partial model can't be taken at face value and should be supported by a supervaluational approach. To this end some definitions will be needed:

DEF. A valuation $I'$ is an extension of a valuation $I$ ($I \subseteq I'$) iff for all $\varphi \in \mathrm{FOR}_{\mathrm{L(M)}}$, if $I(\varphi) = 1$ then $I'(\varphi) = 1$ and if $I(\varphi) = 0$ then $I'(\varphi) = 0$.

DEF. A valuation $I$ is total iff for all relation symbols $R$ of **L(M)**, $I(R) = \langle \bar{R}, \mathrm{E} \setminus \bar{R} \rangle$.

DEF. Let $\mathfrak{M} = \langle \mathrm{E}, I, p_1, p_2 \rangle$ be a model of **L(M)** and $\varphi \in \mathrm{FOR}_{\mathrm{L(M)}}$. A function $I_s : \mathrm{FOR}_{\mathrm{L(M)}} \to \{0,1\}$ is called a supervaluation if $I_s(\varphi) = 1$ iff for all total $I'$ such that $I \subseteq I'$, $I'(\varphi) = 1$ and $I_s(\varphi) = 0$ iff for all total $I'$ such that $I \subseteq I'$, $I'(\varphi) = 0$.

DEF. $\varphi \in \mathrm{FOR}_{\mathrm{L(M)}}$ is valid iff $I(\varphi) = 1$ for all total $I$.

Returning to the last example of $I(x) \notin \bar{A_p}'$, $I(x) \notin \bar{A_q}$, $I(x) \notin (\bar{A_p} \cap \bar{A_q}')$ we can say now that the valuation $I$ may be enlarged in many ways to total valuations. No matter how the process would proceed, however, under any total valuation $I'$, $I'(P(x) \vee \neg P(x)) = 1$.

As we probably want the tautologies to be true under the majority agreement criterion, we are forced to amend it accordingly. The question about the true-value of a sentence $p \vee \neg p$ boils down to the question of whether $I(\mathrm{M}x(A_p(x) \vee \neg A_p(x)) = 1$. That is so only if $|\bar{A_p} \cup \bar{A_p}'| > |\mathrm{E} \setminus (\bar{A_p} \cup \bar{A_p}')|$ and, of course, it is not always the case. Here is a counterexample:

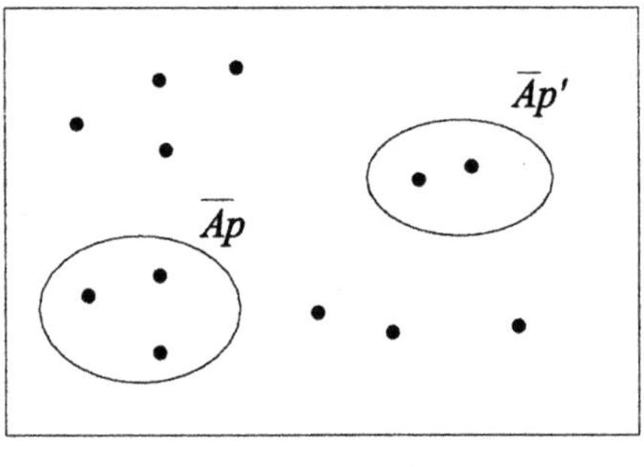


That requirement is fulfilled, however, for any total valuation because in this case $\bar{A_p} \cup \bar{A_p}' = \mathrm{E}$ and $|\mathrm{E} \setminus (\bar{A_p} \cup \bar{A_p}')| = 0$, the consequence being that $I_s(\mathrm{M}x(A_p(x) \vee \neg A_p(x)) = 1$. To be in accord with the intuitive truthfulness of tautologies, I propose to amend the formulation of the majority agreement criterion in the following way:

A sentence $p$ is true in a model $\mathfrak{M}$ if $I(\mathrm{M}x A_p(x)) = 1$ or $I_s(\mathrm{M}x A_p(x)) = 1$;
it is false if $I(\mathrm{M}x A_p(x)) = 0$ or $I_s(\mathrm{M}x A_p(x)) = 0$

Counterintuitive as it may appear at first sight, this formulation may be read as stating that a sentence is true if the majority of a population asserts it or, in the case of compound sentences, would assert it if they only thought of its compounds no matter whether asserting or rejecting them.

## 5. ADEQUACY AND BIVALENCE

So isn't it now the case that supervaluations bring us back to classical logic? In a sense 'yes' and in a sense 'no'; by the supervaluations technique all tautologies are regained. It can also be shown that the criterion meets the minimal test of adequacy stating the equivalence of falsity and truth-of-negation ($\neg p$ is true iff $p$ is false) namely $I(\mathrm{M}x\neg A_p(x)) = 1$ iff $I(\mathrm{M}xA_p(x)) = 0$. It is so because $I(\mathrm{M}x\neg A_p(x)) = 1$ iff $\mid p_1(\neg A_p) \mid > \mid \mathrm{E} \setminus p_1(\neg A_p) \mid$. Since $p_1(\neg A_p) = p_2(A_p)$, it is so iff $\mid p_2(A_p) \mid > \mid \mathrm{E} \setminus p_2(A_p) \mid$ iff $I(\mathrm{M}xA_p(x)) = 0$.

What is not classical is the fact that majority agreement is not bivalent: it is not the case that every sentence is either true or false under this criterion. And here even supervaluations would not help. First of all, for any model whose cardinality is odd there is a total valuation such that $\mid \bar{A}_p \mid = \mid \bar{A}_p{}' \mid$ under it, which renders $I_s(\mathrm{M}xA_p(x))$ undefined. Secondly, and this argument applies for models of arbitrary cardinality, for any partial valuation $I$ under which the sentence $I(\mathrm{M}xA_p(x))$ is undefined there are total valuations ascribing 0 to it as well as those ascribing 1 to it, which is exactly the condition under which $I_s(\mathrm{M}xA_p(x))$ is undefined. We can illustrate this by the example discussed above

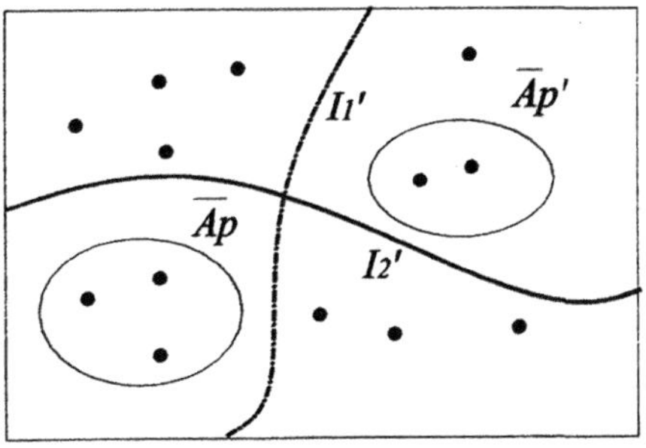


## 6. NONMONOTONICITY

The next nonclassical feature of the majority agreement criterion is its nonmonotonicity. Namely, from the fact that $p \to q$ is true under the criterion and so is $p$ we can not conclude that $q$ is true (from $I(\mathrm{M}x(A_p(x) \to A_q(x)) = 1$ and $I(\mathrm{M}xA_p(x)) = 1$ it does not necessarily follow that $I(\mathrm{M}xA_q(x)) = 1$). It could happen that although the majority of a population asserts $p \to q$ and the majority of the population asserts $p$, but does not assert $q$, as the following example shows.

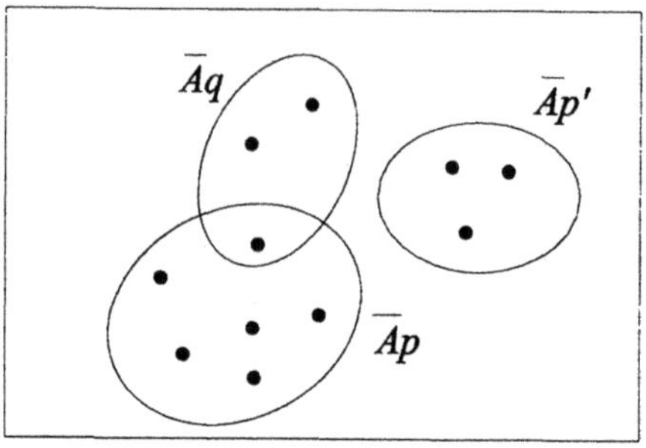


The same applies to the conjunctive version of monotonicity. Namely, from $I(\mathrm{M}xA_p(x)) = 1$ and $I(\mathrm{M}xA_q(x)) = 1$ it does not necessarily follow that $I(\mathrm{M}xA_p(x) \wedge A_q(x)) = 1$.

## 7. CONSENSUS - THE EXTREME CASE

The general quantifier as interpreted in section 3.2. could not serve the purpose of defining the consensus criterion as an extreme case of majority agreement because its behaviour is classical. To this end we would have to add a symbol, $\Pi x$, to the language and stipulate the semantics as follows: $I(\Pi x\psi) = 1$ iff $p_1(\psi) = \mathrm{E}$, $I(\Pi x\psi) = 0$ iff $p_2(\psi) = \mathrm{E}$. Under this constraint only those sentences that are asserted unanimously would be labeled true and only those rejected by everybody would be labeled false. In this way the class of true sentences under the consensus criterion would form a subclass of the true sentences under the majority criterion.

## 8. CONCLUSION

The main advantage of the consensus criterion of truth over the theory of absolute truth was said to be its effectiveness. To clarify what should count as common agreement we can follow two alternative approaches. The first is to state requirements that one must meet in order to be a member of a community to which total agreement is relativised. In this approach, however, the notorious vagueness of the proposed requirements reduces their effectiveness, which was supposed to be the theory's main virtue. The alternative proposal is to assume only those requirements which are not vague and demand majority agreement instead of common agreement. After presenting two approaches to the logical analysis of the majority agreement and consensus criteria I have argued in favour of the second approach on the basis of agreement with our intuitions concerning assertions. What we definitely lose is bivalence, which in the first approach can be preserved under certain assumptions.[5] In both cases, however, truth turns out not to be monotonic.

*Institute of Philosophy, Jagiellonian University*

## NOTES

* During the preparation of this paper the author was supported by *The Ryoichi Sasakawa Young Leaders Fellowship Fund.*

[1] See Poznański, Wundheiler (1934), p. 429.

[2] In fact it is adequate only in the framework of classical models. See Section 7.

[3] $I$ is also called a valuation in what follows.

[4] This definition was published in my paper 'Partial Models for Consensus', in P. 'Weingartner, G. Schurz, G. Dorn, (eds.), *The Role of Pragmatics in Contemporary Philosophy: Contributions of the Austrian Ludwig Wittgenstein Society*, Kirchberg, 1997.

[5] In fact it is adequate only in the framework of classical models. See Section 7.

[6] In models of even cardinality.

## REFERENCES

Barwise, J., Cooper, R.: 1981, 'Generalized Quantifiers and Natural Language', *Linguistics and Philosophy* 4, pp. 159–219.

Mostowski, A.: 1957, 'On a generalisation of quantifiers', *Fundamenta Mathematicae* **44**, pp. 12–36.

Poznański, E., Wundheiler A.: 1934, 'Pojęcie prawdy na terenie fizyki' (The Concept of Truth in Physics), in Pawłowski, T. (ed.), *Logiczna teoria nauki* (The logical Theory of Science), PWN, Warsaw, 1966, pp. 399–448.

AVENIR I. UYEMOV

# THE LVOV-WARSAW SCHOOL AND THE PROBLEM OF A LOGICAL FORMALISM FOR GENERAL SYSTEMS THEORY

The value of philosophy depends not only on the answers it gives to the poignant questions of our life. Methods elaborated by philosophy for sciences able to meet those questions are more important.

The Lvov-Warsaw School is famous mainly because of its contribution to the logic and methodology of science. Ideas which were discussed by the philosophers of the Lvov-Warsaw School are now foundations of scientific theories, namely, Praxeology and Parametric General Systems Theory.

The difficulties in the development of General Systems Theories (GST) have been connected with specific features of their mathematical formalisms. A systems theory becomes a general one if it can be applied to any system. But this is impossible for actual systems theories because of their mathematics.

M. Mesarovic and Y. Takahara were right when they wrote that for real, complicated events the specific language of classical theories which are based on concrete mathematical structures, such as differential equations, arithmetic or abstract algebras, etc., does not allow an adequate description of reality (Mesarovic and Takahara, 1975). Mesarovic and Takahara themselves tackled the problem by choosing a minimal mathematical structure and adding to it new mathematical structures which are necessary for the exploration of various system properties. From our point of view this method has two flaws. The first is the lack of integrity of the mathematical apparatus. The second is even more important. The minimal mathematical structure, according to Mesarovich and Takahara, is a set-theoretical one. The system's concept itself is defined as the type of a set.

However there is a well justified opinion that set-theoretical and system-theoretical approaches are opposite to one another in character (Screider and Scarov 1982, Kantorovic and Plisco, 1983). The main opposition is that of the priority of elements in set theory to the priority of the whole in the system's theory. Let us take a simple example. If we have the row 1, 2, 3, 5, is it correct to ask what numeral is omitted? This question is valid only from the system-theoretical point of view. The set-theoretical approach makes it senseless.

For this reason it is necessary to search for another minimal formal structure for a true GST. This structure must not be less general than the set-theoretical one. We can find it with the help of the categories of philosophy.

Aristotle lists ten categories in his treatise of the same name. The most important of them were singled out in *Metaphysics*: Substance (*Ousia*), Property,



*K. Kijania-Placek and J. Woleński (eds.),*
*The Lvov–Warsaw School and Contemporary Philosophy,* 355–363.
© 1998 *Kluwer Academic Publishers.*

Relation (1089b, 23–24). The Greek '*ousia*' can be interpreted as 'Entity', 'Thing', 'Object'. Therefore we can identify the fundamental triad of Aristotelian categories as the triad 'Thing, Property, Relation.'

Modern mathematical logic reduces these categories to the categories of set and elements. Things are elements, properties are sets of elements, relations are sets of couples, three-tuples and, in general, $n$-tuples. Is this reduction necessary for a mathematical treatment of those categories? The answer depends on the results of philosophical analysis. There was such an analysis in Scholasticism. But in modern philosophy, especially in the categorical systems of Kant and Hegel, the categories of thing, property, relation were not central.

By contrast, in the Lvov-Warsaw School the speculations were mainly connected with the problem of the triad relation thing, property. We do not want to say that there was one true solution of that problem, namely, that of the Lvov-Warsaw School. There were no solutions common for all representatives of that School. Quite the opposite: the Lvov-Warsaw School is characterised by a rich variety of viewpoints. Therefore we can try to choose some of them which are more founded and most appropriate for finding the minimal formal structure for GST. But, if we cannot find a suitable viewpoint, the critical analysis of various positions will be very useful for solving our problem.

We can find the 'Thing-Property-Relation' triad in the works of the founder of the Lvov-Warsaw School, Twardowski. He applied to any thing (*przedmiot*) the Aristotelian opposition of form and matter (Ingarden, 1966). The form of a thing is connected with its property. Any thing is a complicated whole and consists of parts. Every part has a relation to the other parts and to the whole. So we have received the triad: thing, property, relation. But the concepts of this triad are not here completely primary. To understand them it is necessary to know such concepts as form, matter, part, whole and even complexity. Twardowski generalised the concept of 'part.' A part of a thing can be a relation. Such a part is called a formal part of a thing. Relations as formal parts exist between parts of a thing (material or formal) or between parts and whole. A very important idea of Twardowski's, usually associated with Meinong, is the distinction between the problems of things and their parts and the problems of their existence.

Twardowski's student Marian Borowski gave a very detailed analysis of the category 'thing' (*przedmiot*) (Borowski, 1921). He insisted that it is necessary to have a term that denotes everything. Polish has such a term — *przedmiot* which corresponds to *ens vel res* in Latin. Using the results of Scholastic philosophy Borowski listed nine properties of every thing:

A. A thing is *praedicatum universale*. Everything is a thing.
B. A thing is everything about which it is possible to predicate something.

C. Each thing is something. It is not nothing.
D. Each thing has the property of unity.
E. Each thing is distinguishable from another. It has individuality. *A non est non A.*
F. Each thing does not differ from itself. It has the property of identity. $A = A$.
G. Each thing has its own nature, its *quid.*
H. *Complurаritas.* Each thing is one among many.
I. *Connexitas.* Each thing is an element of some integrity.

Borowski added: Each thing is a possible correlate of consciousness. This property is analysed in detail.

Inasmuch as in the definitions listed above we have as definiendum things and as definienens properties we can name, according to Borowski, things as properties. *Vice versa,* it is evident that Borowski's conception implies that properties of things are themselves things. Analogously, relations of things are themselves things. Both properties and relations satisfy all the above clauses in the definition of thing.

Does this mean that the categories thing, property and relation are identical to each other? No, the difference between those categories may have not an ontological but a contextual character, similar to the difference between subject and predicate in grammar. That which acts as a thing in one context may be a property or relation in another. A property, in turn, may be a thing. According to our point of view a property may be a relation as well. A relation may be a thing or a property. This fully corresponds to what takes place in natural languages. When saying something about love, we have in mind love as a thing. Characterising a certain feeling as love, we understand love in the sense of a property. And, finally, love may appear as a relation between people.

Now we can find minimal formal structures which are essentially distinct from the set-theoretical one. Instead of set and element we have thing and property, thing and relation. The first structure we have in Aristotelian logic, the second in the so called 'logic of relations' and in modern predicate logic. But Aristotelian logic and the logic of relations are not sufficiently formal for the satisfaction of modern criteria of logical strictness. The predicate logic is constructed on a set-theoretical base. Properties and relations are reduced to the corresponding classes.

Is it possible to create a rather strict logico-mathematical system on the basis of the understanding of the category of thing developed by Twardowski and Borowski? The author hopes to construct such a system as a formal apparatus for GST. Considering that thing, property, relation are primitive concepts of our formal system, it is called 'The Language of Ternary Description' or

'The Ternary Description Language' (TDL) (Uyemov, 1975, 1979, 1984, 1987, 1995; Leonenko, 1987). We can here give only a very brief characterisation of TDL.

The contextual character of the distinction between the categories thing-property-relation leads to the positional principle of denoting these categories in the formal system. As distinct from predicate logic where things, properties, relations are denoted with the help of different Latin letters, in our formal apparatus the categorial differences are mapped by various places of symbols in relation to the other symbols.

The contextual interpretation of the relationships between the categories under consideration presupposes criteria for distinguishing properties and relations which are essentially different from those used in predicate logic. Instead of a predicate's arity, the distinctive nature of the relationships of properties and relations to things are put in the forefront: attributing certain properties to things, we conserve the thing's identity, establishing a relation we cross a borderline of the thing.

The categories of 'Thing', 'Property' and 'Relation' cannot be defined independently of each other. Each of them is definable through the other in three different ways. Nine various definitions are therefore possible (Uyomov, 1965). Each of them is circular, because it is impossible to avoid circularity when defining categories.

In the construction of TDL we have used not only the triad of thing, property, relation but also another triad : definite, indefinite, arbitrary. This triad has not been exploited theoretically, in spite of its great practical importance. We use that triad at the very beginning of algebra. When we write the simple equation $y = x + 5$ we have an arbitrary number $x$, an indefinite number $y$ and a definite one, 5. The usual treatment of this situation is connected with the use of the categories of set and quantifier. But it is possible to interpret them in an entirely different manner. In this case 'a definite number', 'an indefinite number' and 'an arbitrary number' are names of the three various objects.

With the help of TDL well formed formulae it is possible to express the concept of system and the values of a system's parameters (Uyemov, 1978). At present there is a wide sphere of application of TDL as a kind of deviant logic in parametric GST (Leonenko, 1984; Ujemow, 1989; Uyemov, 1997), in particular for measuring complexity (Mamtchur, Ovtcinnikov and Uyemov, 1989) and integrity (Uyemov, 1996) and in the theory of analogies (Uyemov and Sarayeva, 1992). Of course TDL is not free of criticism. The most essential criticism is connected with categorial basis of TDL. We can find the criticism of that categorial basis in the Lvov-Warsaw School, as well as support for it.

There were two types of such criticism. One line was made by Roman Ingarden, the other of Tadeusz Kotarbiński. Ingarden developed his conception of things, properties and relations in two big volumes entitled *Spór o istnienie świata* (The Controversy About the Existence of the World) in 1947–1948 and in 1961–1962. (Ingarden, 1962, 1961).

According to him genuine reality consists of self dependent, individual, timely determined things. Universalia, e.g. 'triangle', 'man' etc., qualities e.g. 'redness', 'colour' etc. are not, in general, real things (Ingarden, 1962, pp. 354–355). Things are constructive integrities (Ingarden, 1962, p. 373). They have properties. They are substances of properties. But properties are not substances of other properties. They have no properties in turn. All properties of properties are in reality properties of things.

A property of something is not a thing - *non est ens*, but *est entis* (Ingarden, 1962, p. 389). The grounds for this point of view are references to intuition and a fear of *regressus in infinitum* (Ingarden, 1962, p. 392).

Of course, we often speak about properties of properties, e.g. about a blue colour as a dark blue colour. But, according to Ingarden, in such cases we do not have a property of properties. When we speak about the colour of a flower as a blue colour, and after that the blue colour as dark blue, in reality we determine the colour of the flower more precisely. Therefore a property can not have another property but it can have a more precise determination as a property.

The other case of apparent properties of properties is the following: essential-non essential, relative-unrelative, constant-variable, simple-complex, etc. Ingarden insisted that such properties are not properties at all. They are forms or modes of existence of properties. Ingarden narrowed the concept of 'property' by eliminating from it not only forms but also 'constitutive nature' (*konstytutywna natura*), existence of a thing, etc. As a result he had to even reformulate the law of contradiction.

The concept of 'thing' is narrowed as well. It was distinguished from 'totality'. The last is a sum of 'parts' (*całość sumatywna*). Parts are existentially stronger than a sum of them. They can exist outside of a total thing. By contrast, properties are existentially weaker than their thing. Things can exist without a property. They have only been changed in such a case. Therefore it is wrong to identify a self-dependent individual thing with a sum of parts. 'Całość sumatywna' is not a full thing (*nie jest pełnym przedmiotem* (Ingarden, 1962, p. 408)), it is only a scheme of a thing (*jest schematem przedmiotowym*).

The definition of reism given by Kotarbiński is the following: "… every object is a body or a soul (or a combination of these components). Instead of 'a body or a soul' we shall say, more briefly, 'a thing', so that every object is a thing. Now, all those who believe that every object is a thing we call 'reist'; speaking more loosely: the view that every object is a thing we call 'reism' " (Kotarbiński, 1955, p. 488).

The sequences of this thesis are very similar to Ingarden's ideas: "Since every object is a thing, and since, therefore, only things exist, it is not the case that any object is a property, or a relation, or a fact" (Kotarbiński, 1955, p. 489).

Kotarbiński denied the use of properties and relations, even as intentional objects. Sentences like 'John is a tailor', 'Some rocks are crumbly' with an intentional interpretation of the words 'property' and 'relation' are meaningful, but they are false in such sentences as 'There are werewolves' and 'Certain objects are Pegasi'.

As regards expressions like 'The relation of brotherhood is symmetrical' or 'any two objects have common properties', they have no meaning, just as, for instance, the expressions 'there are because', 'certain objects are sees' (Kotarbiński, 1955, p. 491). The difference consists in that the words 'because' and 'sees' are not names, while 'properties' and 'relations', while not names, look like names. They are not names but 'nameoids'.

What is the relation of the outlined theory of a construction of a formal logic-mathematical apparatus? Kotarbiński wrote that it was not necessary to mould special formal-logical means for reism. These are ready made in the 'ontology' of Stanisław Leśniewski (Kotarbiński, 1961, p. 505). Leśniewski rejected J.S. Mill's conception of a singular proposition, in which the subject denotes a certain thing, but the predicate denotes only connotation of that subject. Instead, he understood the true singular proposition as an identification of the thing which was denoted by a subject with the thing which was denoted by a predicate. Certainly it is in accordance with the ideas of Kotarbiński, but it is not a peculiar feature of the reistic viewpoint. That conception is shared by W. Hamilton and R. Lotze, among others. Another coincidence of ontology and reism is more interesting.

In Leśniewski's lectures which were transcribed by J. Słupecki, a singular proposition was understood in such a manner that it can be true only if its subject is a singular name. In other words, any singular proposition which has as a subject an empty name or a general name is false (Słupecki, 1955, p. 14). For example: 'John III Sobieski is the deliverer of Vienna' is a true proposition. 'Hamlet is the hero of a tragedy by Shakespeare' is a false proposition because it has an empty subject. 'The whale is a mammal' is a false proposition, because the subject is a general name. This position is really a reistic one! We can turn a false proposition into a true one by the following transformation:

> 'Hamlet' is the name of the hero of a tragedy by Shakespeare. Here a term 'Hamlet' was employed in *suppositio materialis*, which makes it cease to be an empty name. 'Every whale is a mammal', Here the word 'is' has no meaning of its own, it is only a part of the functor 'every ... is ... (Słupecki, 1955, p. 16).

Is the proposition: 'seniority is a transitive relation' a true one?

Słupecki recognized the difficulty of answering such a question without the help of Kotarbiński. He wrote:

In view of the fact that the decision as to whether any given word is or is not an apparent name lies beyond the scope of logic, it is difficult to analyze the meaning of singular propositions whose subjects are apparent names. For this reason I shall not discuss in more detail singular propositions the subjects of which are names of this type, and I refer the reader to the works of Prof. Kotarbiński (Słupecki, 1955, p. 15).

This comment of Słupecki's is very important. It testifies that reism is not a method of constructing a logical system. It is a method of interpreting some logical systems, for example some parts of Leśniewski's ontology. And this interpretation is rather specific. There is a gap between the current meaning of the word 'is' and that which used in reism and in ontology.

Kazimierz Ajdukiewicz (Ajdukiewicz, 1960) rightly criticised the reduction of all Aristotelian categories to substance (*ousia*). Moreover, it is correct to say that Ingarden and Kotarbiński reduced all categories not to substance in general but to first substances (*ousia proton*).

Let us answer the nominalistic arguments. According to Ingarden, properties have no properties. If we consented to this, the following question would arise: Ingarden wrote very much about properties. Properties are the subject of the discussion. What is its content? Of course the content is properties of the subject, i.e. properties of properties. The property '*non est ens* but *est entis*' is a real property of properties. The negation of any properties of properties leads to self contradiction.

Of course, in some cases *nota notae est nota rei ipsius*. It is right that the colour blue is the colour of a flower. But 'to be a vector' is a property of velocity. Velocity is a property of a tram. Is a vector a property of a tram? The narrowing of the category 'property' as well as the category of 'thing' is a kind of voluntarism. Ingarden liked to use the Latin expression *Sic volo sic jubeo*. Here it can be applied to Ingarden himself. It seems that the limitation of the categories 'property' and 'thing' is made *ad hoc* for saving the tenet of the absence of properties of properties.

But in this case the number of categories is greatly increased. Side by side with the categories 'thing' and 'property' we must speak about 'constitutive natures', 'existence', 'connection between qualitative moments', 'summative totality', 'schemes of a thing' and so on. The categorial system transforms into chaos. Our defence of 'properties of properties' can be applied *mutatis mutandis* to properties of relations.

Let us turn to the works of Kotarbiński. We can argue against his reism in the following way. Categories must be the most general and the most simple concepts which are necessary for explanation of more complicated and particular notions. Accordingly to Kotarbiński every object is a body or a soul. But what is 'a body'? A body is 'something corporeal.' The term 'corporeal' requires an explanation. It means the same as 'temporal, spatial and resistant'. (Kotarbiński, 1955, p. 489). All of these notions require an explanation as well, which Kotarbiński gave with the help of modern physics. Therefore 'thing' is not a category. The categories are: 'gravitating solid' 'a whole consisting of such solids', 'an object of the kind of which such solids consist', 'a whole consisting of elements each of which belongs to one of the preceding three types' (Kotarbiński, 1955, p. 489) In such a way we explain a simple through a complicated, *obscurum per obscurius*. Aristotle could not know what '*ousia*' is because he did not know twentieth century physics. But Kotarbiński did not know the physics of the twenty-first century. Does it mean that he did not know the proper meaning of the term 'thing'?

The arguments which were put forward above against Ingarden and Kotarbiński are arguments against nominalism. Our position is more realistic and, according to the classification of Kotarbiński, more idealistic. The struggle between nominalism and realism has continued through the centuries. We do not pretend that our arguments are decisive and must end that struggle. But we have a right to implement some realistic ideas for a construction of the logico-mathematical formalism of General Systems Theory. And we must be grateful to philosophers of the Lvov-Warsaw School for defending and criticising those ideas.

*Odessa, Ukraine*

## REFERENCES


Ajdukiewicz, K.: 1960, 'Reizm', in *Język i poznanie* (Language and Cognition), vol. I, PWN, Warsaw, pp. 79–101. Eng. trans. as 'A general characterization of the work as a textbook and as a scientific contribution' in (and a review of) T. Kotarbiński, *Gnosiology*, Pergamon Press, Oxford, 1966, pp. 515–536.

Borowski, M.: 1921, *Co to jest przedmiot* (What is an Object), *Przegląd Filozoficzny* **23**.

Ingarden, R.: 1961, *Spór o istnienie świata* (The Controversy About the Existence of the World), t. I, PWN, Warsaw.

Ingarden, R.: 1962, *Spór o istnienie świata* (The Controversy About the Existence of the World), t. II, PWN, Warsaw.

Ingarden, R.: 1966, 'Uwagi o niektórych twierdzeniach ontologicznych w książce Kazimierza Twardowskiego' (Remarks on Some Ontological Statements in K. Twardowski's Book), *Ruch Filozoficzny* **25**, **1–2**, pp. 21–35.

Kantorovic, L.V., Plisco, V.E.: 1983, 'The Systems Ideas in Mathematics', in *Philosophic-Methodological Foundations of Systems Research*, Institute of System Researches, Nauka, Moscow. (In Russian).

Kotarbiński, T.: 1955, 'The Fundamental Ideas of Pansomatism', *Mind* **164**, pp. 488–500.

Kotarbiński, T.: 1961, 'Fazy rozwojowe konkretyzmu' (The Development Stages of Concretism), in *Elementy teorii poznania, logiki formalnej i metodologii nauk* (Elements of Epistemology, Formal logic and Methodology of Sciences), Ossolineum, Wroclaw-Warsaw-Krakow.

Leonenko, L.L., Sarayeva, I.N.: 1984, 'On the Application of Ternary Description Language to Systems Parameter Modelling and Identification of General Systems Laws', *Systems Re-*

*search Methodological Problems*, Yearbook 1984, Nauka, Moscow, pp. 181–193. (In Russian).

Leonenko, L.L.: 1987, 'The Logical Properties of some Calculi of the Language of Ternary Description', *8th Congress of Logic, Methodology and Philosophy of Science. Abstracts*, Vol. 1, pp. 278–280.

Mamtchur, E.A., Ovtcinnikov, N.F., Uyemov, A.I.: 1989, *The Principle of Simplicity and Measures of Complexity*, Nauka, Moscow. (In Russian).

Mesarovic, M.D., Takahara, Y.: 1975, *General Systems Theory. Mathematical Foundation*, Academic Press, New York, San Francisco, London.

Screider, Y.A., Scarov, A.A.: 1982, *The Systems and Models*, Radio and Svjaz, Moscow. (In Russian).

Słupecki, J.: 1955, 'S. Leśniewski's Calculus of Names', *Studia Logica* 3, pp. 7–72.

Ujomov, A.I.: 1965, *Dinge, Eigenshaften und Relationen. Nachwort des Verfassers zur deutchen Ausgabe*, Akademie Verlag, Berlin.

Uyemov, A.I.: 1978, *System Approach and General Systems Theory*, Mysl, Moscow. (In Russian).

Uyemov, A.I.: 1984, 'A Formalism for Parametric General Systems Theory', In *System Research Methodological Problems*, Yearbook 1984, Nauka, Moscow, pp. 154–180. (In Russian).

Uyemov, A.I.: 1987, 'Fundamental Features of the Language of the Ternary Description as a Logical Formalism of Systems Analysis', *8th Congress of Logic, Methodology and Philosophy of Science. Abstracts*, vol. 5, part 1, Section 1–6, Moscow, pp. 337–340.

Ujemov, A.: 1989, 'Formalizacja niedookreślonych modeli systemów za pomocą języka opisu trójkowego' (Formalization of Partially described Models of Systems Using the Ternary Description Language), *Prakseologia* **4 (105)**, pp. 165–174.

Uyemov A.I., Sarayeva I.N.: 1992, *Analogy Method as the Basis for Forecasting Socio — economic Systems Development*, Institute for Market Problems and Economic-Ecological Research, Odessa. (In Russian).

Uyemov A.I.: 1975, 'Logical Apparatus of the Systems Research', *5th Congress of Logic, Methodology and Philosophy of Science. Contributed papers*, London, Ontario, Canada, pp. 129–130.

Uyemov A.I.: 1979, 'The Language of the Ternary Description as an Alternative of the Predicate Calculus', *6th Congress of Logic, Methodology and Philosophy of Science. Abstracts*, Section 5–7, Hannover, pp. 194–198.

Uyemov A.I.: 1995, 'The Language of Ternary Description as a Deviant Logic, part I', in *Boletim da Sociedade Paranaense de Metematica* **15**.

Uyemov A.I.: 1997, 'The Ternary Description Language as a Formalism for the Parametric General Systems Theory', part I, *International Journal of General Systems*. (In print).

Uyemov A.I., Shtakser G.V.: 1996, 'Concerning the Construction of Integrity Measurement Scale'. (Typescript in Russian).

STANISŁAW J. SURMA

# FROM CLOSURE-OPERATIC DEDUCTIVE METHODOLOGY TO NON-STANDARD ALTERNATIVES

## 1. INTRODUCTION

The best known abstract setting for the traditional deductive methodology that has in the course of time become widely accepted as the most essential ingredient of methodological orthodoxy is due to the fundamental work of A. Tarski. In a series of papers of the thirties, especially in papers (1930a), (1930b), (1933) and (1936), Tarski developed a general conceptual set-up which he based on the idea of the consequence operation, or closure operator as it is called in the context of lattice theory today. He took a closure operator **Cn** to be a mapping from $\mathbf{P}(S)$ to $\mathbf{P}(S)$ where $\mathbf{P}(S)$ is the power set of the set $S$ of all sentences of a fixed language and he characterised the operator as reflexive (i.e. $X \subseteq \mathbf{Cn}(X)$); monotonic (i.e. $\mathbf{Cn}(X) \subseteq \mathbf{Cn}(Y)$ whenever $X \subseteq Y$) and idempotent (i.e. $\mathbf{Cn}(\mathbf{Cn}(X)) \subseteq \mathbf{Cn}(X)$) where $X, Y$ are arbitrary subsets of $S$. In his (1935), Tarski re-modelled the **Cn**-based set-up by employing, instead of **Cn**, the idea of what he called a deductive system or the idea of a closure system, as we say today. A closure system **Th** is a subset of $\mathbf{P}(S)$ closed under arbitrary intersections i.e. a subset of $\mathbf{P}(S)$ such that $\bigcap K \subseteq \mathbf{Th}$ for any $K \subseteq \mathbf{Th}$. Various aspects of this Tarskian methodological set-up have been subjected to an extensive investigation in the subsequent years both in Poland and abroad. For more details see Blok and Pigozzi (1989), Cohn (1965), Grzegorczyk (1961), Łoś and Suszko (1958), Pogorzelski (1969) and (1994), Schmidt (1952), Thiele (1956), Weaver (1992) and Wójcicki (1988). In (1952) J. Schmidt proved that, loosely speaking,

$$\mathbf{Cn}(\mathbf{Th}(\mathbf{Cn})) = \mathbf{Cn} \quad \text{and} \quad \mathbf{Th}(\mathbf{Cn}(\mathbf{Th})) = \mathbf{Th}$$

for any **Cn** and **Th**. This establishes, formally, a one-to-one correspondence between all closure operators over the set $S$ on the one hand and all closure systems over $S$ on the other hand. As a move in the same direction one can show that there are other methodological frameworks, each based on a single primitive term, alternative to the Tarskian framework based on **Cn** or on **Th** such as, for instance, **Cons**, **Ln** and **Max** (Cf. Surma, 1980, 1993a, 1993b, 1994, 1995, 1996 and 1997). We refer to these alternatives as non-standard to contrast them with the Tarskian framework, now considered as the standard methodological framework. Below we begin with a brief introduction to these



*K. Kijania-Placek and J. Woleński (eds.),*
*The Lvov–Warsaw School and Contemporary Philosophy*, 365–377.
© 1998 *Kluwer Academic*

non-standard alternatives and then proceed to a more detailed presentation of a normalisation procedure for a uniform description of all these methodological set-ups followed by the application of this procedure to a brief discussion of the problem of saturation within each of these set-ups. On the way we make a number of statements but they are stated without proofs. All these proofs are included in author's (1997). We begin by providing the necessary terminology and related notational conventions.

## 2. BASIC NOTATION

Symbol $S$ denotes the set of all sentences of a fixed object language. It is assumed that $S$ is non-empty. Members of $S$ and subsets of $S$ are symbolised hereafter as $A, B, C, \ldots$, and as $X, Y, Z, \ldots$, respectively, with indices if necessary. For brevity, expressions of the kind $\phi(X \cup \{A_1, A_2, \ldots, A_n\})$ are rendered as $\phi(X, A_1, A_2, \ldots, A_n)$. The set of all subsets of a set $X$ and the set of all supersets of $X$ (within $S$) are denoted and defined as $2^X = \{Y \subseteq S : Y \subseteq X\}$ and $2_X = \{Y \subseteq S : X \subseteq Y\}$, respectively. Clearly, $X \in 2^Y$ iff $Y \in 2_X$; $2_X \subseteq 2_Y$ iff $2^Y \subseteq 2^X$; $X \in 2^X \cap 2_X$; and $2^S \subseteq 2_\emptyset$. If # is an $n$-ary connective and $A_1, A_2, \ldots, A_n \in S$ then $\#A_1 A_2 \ldots A_n \in S$. Thus, whenever a connective # is explicitly specified, $S$ is understood to be closed under this connective. If # is binary, it is customary to write $A\#B$ instead of $\#AB$. In what follows we will be concerned mainly with the standard connectives of negation $\neg$, conditional or implication $\rightarrow$ and biconditional $\equiv$.

## 3. A BRIEF DESCRIPTION OF THREE NON-STANDARD DEDUCTIVE METHODOLOGIES

**Cons** derives from the intuitive idea of consistency and, by the suggested axiomatisation, is characterised as a non-trivial (i.e. $S \notin \mathbf{Cons}$), hereditary ( i.e. $Y \in \mathbf{Cons} \cap 2_X$ implies $X \in \mathbf{Cons}$) and regular (i.e. any $X \in \mathbf{Cons}$ extends to an inclusion-maximal $Y \in \mathbf{Cons}$) property of subsets of $S$. An intuitive motivation for **Ln**, a Lindenbaum operator, comes from the well-known Lindenbaum extension lemma. **Ln** is characterised here as a non-trivial (i.e. $\mathbf{Ln}(S) = \emptyset$), extensive (i.e. $\mathbf{Ln}(X) \subseteq 2_X$), inclusive (i.e. $\mathbf{Ln}(\emptyset) \cap 2_X \subseteq \mathbf{Ln}(X)$) antimonotonic (i.e. $X \subseteq Y$ implies $\mathbf{Ln}(Y) \subseteq \mathbf{Ln}(X)$) and regular (i.e. $X = Y$ for any $X \in \mathbf{Ln}(\emptyset)$ and $Y \in \mathbf{Ln}(X)$) operator from $\mathbf{P}(S)$ to $\mathbf{P}(\mathbf{P}(S))$. **Max**, intuitively a property of maximality, e.g. that of deductive completeness, is characterised as a non-trivial (i.e. $S \notin \mathbf{Max}$) and regular (i.e. $X = Y$ for any $X \in \mathbf{Max}$ and $Y \in \mathbf{Max} \cap 2_X$) family (or property) of subsets of $S$. It has been proved that each of these frameworks can be translated into the standard, closure-operatic framework in a one-to-one and theorem-preserving fashion provided that **Cn** is strongly regular, a term to be defined later in the paper. In other words, it has been proved that

$$\alpha(\beta(\alpha)) = \alpha \quad \text{and} \quad \beta(\alpha(\beta)) = \beta$$

for any $\alpha, \beta \in \{\mathbf{Cn}, \mathbf{Cons}, \mathbf{Ln}, \mathbf{Max}\}$ where each of **Cons**, **Ln**, **Max** is regular. These equations may be of some interest not only for general philosophical reasons but also because they can be treated as providing a proof of the relative consistency of a particular methodological framework, i.e. they can be considered as yielding a consistency proof for a framework on the assumption of the consistency of some other framework involved in the equations. The **Cons**-based methodology and that based on **Ln** seem to be the most interesting alternatives to the **Cn**-based methodology.

## 4. SPECIFICATION OF $\alpha$ BY MEANS OF #

In this paragraph we deal with the interaction between $\alpha$ and # where # is a sentential connective and $\alpha \in \{\mathbf{Cn},\mathbf{Cons},\mathbf{Ln}\}$. Because of lack of space we leave aside **Th** and **Max** as well as some other instances of $\alpha$. When # is fixed and understood then the interaction reduces to a (#)-specification of $\alpha$ and so we are concerned here only with (#)-closure operators, (#)-consistency properties and (#)-Lindenbaum operators. As a further restriction we only instantiate # as negation $\neg$, conditional $\rightarrow$ and biconditional $\equiv$.

### 4.1. *(#)-Closure operators*

The discussion of closure operators, as well as of any other instance of $\alpha$, for that matter, must be carried out in two stages. The first stage was already very briefly reported in the previous paragraph where closure operators were treated as reflexive, monotonic and idempotent operators of a fixed kind. That description was given exclusively in terms of general set theory without the use of any logical constants. For this reason we can refer to that stage as an absolute or language-independent stage. We assumed that $S$, the set of all sentences of the language we are concerned with (or the object-language as it is often called), is a non-empty set of sentences of unspecified structure. At the second stage, to which we now turn, the discussion also assumes some kind of "inner" structure of our object-language. At this stage some members of $S$ can be considered as compound sentences, i.e. as sentences made up of simpler sentences with the help of some specified logical connectives. As a recast of rather well-known facts about the interaction between closure operators and negation we say now that **Cn** is a $(\neg)$-operator iff the condition below is satisfied.

(1) $\mathbf{Cn}(A, \neg A) = S$;
$\mathbf{Cn}(X, A) \cap \mathbf{Cn}(X, \neg A) \subseteq \mathbf{Cn}(X)$.

Among other conditions known to be equivalent to condition (1) we may mention the following one.

(2) $A \in \mathbf{Cn}(X)$ iff $\mathbf{Cn}(X, \neg A) = S$.

We say that **Cn** is a $(\rightarrow)$-operator iff the condition below is satisfied.

(3) $A \rightarrow B \in \mathbf{Cn}(X)$ iff $\mathbf{Cn}(X, B) \subseteq \mathbf{Cn}(X, A)$;
$\mathbf{Cn}(X, A) \cap \mathbf{Cn}(X, A \rightarrow B) \subseteq \mathbf{Cn}(X)$.

We say that **Cn** is a $(\equiv)$-operator iff the condition below is satisfied.

(4) $A \equiv B \in \mathbf{Cn}(X)$ iff $\mathbf{Cn}(X, B) = \mathbf{Cn}(X, A)$;
$(\mathbf{Cn}(X, A) \cap \mathbf{Cn}(X, B) \cap \mathbf{Cn}(X, A \equiv B)) \subseteq \mathbf{Cn}(X)$.

These examples do not prompt any general schema for a uniform definition of a (#)-operator **Cn** that would work in application to any connective. Neither do we expect it to be the case under this heading. We have no option here but to consider each connective individually and work out a suitable definition for it under the guidance of a proof theory adopted explicitly for that connective or, at least, under the guidance of a more or less implicitly assumed and 'intuitively expected' or 'intuitively intended' underpinning.

### 4.2. (#)-*Consistency properties*

Here, too, we build on top of the absolute description of consistency properties, very briefly reported on in the previous paragraph, where consistency properties were characterised as non-trivial, hereditary and, sometimes, also regular properties of subsets of $S$. And we make use of $\neg$, $\rightarrow$ and $\equiv$ to enrich the absolute characterisation by adding to it some relative, language-dependent, characteristics of these properties as well. We say that **Cons** is a $(\neg)$-property iff the condition below is satisfied.

(5) $\{A, \neg A\} \notin \mathbf{Cons}$;
If $X \in \mathbf{Cons}$ then $X \cup \{A\} \in \mathbf{Cons}$ or $X \cup \{\neg A\} \in \mathbf{Cons}$.

We say that **Cons** is a $(\rightarrow)$-property iff the condition below is satisfied.

(6) If $X \cup \{A, A \rightarrow B\} \in \mathbf{Cons}$ then $X \cup \{A, B\} \in \mathbf{Cons}$;
If $X \cup \{B\} \in \mathbf{Cons}$ then $X \cup \{A \rightarrow B\} \in \mathbf{Cons}$;
If $X \in \mathbf{Cons}$ then $X \cup \{A\} \in \mathbf{Cons}$ or $X \cup \{A \rightarrow B\} \in \mathbf{Cons}$.

We say that **Cons** is a $(\equiv)$-property iff the condition below is satisfied.

(7) If $X \cup \{A, A \equiv B\} \in \mathbf{Cons}$ or $X \cup \{B, A \equiv B\} \in \mathbf{Cons}$ then $X \cup \{A, B\} \in \mathbf{Cons}$;
If $X \cup \{A, B\} \in \mathbf{Cons}$ then $X \cup \{A \equiv B\} \in \mathbf{Cons}$;
If $X \in \mathbf{Cons}$ then $X \cup \{A\} \in \mathbf{Cons}$, $X \cup \{B\} \in \mathbf{Cons}$ or $X \cup \{A \equiv B\} \in \mathbf{Cons}$.

### 4.3. (#)-*Lindenbaum operators*

This section is concerned with the problem of axiomatisation of the abstract idea of extension articulated as a primitive operator **Ln** from the powerset $\mathbf{P}(S)$ of some set $S$ to the powerset of the power set $\mathbf{P}(\mathbf{P}(S))$ of that set. One example of $S$ is the set of all sentences of some object-language. In this case an intuitive paradigm example of **Ln** is a correlation, Lindenbaum style, with each set of sentences of some object-language, whether consistent or inconsistent, of the family of all its maximal consistent extensions. With this example in mind, we call the operators Lindenbaum operators. As in the previous cases of **Cn** and **Cons** the discussion of Lindenbaum operators is also carried out in two stages. At the first, absolute, stage, a Lindenbaum operator **Ln** was characterised as a non-trivial, extensive, inclusive, antimonotonic and, possibly, also regular operator from $\mathbf{P}(S)$ to $\mathbf{P}(\mathbf{P}(S))$. At the second, relative, stage the discussion involves also some of the properties of Lindenbaum operators that depend on the 'structure' of $S$. Here we begin by saying that **Ln** is a $(\neg)$-operator iff the following condition is satisfied:

(8) $\mathbf{Ln}(A) \cap \mathbf{Ln}(\neg A) = \emptyset$;
$\mathbf{Ln}(A) \cup \mathbf{Ln}(\neg A) = \mathbf{Ln}(\emptyset)$.

Condition (8) is equivalent to:

(9) $\mathbf{Ln}(X, A) \cap \mathbf{Ln}(X, \neg A) = \emptyset$;
$\mathbf{Ln}(X) \subseteq \mathbf{Ln}(X, A) \cup \mathbf{Ln}(X, \neg A)$.

We say that **Ln** is a $(\rightarrow)$-operator iff the condition below is satisfied.

(10) $\mathbf{Ln}(A, A \rightarrow B) \subseteq \mathbf{Ln}(A, B)$;
$\mathbf{Ln}(B) \subseteq \mathbf{Ln}(A \rightarrow B)$;
$\mathbf{Ln}(A) \cup \mathbf{Ln}(A \rightarrow B) = \mathbf{Ln}(\emptyset)$.

Condition (10) is equivalent to:

(11) $\mathbf{Ln}(X, A, A \rightarrow B) \subseteq \mathbf{Ln}(X, A, B)$;
$\mathbf{Ln}(X, B) \subseteq \mathbf{Ln}(X, A \rightarrow B)$;
$\mathbf{Ln}(X) \subseteq \mathbf{Ln}(X, A) \cup \mathbf{Ln}(X, A \rightarrow B)$.

We say that **Ln** is a $(\equiv)$-operator iff the condition below is satisfied.

(12) $\mathbf{Ln}(A, A \equiv B) \cup \mathbf{Ln}(B, A \equiv B) \subseteq \mathbf{Ln}(A, B)$;
$\mathbf{Ln}(A, B) \subseteq \mathbf{Ln}(A, B, A \equiv B)$;
$\mathbf{Ln}(A) \cup \mathbf{Ln}(B) \cup \mathbf{Ln}(A \equiv B) = \mathbf{Ln}(\emptyset)$.

Conditions (12) is equivalent to:

(13) $\mathbf{Ln}(X, A, A \equiv B) \cup \mathbf{Ln}(X, B, A \equiv B) \subseteq \mathbf{Ln}(X, A, B)$;
$\mathbf{Ln}(X, A, B) \subseteq \mathbf{Ln}(X, A, B, A \equiv B)$;
$\mathbf{Ln}(X) \subseteq \mathbf{Ln}(X, A) \cup \mathbf{Ln}(X, B) \cup \mathbf{Ln}(X, A \equiv B)$.

## 5. (#)-NORMALISATION OF $\alpha$

Here we define '$\alpha$ is (#)-normal' for any $\alpha$ and we prove — for all those # for which we have a definition of '$\alpha$ is a (#)-operator' — that $\alpha$ is (#)-normal iff $\alpha$ is a (#)-operator. Our (meta-linguistic) normalisation procedure is somewhat related to but simpler than that based on the well-known (object-language) disjunctive-conjunctive normal form for #. It seems simpler because it makes disjunction and conjunction dispensable. Typically, the latter procedure is worked out from within a fixed object language and uses the object-language concepts of conjunction $\wedge$, disjunction $\vee$ and negation $\neg$ as defining ingredients. Unlike disjunctive-conjunctive normalisation, however, the normalisation procedure described in this paragraph involves neither disjunction $\vee$ nor conjunction $\wedge$. They are both conveniently replaced by their corresponding closure-operatic counterparts. Negation $\neg$ remains the only object-language component that is used both in the construction of our definition and in the definition of the object-language disjunctive-conjunctive normal form procedure. Another advantage of this procedure is that it provides a unified conceptual and terminological schema which helps generate, uniformly, arbitrary (#)-closure operators. In accordance with our restrictions we must now consider three instances of $\alpha$, namely, **Cn**, **Cons** and **Ln**.

### 5.1. (#)-*Normal closure operators*

To explain the general idea behind the procedure we exemplify it first using the standard sentential connectives of $\rightarrow$ and $\equiv$. We say that **Cn** is ($\rightarrow$)-normal iff $\mathbf{Cn}(X, A \rightarrow B) = \mathbf{Cn}(X, A, B) \cap \mathbf{Cn}(X, \neg A, B) \cap \mathbf{Cn}(X, \neg A, \neg B)$. It is clear that this definition depends on negation $\neg$ and for this reason it only applies to ($\neg$)-closure operators. One can prove that for any ($\neg$)-closure operator **Cn** the conditions below are equivalent.

(i) **Cn** is a ($\rightarrow$)-operator.
(ii) **Cn** is ($\rightarrow$)-normal.
(iii) The fact that $A \rightarrow B \in \mathbf{Cn}(X)$ is equivalent to the fact that $\mathbf{Cn}(X, A, \neg B) = S$.

We say that **Cn** is ($\equiv$)-normal iff

$$\mathbf{Cn}(X, A \equiv B) = \mathbf{Cn}(X, A, B) \cap \mathbf{Cn}(X, \neg A, \neg B).$$

We can prove that for any ($\neg$)-closure operator **Cn** the conditions below are equivalent.

(i) **Cn** is a ($\equiv$)-operator.
(ii) **Cn** is ($\equiv$)-normal.

(iii) The fact that $A \equiv B \in \mathbf{Cn}(X)$ is equivalent to the fact that $\mathbf{Cn}(X, \neg A, B) \cap \mathbf{Cn}(X, A, \neg B) = S$.

In contrast to the definitions in the last paragraph of a $(\rightarrow)$-closure operator and of a $(\equiv)$-closure operator our definition of a $(\rightarrow)$-normal closure operator and that of a $(\equiv)$-normal closure operator generalise easily and uniformly to an arbitrary sentential connective. To this end we assume hereafter that the alphabet of the object language contains two sentential connectives, negation $\neg$, as a constant connective, and an $n$-ary connective #, as a variable connective. We also tacitly assume that the object language is given the classical, two-valued semantic interpretation. Accordingly, let $f$ be a valuation function (i.e. a homomorphism) from the formula algebra $(S, \neg, \#)$ to the reduct $(\{0,1\}, \neg, \#)$ of the Boolean algebra over the set $\{0,1\}$. This means that if $\neg$ and $\#$ are $f$-correlated with $\neg$ and #, respectively, then $f(\neg A) = \neg f(A) = 1 - f(A)$ and $f(\#A_1A_2\ldots A_n) = \#(f(A_1), f(A_2), \ldots, f(A_n))$. For any $i \leq n$, we define

$$A_i[f] = A_i \quad \text{iff} \quad f(A_i) = 1 \quad \text{and} \quad A_i[f] = \neg A_i \quad \text{iff} \quad f(A_i) = 0.$$

We say that $\mathbf{Cn}$ is (#)-normal iff

$$\mathbf{Cn}(X, \#A_1A_2\ldots A_n) = \bigcap\{\mathbf{Cn}(X, A_1^{[f_j]}, A_2^{[f_j]}, \ldots, A_n^{[f_j]}) : j \leq 2^n \text{ and } f_j(\#A_1A_2\ldots A_n) = 1\}.$$

We have the following theorem: If $\mathbf{Cn}$ is a $(\neg)$-closure operator then the conditions below are equivalent.

(i) $\mathbf{Cn}$ is (#)-normal.

(ii) The fact that $\#A_1A_2\ldots A_n \in \mathbf{Cn}(X)$ is equivalent to the fact that $\bigcap\{\mathbf{Cn}(X, A_1^{[f_j]}, A_2^{[f_j]}, \ldots, A_n^{[f_j]}) : j \leq 2^n$ such that $f_j(\#A_1A_2\ldots A_n) = 0\} = S$.

As we said earlier, the introduction of the normalisation procedure for (#)-closure operators has been motivated by the desire to simplify and possibly even mechanise our dealing with arbitrary (#)-closure operators as discussed in the previous paragraph in a rather non-effective way. And, as far as we see it in relation to closure operators, simplify and mechanise it does.

## 5.2. (#)-*Normal consistency properties*

This section provides a unifying conceptual and terminological schema for the generation of all (#)-consistency properties based on a procedure similar to that applied to normal (#)-closure operators in the previous section. As in the case of normal (#)-closure operators we first explain the intuitions behind the normalisation procedure using the standard connectives of $\rightarrow$ and $\equiv$. And as in the case of normal (#)-closure operators the normalisation resembles the well known disjunctive-conjunctive normal form procedure. Below we use a straightforward concept which we now define. We say that $X$ is **Cons**-maximal iff $X \in$ **Cons** and $Y = X$ for any $Y \in$ **Cons**$\cap 2_X$ i.e. iff $X$ is an inclusion-maximal member of **Cons**. We say that **Cons** is $(\rightarrow)$-normal iff the fact that $X \cup \{A \rightarrow B\} \in$**Cons** is equivalent to the fact that $X \cup \{A, B\} \in$**Cons**, $X \cup \{\neg A, B\} \in$ **Cons** or $X \cup \{\neg A, \neg B\} \in$ **Cons** for any **Cons**-maximal $X$. We can prove that for any $(\neg)$-consistency property **Cons** the conditions below are equivalent.

(i) **Cons** is a $(\rightarrow)$-property.
(ii) **Cons** is $(\rightarrow)$-normal.
(iii) The fact that $X \cup \{A \rightarrow B\} \in$ **Cons** is equivalent to the fact that $X \cup \{A, \neg B\} \notin$ **Cons** for any **Cons**-maximal $X$.

We say that **Cons** is $(\equiv)$-normal iff the fact that $X \cup \{A \equiv B\} \in$ **Cons** is equivalent to the fact that $X \cup \{A, B\} \in$ **Cons** or $X \cup \{\neg A, \neg B\} \in$ **Cons** for any **Cons**-maximal $X$. We can prove that for any $(\neg)$-consistency property **Cons** the conditions below are equivalent.

(i) **Cons** is a $(\equiv)$-property.
(ii) **Cons** is $(\equiv)$-normal.
(iii) The fact that $X \cup \{A \equiv B\} \in$ **Cons** is equivalent to the fact that $X \cup \{\neg A, B\} \notin$ **Cons** and $X \cup \{A, \neg B\} \notin$ **Cons** for any **Cons**-maximal $X$.

It is worthwhile mentioning that conditions (i) and (ii) of the last two theorems continue to be equivalent to each other after the clause "for any **Cons**-maximal $X$", appearing in the definitions of "**Cons** is $(\rightarrow)$-normal" and "**Cons** is $(\equiv)$-normal" is dropped. The idea of normalisation, as illustrated above, easily generalises to an arbitrary sentential connective. In order to describe the generalisation we can use the same setting as in the case of (#)-normal closure operators. We say that **Cons** is (#)-normal iff, for any **Cons**-maximal X,

> the fact that $X \cup \{\#A_1 A_2 \dots A_n\} \in$ **Cons** is equivalent to the fact that $X \cup \{A_1^{[f_j]}, A_2^{[f_j]}, \dots, A_n^{[f_j]}\} \in$ **Cons** for some $j \leq 2^n$ such that $f_j(\#A_1 A_2 \dots A_n) = 1$.

We can prove that if **Cons** is a $(\neg)$-consistency property then the conditions below are equivalent.

(i) **Cons** is (#)-normal.
(ii) The fact that $X \cup \{\#A_1 A_2 \dots A_n\} \in$ **Cons** is equivalent to the fact that $X \cup \{A_1^{[f_j]}, A_2^{[f_j]}, \dots, A_n^{[f_j]}\} \notin$ **Cons** for any $j \leq 2^n$ such that $f_j(\#A_1 A_2 \dots A_n) = 0$ and any **Cons**-maximal $X$.

### 5.3. (#)-*Normal Lindenbaum operators*

This section provides a unifying conceptual and terminological schema for the generation of all (#)-Lindenbaum operators based on a procedure similar to that used in application to normal (#)-closure operators and (#)-consistency properties. As in the case of normal (#)-closure operators and (#)-normal consistency properties we first explain the intuitions behind the normalisation procedure using the standard sentential connectives of $\rightarrow$ and $\equiv$.

We say that **Ln** is ($\rightarrow$)-normal iff the fact that $Y \in \mathbf{Ln}(X, A \rightarrow B)$ is equivalent to the fact that $Y \in \mathbf{Ln}(X, A, B) \cup \mathbf{Ln}(X, \neg A, B) \cup \mathbf{Ln}(X, \neg A, \neg B)$ for any $Y \in \mathbf{Ln}(X)$. We can prove that for any ($\neg$)-Lindenbaum operator **Ln** the conditions below are equivalent.

(i) **Ln** is a ($\rightarrow$)-operator.
(ii) **Ln** is ($\rightarrow$)-normal.
(iii) The fact that $Y \in \mathbf{Ln}(X, A \rightarrow B)$ is equivalent to the fact that $Y \notin \mathbf{Ln}(X, A, \neg B)$ for any $Y \in \mathbf{Ln}(X)$.

We say that **Ln** is ($\equiv$)-normal iff the fact that $Y \in \mathbf{Ln}(X, A \equiv B)$ is equivalent to the fact that $Y \in \mathbf{Ln}(X, A, B) \cup \mathbf{Ln}(X, \neg A, \neg B)$ for any $Y \in \mathbf{Ln}(X)$. We can prove that for any ($\neg$)-Lindenbaum operator **Ln** the conditions below are equivalent.

(i) **Ln** is a ($\equiv$)-operator.
(ii) **Ln** is ($\equiv$)-normal.
(iii) The fact that $Y \in \mathbf{Ln}(X, A \equiv B)$ is equivalent to the fact that $Y \notin \mathbf{Ln}(X, A, \neg B) \cup \mathbf{Ln}(X, \neg A, B)$ for any $Y \in \mathbf{Ln}(X)$.

The above discussion, aimed at providing an intuitive explanation of the idea of normalisation of **Ln**, can now be generalised to an arbitrary sentential connective. In order to describe the generalisation we use the same setting as in the case of (#)-normal closure operators and (#)-normal consistency properties.

We say that **Ln** is (#)-normal iff the fact that $Y \in \mathbf{Ln}(X, \#A_1A_2 \ldots A_n)$ is equivalent to the fact that $Y \in \bigcup\{\mathbf{Ln}(X, A_1^{[f_j]}, A_2^{[f_j]}, \ldots, A_n^{[f_j]}) : j \leq 2^n$ and $f_j(\#A_1A_2 \ldots A_n) = 1\}$ for any $Y \in \mathbf{Ln}(X)$. We can prove that for any ($\neg$)-Lindenbaum operator **Ln** the conditions below are equivalent.

(i) **Ln** is (#)-normal.
(ii) The fact that $Y \in \mathbf{Ln}(X, \#A_1A_2 \ldots A_n)$ is equivalent to the fact that $Y \notin \bigcup\{\mathbf{Ln}(X, A_1^{[f_j]}, A_2^{[f_j]}, \ldots, A_n^{[f_j]}) : j \leq 2^n$ and $f_j(\#A_1A_2 \ldots A_n) = 0\}$ for any $Y \in \mathbf{Ln}(X)$.

## 6. (#)-SATURATION OF $\alpha$

Here we begin by setting some more terminology. We say that $X$ is (#)-complete iff the fact that $(\# A_1 A_2 \ldots A_n) \in X$ is equivalent to the fact that

$$\#^*(A_1 \in X)(A_2 \in X) \ldots (A_n \in X),$$

for any $A_1, A_2, \ldots, A_n$, where #* is the meta-linguistic correlate of the object-language connective #. However, to simplify notation in this respect we may symbolise the meta-linguistic correlate #* of # by the same symbol # with the star * omitted. In normal circumstances this does not lead to any confusion. The examples below help to illustrate the above definition. The clause below is the defining condition of 'X is ($\rightarrow$)-complete'.

> The fact that $(A \rightarrow B) \in X$ is equivalent to the fact that
> $(A \in X) \rightarrow (B \in X)$
> i.e., loosely speaking, is equivalent to the fact that
> $(A \in X)$ implies that $(B \in X)$.

The defining condition of 'X is ($\neg$)-complete' is the clause below.

> The fact that $(\neg A) \in X$ is equivalent to the fact that $\neg(A \in X)$
> i.e., loosely speaking, is equivalent to the fact that $(A \notin X)$.

It may be remarked that, from an intuitive point of view, the above definition of the concept of connective-completeness is a statement about the co-extensiveness of two characterisations of a logical constant, an 'inside-a-system' characterisation of the constant, on the one hand, and its 'outside-the-system' characterisation, on the other hand. Indeed, intuitively, a set $X$ is complete with respect to a logical constant iff the constant does in $X$ precisely what it is seen doing there. Connective-completeness plays some role in our conceptualisation of the problem of deductive methodology. Next we define the concept of a regular operator $\alpha$ (or a regular property $\alpha$ as the case may be) as well as that of a $\alpha$-maximal set where $\alpha$ is instantiated as **Cn**, **Cons** or as **Ln**.

### 6.1. *The case of* **Cn**

We say that $X$ is (**Cn**, $A$)-maximal iff $A \notin \mathbf{Cn}(X)$ and $A \in \mathbf{Cn}(X, B)$ for any $B \notin X$. $X$ is **Cn**-maximal iff $X$ is (**Cn**, $A$)-maximal for some $A$. We say that an operator **Cn** is regular iff, for any $A, X$, the fact that $A \notin \mathbf{Cn}(X)$ implies that there is a $Y \in 2_X$ such that $A \notin \mathbf{Cn}(Y)$ and $A \in \mathbf{Cn}(Y, B)$ for any $B \notin Y$, i.e. iff, for any $A$, every $A$-omitting set can be extended to a set which has no $A$-omitting proper superset. This concept of regularity of **Cn**, but not its name, is implicit in H. Thiele's paper (1956).

We say that $X$ is absolutely **Cn**-maximal iff $\mathbf{Cn}(X) \neq S$ and $\mathbf{Cn}(X, A) = S$ for any $A \notin X$. We say that an operator **Cn** is absolutely regular iff, for any $X$, the fact that $\mathbf{Cn}(X) \neq S$ implies that there is $Y \in 2_X$ such that $\mathbf{Cn}(Y) \neq S$ and $\mathbf{Cn}(Y, A) = S$ for any $A \notin Y$. This concept, although not the name, is implicit in the well-known extension theorem due to A. Lindenbaum (cf. Tarski, 1930a, 1930b, 1935).

We say that $X$ is strongly $(\mathbf{Cn}, A)$-maximal iff $A \notin \mathbf{Cn}(X)$ and $\mathbf{Cn}(X, B) = S$ for any $B \notin X$. $X$ is strongly **Cn**-maximal iff $X$ is strongly $(\mathbf{Cn}, A)$-maximal for some $A$. We say that an operator **Cn** is strongly regular iff, for any $A$ and $X$, the fact that $A \notin \mathbf{Cn}(X)$ implies that there is $Y \in 2_X$ such that $A \notin \mathbf{Cn}(Y)$ and $\mathbf{Cn}(Y, B) = S$ for any $B \notin Y$. This concept is implicit in A. Tarski's works (1930a), (1930b) and (1935).

### 6.2. *The case of* **Cons**

**Cons** is regular iff for any $X \in \mathbf{Cons}$ there is $Y \in \mathbf{Cons} \cap 2_X$ such that $Z = Y$ for any $Z \in \mathbf{Cons} \cap 2_Y$ i.e. iff all members of **Cons** are inclusion-maximalisable.

### 6.3. *The case of* **Ln**

Here we only need the following definition already provided in the third paragraph: **Ln** is regular iff $X = Y$ for any $X \in \mathbf{Ln}(\emptyset)$ and any $Y \in \mathbf{Ln}(X)$. Now to the key definition of this paragraph. We say that $\alpha$ is (#)-saturated iff $X$ is (#)-complete for any $X$ such that $\Gamma(\alpha, X)$. Clearly, for each particular instance of $\alpha$, the condition $\Gamma(\alpha, X)$ needs to be specified. This is done as follows. '$\Gamma(\mathbf{Cn}, X)$' is specified as '$X$ is **Cn**-maximal'; '$\Gamma(\mathbf{Cons}, X)$' is specified as '$X$ is **Cons**-maximal', as already defined in the previous section; and, finally, '$\Gamma(\mathbf{Ln}, X)$' becomes simply '$X \in \mathbf{Ln}(\emptyset)$'. One can prove that for any $\alpha \in \{\mathbf{Cn}, \mathbf{Cons}, \mathbf{Ln}\}$ and any # if $\alpha$ is regular then the conditions below are equivalent.

(22) $\alpha$ is (#)-normal.
(23) $\alpha$ is (#)-saturated.

## 7. A FINAL COMMENT

Of the metalogical set-ups mentioned above the one based on **Cn** is, no doubt, the most obvious. Virtually all the known frameworks for metalogic are, at the very core, closure-operatic, the best known among them being Tarski's general consequence theory. The next framework, mentioned above, is that based on **Th**. The treatment of the set-theoretic component of this program is motivated by Tarski's theory of deductive systems. For more recent references cf.

Cohn (1965), Weaver (1992), and Pollard and Martin, (1994). A consistency-theoretic framework for metalogic, based on **Cons** as a primitive idea, is not only possible but also seems attractive on intuitive grounds. However, an implementation of this program inevitably requires that the definition of consistency be freed from negation-dependence and, indeed, from dependence on any other logical constant. If logical constants are to be explained in terms of consistency then they cannot appear in the general definition of the latter. And, indeed, the equivalence of the consistency-theoretic framework with that based on the idea of a closure operator can be established already at the set-theoretic level, i.e. in the absence of any logical constant. Frameworks based on the idea of maximality are also possible. Yet, implementations of this idea are affected by the fact that it can be categorised in more than one way. Sometimes it is categorised as a function which maps a set to the family of all its maximal extensions of a kind and sometimes as a family of maximally extended sets. Of the two options **Max** and **Ln**, the **Ln**-based methodology seems to be more appealing than **Max**. It may be worthwhile noting that the set-theoretic or absolute component of both the **Cons**-based methodology and that based on **Max** was inspired by R. Suszko as early as in the sixties (cf. Żandarowska, 1966).

*The University of Auckland, New Zealand*

## REFERENCES

Blok, W.J., Pigozzi, D.: 1989, *Algebraizable Logics*, Memoirs of the American Mathematical Society, **77**.

Cohn, P.M.: 1965, *Universal Algebra*, Harper and Row, New York.

Grzegorczyk, A.: 1961, *Zarys logiki matematycznej*, PWN, Warsaw, (An Outline of Mathematical Logic), Kluwer Academic Publishers, Dordrecht, 1974.

Łoś, J., Suszko, R.: 1958, 'Remarks on Sentential Logics', *Proc. Kon. Nederl. Akad. van Wetenschappen, Series A* **61**, pp. 177–183.

Pogorzelski, W.A.: 1969, *Klasyczny rachunek zdań. Zarys teorii* (Classical Sentential Calculus. An Outline of the Theory), PWN, Warsaw.

Pogorzelski, W.A.: 1994, *Notions and theorems of elementary formal logic*, University of Warsaw, Białystok Branch Publishers, Białystok.

Pollard, S., Martin, M.: 1994, 'Contractions of closure systems', *Notre Dame Journal of Formal Logic* **35**, No 1, pp. 108–115.

Schmidt, J.: 1952, 'Über die Rolle den transfiniten Schlussweisen in einer allgemeinen Idealtheorie', *Mathematische Nachrichten* **7**, pp. 165–182.

Surma, S.J.: 1980, 'The Growth of Logic out of the Foundational Research in Mathematics', in E. Agazzi (ed.), *Modern Logic — A Survey*, D. Reidel Publishing Company, Dordrech, pp. 15–33.

Surma, S.J.: 1993a, 'Remarks on the consistency-theoretic versus consequence-theoretic approaches to metalogic. Abstract', *The Journal of Symbolic Logic* **58**, 1482.

Surma, S.J.: 1993b, 'W duchu Tarskiego: O alternatywach teorio-dowodowej metalogiki' (In the Spirit of Tarski: On Alternatives to the Proof-theoretic Metalogic), *Filozofia Nauki* **1**, pp. 49–65.

Surma, S.J.: 1994, 'Alternatives to the Consequence-theoretic Approach to Metalogic', in Proceedings of the IX Latin American Symposium on Mathematical Logic, August 3 — 8, 1992 *Notas de Logica Matematica* **39**, Universidad del Sur, Bahia Blanca, Argentina, pp. 1–30.

Surma, S.J.: 1995, 'On closure operators one-to-one associated with fixed object languages. Abstract', *The Bulletin of Symbolic Logic* **1(3)**, p. 358.

Surma, S.J.: 1996, 'Towards an abstract theory of Lindenbaum operators. Abstract', *The Bulletin of Symbolic Logic* **2(1)**, pp. 119–120.

Surma, S.J.: 1997, 'An Investigation into the Relationship between Closure Operators, Closure Systems, Consistency, Lindenbaum Operators, Families of Maximal Sets and Separation', unpublished.

Tarski, A.: 1930a, 'Über einige fundamentale Begriffe der Metamathematik', *Comptes Rendus des séances de la Société des Sciences et des Lettres de Varsovie* **Cl. III, 23**, pp. 22–29; Eng. trans. as 'On Some Fundamental Concepts of Metamathematics', in Tarski (1983), pp. 30–37.

Tarski, A.: 1930b, 'Fundamentale Begriffe der Methodologie der deduktiven Wissenschaften', *Monatshefte für Mathematik und Physik* **37**, pp. 361–404; Eng. trans. as 'Fundamental Concepts of the Methodology of the Deductive Sciences', in Tarski (1983), pp. 60–109.

Tarski, A.: 1933, *Pojęcie prawdy w językach nauk dedukcyjnych*, Tow. Naukowe Warszawskie, Warsaw. Eng. trans. as 'The Concept of Truth in Formalized Languages' in Tarski (1983), pp. 152–278.

Tarski, A.: 1935, 'Grundzüge des Systemenkalküls', *Fundamenta Mathematicae* **25**, pp. 503–526; Eng. trans. as 'Foundations of the Calculus of Systems', in Tarski (1983), pp. 342–383.

Tarski, A.: 1936, 'Über den Begriff der logischen Folderung', *Actes du Congrès International de Philosophie Scientifique* **7**, (*Actualités Scientifiques et Industrielles* **394**), pp. 1–11; Eng. trans. as 'On the Concept of Logical Consequence', in Tarski (1983), pp. 409–420.

Tarski, A.: 1983, *Logic, Semantics, Metamathematics*, Second Edition, Hackett, Indianapolis, Indiana, USA.

Thiele, H.: 1956, 'Eine Axiomatisierung der zweiwertigen Prädikatenkalküle der ersten Stuffe, welche die Implikation enhalten', *Zeitschrift für Mathematische Logik und Grundlagen der Mathematik* **2**, pp. 93–106xsy.

Weaver, G.: 1992, 'Unifying some modifications of the Henkin construction', *Notre Dame Journal of Formal Logic* **33**, No. 3, pp. 450–460.

Wójcicki, R.: 1988, *Theory of Logical Calculi*, Kluwer Academic Publishers, Dordrecht.

Żandarowska, W.: 1966, 'O związkach pomiędzy wynikaniem, sprzecznością i zupełnością' (On Certain Connections between Consequence, Inconsistency and Completeness), *Studia Logica* **18**, pp. 165–178.

PAUL WEINGARTNER

# FORGOTTEN AND NEGLECTED SOLUTIONS OF PROBLEMS IN PHILOSOPHICAL LOGIC

The purpose of this paper is to show that there are a number of important solutions (or partial solutions) to philosophical and logical problems which are very widely unknown or forgotten. I shall show this by giving examples from the history of philosophy. The logical reconstruction of the philosophical or logical text is mine. The amount of interpretation and construction necessary differs from text to text. Thus for instance in the first two examples (Aristotle's denial of a first genus and Paulus Venetus' proposed solution of the Liar) not much more than correct formulation in symbolic language was required because the text itself is rather complete. On the other hand interpretation and reconstruction are necessary for the third and fourth examples, similarly for the last two examples.

## 1. ARISTOTLE: IS BEING A MOST COMPREHENSIVE CLASS (SET) OR PREDICATE?

This question is not identical with the question 'What is being?' But the latter is very wide and vague and the history of ideas shows us that such wide and perennial questions can get interesting partial answers only if the original question is split into more precise and detailed subquestions. One of the subquestions which proved to be interesting and rewarding was the question whether being is a highest genus. It was Aristotle who already gave an answer to this question, an answer which differs fundamentally from the theories of Parmenides and Meinong who held that there was a most comprehensive class or predicate: It was 'being' for Parmenides and 'Gegenstand' for Meinong. Aristotle made two main assumptions concerning being:

- A1 The term 'being' has different meanings; there are mainly four: being in the sense of the 10 categories, of accident, of potentiality and actuality and of true (is the case) and false (is not the case).
- A2 Being is not a most comprehensive (highest) genus.

In fact Aristotle tried to prove A2 by the following argument:

1. For every genus there is a *differentia specifica* of it.
2. If $y$ is the *differentia specifica* of $x$ then it is not the case that: $y$ is a (subset of) $x$.



*K. Kijania-Placek and J. Woleński (eds.),*
*The Lvov–Warsaw School and Contemporary Philosophy,* 379–393.
© 1998 *Kluwer Academic Publishers.*

3. Assume that there is a genus (being) such that everything is properly included in it. Then the difference is (properly) included in it (by this assumption) and not included in it (by 1. and 2.).[1]

This argument can be translated into the following formal proof:

| | | |
|---|---|---|
| 1. | $\forall x(Gx \rightarrow \exists y(y \text{ Diff } x))$ | |
| 2. | $\forall x \forall y(y \text{ Diff } x \rightarrow \neg(y \subset x))$ | |
| $\rightarrow$ 3. | $\exists x \forall y(G(x) \wedge (y \subset x))$ | Assumption of the highest genus |
| 4. | $\forall y(Ga \wedge (y \subset a))$ | Existential Instantiation 3. |
| 5. | $Ga$ | from 4 |
| 6. | $\forall y(y \subset a)$ | from 4 |
| 7. | $Ga \rightarrow \exists y(y \text{ Diff } a)$ | Universal Instantiation 1. |
| 8. | $\exists y(y \text{ Diff } a)$ | M.P. 5, 7 |
| 9. | $(b \text{ Diff } a)$ | Existential Instantiation 8. |
| 10. | $(b \text{ Diff } a) \rightarrow \neg(b \subset a)$ | Universal Instantiation 2. |
| 11. | $\neg(b \subset a)$ | M.P. 9, 10 |
| 12. | $b \subset a$ | Universal Instantiation 6. |

13. Assumption 3. $\rightarrow ((b \subset a) \wedge \neg(b \subset a))$

For Aristotle premises 1 and 2 are true. They are in fact part of his doctrine of genus, species and differentia. And it is plain that they — together with assumption 3 — must lead to a contradiction.[2] We may doubt assumption 1 or 2 or both. But what is interesting is that Aristotle was convinced that being cannot be a highest genus, cannot be an all-comprehensive class and this he tried to prove in some way. With this conviction he anticipated a fundamental insight which was proved in a much more sophisticated way 2300 years later by mathematics: that classes or sets which embrace everything are contradictory.

## 2. PAULUS VENETUS: A SOLUTION OF THE LIAR

Before Paulus Venetus offers his own solution he discusses fourteen other proposals given by medieval logicians and philosophers. Among them there are some very interesting ones which have been rediscovered (or reinvented) independently in the twentieth century (as for instance the distinction of levels of languages or the so called zig-zag method mentioned by Russell). The basic idea of Paulus Venetus is his doctrine of the twofold function of a sentence: every (indicative) sentence (1) means its content, i.e. its proposition and (2) says of itself that it is so as it says, i.e. that it is true. The first function has been known since the Greeks and the Stoics pointed it out explicitly. The second is that which is usually neglected and therefore may lead to the Liar. In any case the second function has to be incorporated in an adequate truth condition because it guarantees the normal (positive) usage of language (of indicative sentences). Thus when I am saying to my friend (in an art exhibition): 'This picture is very beautiful' he interprets this sentence as if I would have added: it is so, or I think it is so. But this normal usage may be violated when I am using some ironical voice and some smile in order to indicate the contrary. These are exceptions to the normal (positive) usage of language.

The following comparison shows on the left side how the antinomy arises and on the right side how it is solved by Paulus Venetus by revising the truth condition where he incorporates function (2). Function (1) is expressed by Paulus Venetus as: the sentence *A* means the proposition *p*. Consequently the liar sentence is expressed by: the sentence *A* means that *A* is false. The second function is expressed as: the sentence *A* says of *A* that *A* is true, thus what the sentence *A* says of itself is: *A* is true.

| *How the liar arises:* | *How the liar is solved:* |
|---|---|
| 1. *A* means that *A* is false<br>(assumption of the liar) | 1.' *A* means that *A* is false<br>(assumption of the liar) |
| 2. If *A* means that *p* then:<br>*A* is true iff *p* | 2.' If *A* means that *p* then:<br>*A* is true iff *p* and *A* is true. |
| 3. If *A* means that *A* is false then:<br>*A* is true iff *A* is false<br>(from 2. by substitution) | 3.' If *A* means that *A* is false then:<br>*A* is true iff *A* is false and *A* is true<br>(from 2.' by substitution) |
| 4. *A* is true iff *A* is false<br>(from 1. and 3. by M.P.) | 4.' *A* is true iff *A* is false and *A* is true<br>(from 1.' and 3.' by M.P.) |
| | 5.' *A* is false<br>(from 4.' since the right side must be false) |

The essential point of Paulus Venetus is his revision of the truth-condition 2.[3] The unrevised version 2 is very much like Tarski's truth condition except the antecedent-condition which includes meanings. Tarski avoids them by distinguishing the metalinguistic name of the sentence from the translation of the sentence into the metalanguage.

## 3. FROM ARISTOTLE TO DE MORGAN: NON MODAL INTENSIONS

Aristotle distinguishes eight meanings of $a$ is *in* $b$, two of them are as follows:

Again, those divisions into which the form, apart from quantity, can be divided, are also called parts of the form. Hence species are called parts of their genus.

This passage is concerned with extension, the next one with intension:

The elements in the definition of each thing are also called parts of the whole. Hence the genus is even called a part of the species, whereas in another sense the species is part of the genus.[4]

This kind of non modal intensional inclusion was used by many philosophers like Boethius, Aquinas, Leibniz and De Morgan:

The logicians who have *recently* introduced the distinction of extension and comprehension, have altogether missed this opposition of the quantities, and have imagined that the quantities remain the same [ ... ] In this the logicians have abandoned both Aristotle and the laws of thought from which he drew the few clear words of his dictum: "the genus is said to be part of the species; but in another point of view ($\alpha\lambda\lambda\omega\varsigma$) the species is part of the genus". *All animal* is in *man*, notion in notion: *all man* is in *animal*, class in class. In the first, all the notion *animal* part of the notion *man*: in the second, all the class *man* part of the class *animal*. Here is the opposition of the quantities.[5]

There is hardly a problem of how to interpret the traditional concept of inclusion, if it is expressed by saying: species is contained (included) in genus like a part is in its whole. What is meant is the standard class-inclusion in the form:

$$a \subseteq b \leftrightarrow (\forall x(x \in a \rightarrow x \in b))$$

Since $\{x : x \in a\}$ may be interpreted as the extension of $a$, it is clear that $a \subseteq b$ and $a \subset b$ are called *extensional inclusions*.

There is, however, a problem of how to interpret the type of inclusion which is expressed by saying: genus is contained in species, insofar as the set of properties (characteristics) of the genus are included in the set of properties (characteristics) of the species; or since the genus-term is a part of the whole definiens by which species is defined. Or more generally and independent of an interpretation of the tradition: How can one define the kind of inclusion which is meant when I say: the set of properties which are common to all elements of class $a$ are included in the set of properties which are common to all elements of class $b$? It was clear to me from the very beginning that an adequate answer to this question will provide a key-concept for a new interpretation of intensions in the tradition. The solution to this question is the following definition of intensional inclusion: [6]

$a$ is intensionally included in $b$ (symbolically: $a \sqsubseteq b$) iff for all $Z$: if $a$ is included in $Z$ then $b$ is included in $Z$. Symbolically:

$$a \sqsubseteq b \leftrightarrow \forall_Z(a \subseteq Z \rightarrow b \subseteq Z)$$

The concept of intensional inclusion just defined is the dual concept to that of the familiar extensional inclusion $\subseteq$ usually called set- or class-inclusion.

That the given definition is in fact an adequate interpretation of the kind of inclusion involved in the principle of the inverse relation between extension and intension is proved by the fact that this principle follows as a theorem from the dual theory, i.e., $a \subseteq b \leftrightarrow b \sqsubseteq a$ is a theorem of the theory of (virtual) classes if it is extended by the above definition. It says: $a$ is extensionally included in $b$ iff $b$ is intensionally included in $a$. In order not to shift into serious Second Order one has to make some suitable restrictions on $Z$ in the above definition, similar to the restricted usage of $F$ in an identity axiom like $x = y \rightarrow (Fx \leftrightarrow Fy)$.

One may further extend duality by introducing a definition for intensional intersection like: $a \sqcap b = c \leftrightarrow \forall_Z((a \subseteq Z \wedge b \subseteq Z) \leftrightarrow c \subseteq Z)$. Adding these definitions of intensional inclusion and intersection to the standard theory of (virtual) classes leads to a dual theory of (virtual) classes which is both extensional and intensional. Since the extension is conservative and the definitions are noncreative (provided one does not add more) the dual part which is intensional is already implicitly contained in the standard theory we know. It only was not made explicit and so a reasonable and simple way of having some kind of intension which offers an adequate interpretation of traditional doctrines has not been used.

## 4. RELEVANCE IMPLICIT IN SYLLOGISTICS

The syllogistic modes of Aristotle's syllogistics allow three terms in the premises (subject-, predicate-, middle-term) and two of them in the conclusion (subject- and predicate term). This is an interesting relevance restriction for not introducing new predicates into the conclusion. The principle may be applied also in propositional logic concerning propositional variables. The two relevance principles can be defined as follows:

4.1 A valid implicational formula $\alpha \rightarrow \beta$ (inference $\alpha \vdash \beta$) is $A_0$-relevant iff there is no propositional variable which occurs in $\beta$ but not in $\alpha$.

4.2 A valid implicational formula $\alpha \rightarrow \beta$ (inference $\alpha \vdash \beta$) is $A_1$-relevant iff there is no propositional variable and no predicate which occurs in $\beta$ but not in $\alpha$.

The latter relevance restriction can be extended from predicates to special operators like epistemic operators, value operators or deontic operators. Thus extended this relevance criterion rules out most of the paradoxes in the areas of Deontic Logic, Value Theory, Epistemic Logic, and the Theories of Confirmation, Explanation, Law Statements, Disposition Predicates.

### 4.1. *Examples of $A_1$-relevant and irrelevant (valid) formulas*

$p \to p, p \to \neg\neg p$,
laws of commutation, association, distribution, De Morgan, modus ponens, modus tollens,
hypothetical syllogism in the form $((p \to q) \wedge (q \to r)) \to (p \to r)$,
disjunctive syllogism, $(p \wedge q) \to p$ (simplification),
$(p \wedge p) \leftrightarrow p, ((p \vee q) \to r) \to (p \to r), (p \vee (p \wedge q)) \to p$,
$(p \to q) \leftrightarrow (\neg q \to \neg p), ((p \to r) \wedge (q \to r)) \to ((p \vee q) \to r)$,
$((p \to q) \to p) \to p$.
$\forall x Fx \to \exists x Fx, \forall x Fx \to Fa, Fa \to \exists x Fx, ((a = b) \wedge Fa) \to Fb$,
$\forall x(Fx \wedge Gx) \to \forall x Fx, (\forall x(Fx \to Gx) \wedge Fa) \to Ga$,
$\forall x(Fx \to Gx) \to (\forall x Fx \to \forall x Gx) \ldots$ etc.

Valid formulas which are not $A_1$-relevant are:
$(p \wedge \neg p) \to q, \neg p \to (p \to q)$,
$p \to (p \vee q), \neg p \to (p \to (q \wedge \neg q)), (p \to q) \to ((p \wedge r) \to q)$.
$\forall x Fx \to \forall x(Fx \vee Gx), \forall x(Fx \to Gx) \to (\forall x(Fx \wedge Hx) \to Gx)$,
$(a = b) \to (Fa \to Fb)$.

### 4.2. *Properties of $A_0$-Relevance*

In two other papers[7] the following properties of $A_0$-relevance have been proved:

(1) $A_0$-relevant implication is transitive, i.e., if $\alpha \to \beta$ and $\beta \to \gamma$ are both $A_0$-relevant and valid then so is $\alpha \to \gamma$.
(2) The set of all valid implicational formulas satisfying $A_0$-relevance is closed under substitution, whereas for valid implicational formulas which are not $A_0$-relevant this is not the case.
(3) The set of all valid implicational formulas satisfying $A_0$-relevance is representable by finite matrices (truth-tables).
(4) The set of all valid implicational formulas satisfying $A_0$-relevance is not closed under modus ponens (with respect to $A_0$-relevance).
(5) $A_0$-relevant implication is not $A_0$-relevance preserving.
(6) If $\alpha \to \beta$ is $A_0$-relevant then all four cases are possible concerning the $A_0$-relevance of $\alpha$ and $\beta$ (provided $\alpha$ and $\beta$ are implicational formulas): Both, neither or one of both may be $A_0$-relevant.
(7) $A_0$-relevant implication is not closed under contraposition (for instance, $(p \wedge q) \to p$ is $A_0$-relevant but $\neg p \to \neg(p \wedge p)$ is not).

### 4.3. *Some Applications*

The valid but $A_1$-irrelevant formula $p \to (p \vee q)$ or the respective inference $p \vdash (p \vee q)$ is the culprit in a number of paradoxes: the Ross Paradox, the Paradox of Free Choice (the analogous paradoxes can be constructed with value-operators instead of deontic operators); Hesse's Confirmation Paradox, Goodman's Paradox, the Paradox of Irrelevant Confirmans-Weakening. The valid but $A_1$-irrelevant formula $\neg p \rightarrow (p \rightarrow q)$ viz. the inference $\neg p \vdash (p \rightarrow q)$ is responsible for the following paradoxes: the Paradox of Derived Obligation, the Paradox of Commitment (analogous paradoxes in value theory if the deontic operators are replaced by value operators) and the Paradox of Disposition Predicates. Applications of that sort can be continued.[8]

The $A_1$-criterion is not sufficient for more complicated kinds of paradoxes. For instance it allows tautologies in the conclusion, provided the propositional variables or predicates occurring in them are those used in the premises. A stronger criterion offers a solution to a number of further paradoxes and — if supplemented by splitting the consequence into smallest consequence elements — to the problem of verisimilitude.[9]

## 5. WEAKER LOGICS FOR THE INTERPRETATION OF PHILOSOPHICAL TEXTS

It might not be too well-known that Aristotle did not treat the principle of non-contradiction as equivalent with that of the *tertium non datur*. First he treats them in different places in his metaphysics. Second he knew that the *tertium non datur* leads to certain difficulties when applied to future contingencies (as is clear from *De Interpretatione*). Third he points out that the validity of the *tertium non datur* is dependent in a non-trivial sense on the definiteness of concepts involved.[10] Fourth, as Bernays observed, the *tertium non datur* is not needed to build up the syllogistic forms. As is well known, Brouwer rejected the *tertium non datur* as a valid principle for mathematics. The main reasons are problematic existence proofs accepted in classical mathematics. Intuitionistically a statement $\exists x Fx$ can only be established if one provides a construction of a natural number $k$ and a proof of $F(k)$. On the other hand, in order to prove $\neg\exists x Fx$, for each individual $n$ a proof of $\neg Fn$ would be required. That is, the translation of $\exists x Fx \vee \neg\exists x Fx$ is this: Either there is a construction of a natural number $k$ and a proof of $F(k)$ or there is a uniform proof that shows the falsity of $F(n)$ for each $n$. But clearly it might be that neither of the above alternatives is true.

If the range of $x$ in $Fx$ is finite then the validity of $\exists x Fx \vee \neg\exists x Fx$ has never been questioned because then $\forall x \neg Fx$ is nothing but a conjunction of a finite number of decidable statements. But if the range of $x$ in $\forall x \neg Fx$ is infinite one cannot give for each individual $n$ a proof of $\neg F(n)$ because the process with respect to an infinite domain can never be finished.

5.1.

Is it reasonable to use intuitionistic logic when interpreting philosophical texts, especially when they are connected with existence proofs? Elsewhere[11] I have defended the view that for the interpretation of philosophical texts the following rule seems adequate: Take (interpret) every part of the text but use logical principles which are as weak as possible. The rule to the contrary (strong logical principles and only selected parts of the text) is very widespread. If the first method is taken then philosophical texts which contain existence proofs (like those for the existence of God) should be interpreted — at least as a trial — by intuitionistic logic, even if such a proof would not end up with the existence but only with the non-absurdity of the existence. As an example I shall interpret Anselm's Ontological argument with intuitionistic logic.

Anselm's Ontological Argument consists of two parts. The first is concerned with the assumption that every man has in his mind the idea of a most perfect being in the sense of *id quo maius cogitari non potest*. One may very well doubt already this assumption.[12] But it seems to me a serious mistake if this first part is ignored in interpretations of Anselm's Argument, as is frequently done. In any case, interpretation with intuitionistic means makes it necessary to use this first part. Moreover the question of the definiteness of the predicate *id quo maius cogitari non potest* will enter again later.

The second part has the form of an indirect existence proof: Assume that it is not the case that such a most perfect being (in the sense of *id quo maius cogitari non potest*) exists in reality (but only in mind). Then it wouldn't be *id quo maius cogitari non potest* (i.e. most perfect) contradicting the assumption; since to exist both in the mind and in reality is more perfect than just to exist in the mind. i.e. to be *id quo maius cogitari non potest* (most perfect) implies existence in the mind and in reality.

### 5.2. *Anselm's Ontological Argument interpreted with Intuitionistic Logic*

Abbreviating *id quo maius cogitari non potest* by the predicate $QM$ we get the following *tertium non datur*: $\exists x QMx \vee \neg\exists x QMx$. Now according to intuitionistic mathematics this has to be interpreted as: Either there is a conceptual construction of an object $k$ and a proof of $QM(k)$ or there is a uniform proof that shows the falsity of $QM(n)$ for each object $n$. Clearly there is neither of both if $x$ is a $QM$. It is understandable that it would be hard to get a proof of $QM(k)$. But also to show the falsity of $QM(n)$ for each $n$ would mean something hardly possible. Although classically such a proof would not mean intuitionistically to show for each $n$ that it belongs to the class *id quo maius cogitare potest*, i.e., to the complement of $QM$, whatever that complement class might be, perhaps an all-embracing universal class and therefore suspect of inconsistency. The situation may seem different at first sight if we change the predicate $QM$ to the predicate '*id quo maius cogitari non potest* which exists in reality' ($QMR$). In this case we have two different 'negations' or complements: (a) ($\overline{QM}R$, i.e. *id quo maius cogitari potest which exists in reality*. This set can be finite such that a proof of $\forall x\overline{QM}Rx$ might be possible, even intuitionistically. But this set is not the one Anselm has in mind. (b) $QM$ , i.e. *id quo maius cogitari non potest which does not exist in reality*. This set is the one to be used in *a tertium non datur* of the form $\exists xQMRx \vee \neg\exists xQMRx$. And here we are in a similar situation as above. There is hardly a proof of $QMR(k)$ and there is hardly one of the falsity of $QMRn$ (with respect to $R$) for each $n$, i.e. one which shows for each $n$ that it belongs to $QM\overline{R}$.

One important reason for the similarity of this situation to the one in intuitionistic mathematics seems to be that the concept of a $QM$ or of a $QMR$ is rather indefinite and that the range of the variable $x$ (over natural and supernatural beings?) is also quite unclear.[13]

Let me finally show how Anselm's Ontological Argument would look if only intuitionistically valid steps are used. For this let $QM$, $QMM$ and $QMR$ be the following predicates respectively: *id quo maius cogitari non potest*, '*id quo maius cogitari non potest* which is in the mind', '*id quo maius cogitari non potest* which exists in reality'.

| | |
|---|---|
| 1. $QMMa$ | |
| 2. $\forall x(QMMx \rightarrow QMx)$ | |
| 3. $\forall x((QMMx \wedge \neg QMRx) \rightarrow \neg QMx)$ | |
| 4. $\neg\exists xQMRx$ | Assumption for indirect proof |
| 5. $QMMa \rightarrow QMa$ | from 2 (instantiation) |
| 6. $QMa$ | from 1, 5 M. P. |
| 7. $\forall x\neg QMRx$ | from 4 |
| 8. $\neg QMRa$ | from 7 (instantiation) |
| 9. $QMMa \wedge \neg QMRa$ | from 1 and 8 |
| 10. $(QMMa \wedge \neg QMRa) \rightarrow \neg QMa$ | from 3 (instantiation) |
| 11. $\neg QMa$ | from 9 , 10 M. P. |
| 12. $QMa \wedge \neg QMa$ | from 6, 11 |
| 13. $\neg\exists xQMRx \rightarrow (QMa \wedge \neg QMa)$ | Conditional indirect proof |
| 14. $\neg(QMa \wedge \neg QMa)$ | Principle of Non-Contradiction |
| 15. $\neg\neg\exists xQMRx$ | from 13, 14 M. T. |

Premise 1 is Anselm's assumption that a certain $a$ is such an '*id quo maius cogitari non potest* which is in the mind'. Observe that $\exists xQMMx$ would not be sufficient intuitionistically as a starting point, since such an existential statement is only intuitionistically sound if a construction of a concrete object $a$ and a proof of QMMa can be given. But I think that what Anselm claims is in fact $QMMa$, since everybody has in his own mind the concept of a $QM$. Premise 2 is rather trivial, even if $QMa$ and $\exists xQMx$ can be derived from premises 1 and 2, since the predicate $QM$ is (here) understood in such a way as to leave completely open the question of whether $x$ is a $QM$ which exists in reality or which does not exist in reality.

Premise 3 says that an $x$ which is a $QM$ which is in the mind but is (at the same time) a $QM$ which is not in reality is in fact not a $QM$ (at all). This is one of the essential points in Anselm's argument. In a more detailed translation of the argument one would certainly need a justification for premise 3 by the further premise that $x$ being $QMM$ and $QMR$ is greater (i.e. the only candidate for being $QM$) than $x$ being $QMM$ and not $QMR$. But these details are not essential for what is intended to be shown here. As long as premise 3 is a consequence of these more sophisticated premises it can be used in the proof above. Since my interpretation is one of Chapter 2 of the *Proslogion* (not of Chapter 3 where modalities come in) the essential statement of the text for premise 3 is: *Si ergo id quo maius cogitari non potest, est in solo intellectu: id ipsum quo maius cogitari non potest, est quo maius cogitari potest.* Therefore it would seem that having a $QMx$ is a non-$QMx$ (i.e. $QMx \rightarrow \neg QMx$) in the consequent part is a translation more close to the text. This can be admitted and premise 3 can be changed into: $\forall x((QMMx \wedge \neg QMRx) \rightarrow (QMx \rightarrow \neg QMx))$. As can easily be seen, the new premise leads together with the others to the same result since $QMa$ is obtained from premises 1 and 2. Observe further that premise 3 is not intuitionistically equivalent to $\forall x(QMx \rightarrow (QMMx \rightarrow QMRx))$ although it would be so classically. This observation is important since the above result by contraposition seems much too strong a claim and not justified by Anselm's text. It shows that there are non-trivial differences when Anselm's argument is interpreted by intuitionistic logic rather than classical logic. Premise 4 is the negative assumption for the indirect proof. For the steps in the argument observe the following: Universal Instantiation, *modus ponens* and *modus tollens* are intuitionistically valid. The step from 4 to 7 is sound since $\neg\exists xFx$ and $\forall x\neg Fx$ are intuitionistically equivalent.

The conclusion (15) is rather essential. Observe that it would be not allowed intuitionistically to pass from (15) to $\exists xQMRx$ — as in classical logic — since $\neg\neg p \rightarrow p$ is invalid in intuitionistic logic. Therefore the argument ends up not with the existence of the '*id quo maius cogitari non potest* which exists in reality'. In intuitionistic terms it ends up with the absurdity of the absurdity, i.e. the non-contradictoriness of the $QMR$. Where classically one proves the existence (in an indirect proof) one can prove intuitionistically only the non-contradictoriness (the consistency). This is just an instance of a theorem of

Glivenko which says that if $p$ can be proved classically the absurdity of the absurdity (i.e. the non-contradictoriness) of $p$ can be proved intuitionistically.

## 6. IS THE PRACTICAL SYLLOGISM LOGICALLY VALID?

The following argument is called by von Wright his final formulation of the practical syllogism:[14]

(1) From now on $A$ intends to bring about $X$ at time $t$.
(2) From now on $A$ considers that unless he does $X$ no later than at time $t'$, he cannot bring about $X$ at time $t$.

---

(3) Therefore, no later than when he thinks time $t'$ has arrived, $A$ sets himself to do $Y$, unless he forgets about time or is prevented.

The symbols $A$, $X$, $Y$ and $t$ in the above argument can be regarded as variables or place-holders, i.e., the argument is understood as a general argument-form. The question whether this argument (or argument-form) is logically valid has been a matter of dispute from the time it was first proposed.[15]

But in fact the answer is very simple: The Practical Syllogism is not logically valid because there is no logically valid argument form of either (two valued) propositional- or first order predicate calculus of which the Practical Syllogism would be an instance.[16]

Some authors who commented on the Practical Syllogism spelled out clearly that it is logically invalid.[17] Others are more or less ambiguous about the matter.[18] How to change the Practical Syllogism into a logically valid argument? My proposal is to change the Practical Syllogism into an argument which is a specific form of the general structure of teleological arguments which runs thus.

### 6.1. *General Structure of Teleological Arguments*

1. For all $x$: If
   a) $p$ is a goal for $x$ with respect to the field $f$
   b) and $x$ beliefs (or knows) that $q$ is a necessary condition for $p$ with respect to $f$
   c) and $x$ is able to bring it about that $q$
   d) and $x$ prefers $p$ over non-$q$ with respect to $f$
   then: $x$ acts in such a way that $q$ obtains.
2. a) $p$ is a goal for person A with respect to (special) $f$

b) and A beliefs (or knows) that $q$ is a necessary condition for $p$ with respect to $f$

c) and A is able to bring it about that $q$

d) and A prefers $p$ over non-$q$ with respect to $f$

3. Therefore: A acts in such a way that $q$ obtains.

That this argument is a logically valid argument is easily seen by the fact that it is an instance of modus ponens + universal instantiation. Moreover: The premises and the conclusions are empirical statements, i.e. statements which are factually (not logically) true or false. Whether the first premise (1) is universally true is questionable. It may have exceptions such that the universal statement is false (though the conclusion still follows from the premises). If the first premise is interpreted as a statistical law (a law which allows exceptions but is true for most cases) one might say that the conclusion is not fully, but highly significantly explained. Whether probability is applicable has to be decided in the special case (or area). The argument is a teleological argument in the narrower sense since it contains not only value-statements in general in the premises but value statements the values of which are goals, ends, purposes. The conclusion on the other hand does not contain a value-statement. Thus turned into a jargon it "explains facts with the help of values" or when applied to human actions: it explains actions with the help of motives or goals.

### 6.2. *Proposed Change for the Practical Syllogism*

The proposed change of the Practical Syllogism becomes clear if we compare the gist of both arguments (that of von Wright and the general one given above) in its most simple form.

(1) $A$ intends $p$
(2) $A$ considers $q$ necessary for $p$

---

(3) Therefore: $A$ sets himself to do $q$

(1') $\forall x((x \text{ intends } p \land x \text{ considers } q \text{ necessaryfor } p) \rightarrow x \text{ sets himself to do } q)$
(2') $A$ intends $p \land A$ considers $q$ necessary for $p$

---

(3') Therefore: $A$ sets himself to do $q$

Whereas the first argument is logically invalid the second is valid. The main difference is that in the second there is a law-like statement which contains already the connection between premises and conclusion of the first argument as a universal implication. The premises and the conclusion of the second argument are understood to be true or false empirical statements. Nothing is claimed to be analytic; on the contrary, law-like statements like the first premise have to be tested empirically. It then will turn out, especially if one considers all the counterexamples which have been discussed in connection with Practical Syllogism, that the first premise (the kind of law-statement) has to be formulated in a much more sophisticated way. A large group of counterexamples are ruled out by the formulation of the first premise proposed in the general structure of the teleological argument given in 6.1, for others further revision might be necessary.

*University of Salzburg and Institut für Wissenschaftstheorie,*
*International Research Center, Salzburg*

## NOTES

[1] Metaphysics, 998b22-27.

[2] Bocheński in his (1961), p. 54, has stated Aristotle's argument with essentially the same three premises as above, but without symbolizing them and without giving a formal proof. He then says that the argument is fallacious because "*b* 'is' *a*" (where *a* is the highest genus) is meaningless. He seems to think that there should be some type-theoretical distinction. But when symbolized as above no such problem seems to arise, i.e. the contradiction is caused by the incompatibility of the three premises.

[3] The above picture is, if I am right, the correct gist of Bocheński's interpretation in his (1961), p. 250, which however includes at least redundancies if not misleading parts. I have elaborated Paulus Venetus' idea and two new types of truth conditions for solving the liar in more detail in my forthcoming book (1997), Ch. 7.

[4] Aristotle (Metaphysics) 1023b17 and 1023b23.

[5] De Morgan (1966).

[6] Cf. Weingartner (1974) and (1976, 1) Ch. 3.46.

[7] Weingartner (1985) and Weingartner (1994). Some of the proofs in the latter essay are due to Andrzej Wroński.

[8] For details see Weingartner-Schurz (1986).

[9] This stronger criterion was proposed first in Schurz-Weingartner (1987) and has been developed further in Weingartner (1988) and Schurz (1991).

[10] Cf. (Met) 1012a20-24.

[11] Weingartner (1996).

[12] Cf. for instance Thomas Aquinas (STh) I, 2, 1 ad 2.

[13] Recall Aristotle's view that the application of the *tertium non datur* needs definite concepts, cf. note 10.

[14] G. H. von Wright (1971), p. 96.

[15] G. H. von Wright (1963). Many authors including von Wright cite also passages from Aristotle, especially that of *Nichomachean Ethics* 1147a25-33 and of *De Motu Animalium* 701a4-33. As R. Stranzinger recently pointed out in his (1984) they seem, however, to have overlooked so far the following interesting passages in the *Nicomachean Ethics*: 1141a and 1140a, b. For a discussion of the Aristotelian views on Practical Syllogism see also J. Hintikka (1974).

[16] Moreover nobody has proposed an extended or new logic — as far as I know — in which such an inference is a general valid form. To try to solve the problem with strong axioms which are then called 'analytically true' does not seem to be a solution but rather a *petitio principii*.

[17] One is N. Rescher in his essay (1966). This essay contains some interesting counterexamples to the Practical Syllogism (formulated with the help of Rescher's task-theses). Cf. p. 129, and a lot of related interesting topics (for instance about practical inconsistencies). Another is J. Hintikka in (1974), especially p. 97 ff. though he is concerned mainly with the Aristotelian form of Practical inference. However Rescher's proposal to revise the Practical Syllogism is a completely different one, since he interprets it as a normative argument with a norm (task-thesis) as the conclusion. Whereas in my proposal the conclusion of the Practical Syllogism is an indicative (factual) sentence and only the premises contain values (goals, motives).

[18] So for instance Tuomela says in his interesting work (1974), p. 20f.: "We shall accordingly below accept the practical syllogism as valid for conceptual reasons." It is not clear, however, what the phrase "valid for conceptual reasons" really means. Especially since two pages later he gives an "amended version" (ibid., p. 23) of the Practical Syllogism which is an essentially different argument. In fact it is of the general structure given below which I proposed in (1984), Ch. 4.5 (cf. also (1971), p. 109f.) as a general form of a teleological argument. Cf. also Tuomela (1977), p. 177ff. Another ambiguous and obscure statement is given by Weinberger when he says: "P1 (Practical Inference) und teleologische Überlegung ist Schließen zwar nicht im Sinne von wahrheitskonservierender Deduktion, aber im Sinne formaler Schlüssigkeit". Cf. (199), p. 28. In R. Bubner's chapter on Practical Inference (cf. his (1982)) it is not clear what kind of sentences are involved in his "Praktischer Schluss". Be cause of this ambiguity it is not clear what he denies if he denies the correctness of the Practical Syllogism.

## REFERENCES


Aristotle: 1958, *Metaphysics*, intr. comm. W.D. Ross, Oxford.

Bocheński, J. M.: 1961, *A History of Formal Logic*, University of Notre Dame Press, Notre Dame, IN.

Bubner, R.: 1982, 'Handlung, Sprache, Vernunft', Frankfurt.

De Morgan, A.: 1966, 'Syllabus of a Proposed System of Logic', in De Morgan: *On the Syllogism and other Logical Writings*, P. Heath (ed.), Routledge & Kegan Paul, London, pp. 147–207.

Hintikka, J.: 1974, 'Practical vs. Theoretical Reason — An Ambiguous Legacy', in Körner, St. (ed.): *Practical Reason*, Dordrecht, pp. 83–102.

Rescher, N.: 1966, 'Practical Reasoning and Values', *The Philosophical Quarterly* **16**, pp. 121–136.

Schurz, G.: 1991, 'Relevant Deduction', *Erkenntnis* **35**, pp. 391–437.

Schurz, G., Weingartner, P.: 1987, 'Verisimilitude Defined by Relevant Consequence- Elements. A New Reconstruction of Popper's Original Idea', in Kuipers, T. (ed.): *What is Closer-to-the-Truth?*, Rodopi, Amsterdam, pp. 47–77.

Stranzinger, R.: 1984, 'Anmerkungen zum Praktischen Syllogismus', in Fischer, M., Schreiner, M., Hock, E. (eds.): *Hermeneutik und Strukturtheorie*, separate issue of *Archiv für Rechts- und Sozialphilosophie*, pp. 111–120.

Thomas Aquinas: STh 1952, *Summa Theologica*. Caramello, P. (ed.), Marietti-Rome.

Tuomela, R.: 1974, *Human Action and its Explanation*, Reports from the Institute of Philosophy, University of Helsinki, No. 2.

Tuomela, R.: 1977, *Human Action and its Explanation*, Reidel, Dordrecht.

von Wright, H.: 1963, 'Practical Inference', in *The Philosophical Review* **72**, pp. 159–179.

von Wright, H.: 1971, *Explanation and Understanding*, Cornell University Press, Ithaca, N.Y.
Weinberger, O.: 1979, 'Handeln und Schließen, Überlegungen zum Begriff der praktischen Inferenz', *Philosophica* **23**, pp. 5–36.
Weingartner, P.: 1971, *Wissenschaftstheorie I*, Frommann-Holzboog, Stuttgart-Bad Cannstatt (2nd Ed. 1978).
Weingartner, P.: 1974 'On the Characterization of Entities by Means of Individuals and Properties', *Journal of Philosophical Logic* **3**, pp. 323–336.
Weingartner, P.: 1976, *Wissenschaftstheorie II, I. Grundlagenprobleme der Logik und Mathematik*, Frommann-Holzboog, Stuttgart-Bad Cannstatt.
Weingartner, P.: 1982, 'Conditions of Rationality for the Concepts Belief, Knowledge and Assumption', *Dialectica* **36**, pp. 243–263.
Weingartner, P.: 1984, 'On the Introduction of Teleological Arguments into Scientific Discourse', in *La Sicence Face aux Attentes de l'Homme Contemporain. Colloque de l'Acadèmie Internationale de Philosophie des Science*, Sevilla 1983, Archives de l'Institut Internationale des Sciences Théoriques **26**, Bruxelles, pp. 196-218.
Weingartner, P.: 1985, 'A Simple Relevance-Criterion for Natural Language and its Semantics', in Dorn, G., Weingartner, P. (eds.): *Foundations of Logic and Linguistics: Problems and the Solutions*, Plenum Press, New York, pp. 563–575.
Weingartner, P.: 1988, 'Remarks on the Consequence-Class of Theories', in Scheibe, E. (ed.): *The Role of Experience in Science*, de Gruyter, Berlin, pp. 161–180.
Weingartner, P.: 1994, 'Can There be Reasons for Putting Limitations on Classical Logic?', in Humphries, P. (ed.): *Patrick Suppes, Scientific Philosopher*, vol. 3, Kluwer Academic Publishers, Dordrecht, pp. 89–124.
Weingartner, P.: 1996, 'Kann sich das natürliche Gesetz ändern?', Kommentar zu Thomas von Aquin, Summa Theologica I-II, 94, 5, in Löffler, W., Runggaldier, E. (eds.): *Dialog und System. St. Augustin*, Akademia Verlag, St. Augustin, pp. 197–230.
Weingartner, P.: 1997, 'Basic Questions on Truth', (forthcoming).
Weingartner, P., Schurz, G.: 1986, 'Paradoxes Solved by Two Simple Relevance Criteria', *Logique et Analyse* **113**, pp. 3–40.

# INDEX OF NAMES

190. W. Callebaut and R. Pinxten, *Evolutionary Epistemology*. A Multiparadigm Program, with a Complete Evolutionary Epistemology Bibliograph. 1987 ISBN 90-277-2582-9
191. J. Kmita, *Problems in Historical Epistemology*. 1988 ISBN 90-277-2199-8
192. J. H. Fetzer (ed.), *Probability and Causality*. Essays in Honor of Wesley C. Salmon, with an Annotated Bibliography. 1988 ISBN 90-277-2607-8; Pb 1-5560-8052-2
193. A. Donovan, L. Laudan and R. Laudan (eds.), *Scrutinizing Science*. Empirical Studies of Scientific Change. 1988 ISBN 90-277-2608-6
194. H.R. Otto and J.A. Tuedio (eds.), *Perspectives on Mind*. 1988 ISBN 90-277-2640-X
195. D. Batens and J.P. van Bendegem (eds.), *Theory and Experiment*. Recent Insights and New Perspectives on Their Relation. 1988 ISBN 90-277-2645-0
196. J. Österberg, *Self and Others*. A Study of Ethical Egoism. 1988 ISBN 90-277-2648-5
197. D.H. Helman (ed.), *Analogical Reasoning*. Perspectives of Artificial Intelligence, Cognitive Science, and Philosophy. 1988 ISBN 90-277-2711-2
198. J. Wolenski, *Logic and Philosophy in the Lvov-Warsaw School*. 1989 ISBN 90-277-2749-X
199. R. Wójcicki, *Theory of Logical Calculi*. Basic Theory of Consequence Operations. 1988 ISBN 90-277-2785-6
200. J. Hintikka and M.B. Hintikka, *The Logic of Epistemology and the Epistemology of Logic*. Selected Essays. 1989 ISBN 0-7923-0040-8; Pb 0-7923-0041-6
201. E. Agazzi (ed.), *Probability in the Sciences*. 1988 ISBN 90-277-2808-9
202. M. Meyer (ed.), *From Metaphysics to Rhetoric*. 1989 ISBN 90-277-2814-3
203. R.L. Tieszen, *Mathematical Intuition*. Phenomenology and Mathematical Knowledge. 1989 ISBN 0-7923-0131-5
204. A. Melnick, *Space, Time, and Thought in Kant*. 1989 ISBN 0-7923-0135-8
205. D.W. Smith, *The Circle of Acquaintance*. Perception, Consciousness, and Empathy. 1989 ISBN 0-7923-0252-4
206. M.H. Salmon (ed.), *The Philosophy of Logical Mechanism*. Essays in Honor of Arthur W. Burks. With his Responses, and with a Bibliography of Burk's Work. 1990 ISBN 0-7923-0325-3
207. M. Kusch, *Language as Calculus vs. Language as Universal Medium*. A Study in Husserl, Heidegger, and Gadamer. 1989 ISBN 0-7923-0333-4
208. T.C. Meyering, *Historical Roots of Cognitive Science*. The Rise of a Cognitive Theory of Perception from Antiquity to the Nineteenth Century. 1989 ISBN 0-7923-0349-0
209. P. Kosso, *Observability and Observation in Physical Science*. 1989 ISBN 0-7923-0389-X
210. J. Kmita, *Essays on the Theory of Scientific Cognition*. 1990 ISBN 0-7923-0441-1
211. W. Sieg (ed.), *Acting and Reflecting*. The Interdisciplinary Turn in Philosophy. 1990 ISBN 0-7923-0512-4
212. J. Karpiński, *Causality in Sociological Research*. 1990 ISBN 0-7923-0546-9
213. H.A. Lewis (ed.), *Peter Geach: Philosophical Encounters*. 1991 ISBN 0-7923-0823-9
214. M. Ter Hark, *Beyond the Inner and the Outer*. Wittgenstein's Philosophy of Psychology. 1990 ISBN 0-7923-0850-6
215. M. Gosselin, *Nominalism and Contemporary Nominalism*. Ontological and Epistemological Implications of the Work of W.V.O. Quine and of N. Goodman. 1990 ISBN 0-7923-0904-9
216. J.H. Fetzer, D. Shatz and G. Schlesinger (eds.), *Definitions and Definability*. Philosophical Perspectives. 1991 ISBN 0-7923-1046-2
217. E. Agazzi and A. Cordero (eds.), *Philosophy and the Origin and Evolution of the Universe*. 1991 ISBN 0-7923-1322-4
218. M. Kusch, *Foucault's Strata and Fields*. An Investigation into Archaeological and Genealogical Science Studies. 1991 ISBN 0-7923-1462-X

219. C.J. Posy, *Kant's Philosophy of Mathematics*. Modern Essays. 1992 ISBN 0-7923-1495-6
220. G. Van de Vijver, *New Perspectives on Cybernetics*. Self-Organization, Autonomy and Connectionism. 1992 ISBN 0-7923-1519-7
221. J.C. Nyíri, *Tradition and Individuality*. Essays. 1992 ISBN 0-7923-1566-9
222. R. Howell, *Kant's Transcendental Deduction*. An Analysis of Main Themes in His Critical Philosophy. 1992 ISBN 0-7923-1571-5
223. A. García de la Sienra, *The Logical Foundations of the Marxian Theory of Value*. 1992 ISBN 0-7923-1778-5
224. D.S. Shwayder, *Statement and Referent*. An Inquiry into the Foundations of Our Conceptual Order. 1992 ISBN 0-7923-1803-X
225. M. Rosen, *Problems of the Hegelian Dialectic*. Dialectic Reconstructed as a Logic of Human Reality. 1993 ISBN 0-7923-2047-6
226. P. Suppes, *Models and Methods in the Philosophy of Science: Selected Essays*. 1993 ISBN 0-7923-2211-8
227. R. M. Dancy (ed.), *Kant and Critique: New Essays in Honor of W. H. Werkmeister*. 1993 ISBN 0-7923-2244-4
228. J. Woleński (ed.), *Philosophical Logic in Poland*. 1993 ISBN 0-7923-2293-2
229. M. De Rijke (ed.), *Diamonds and Defaults*. Studies in Pure and Applied Intensional Logic. 1993 ISBN 0-7923-2342-4
230. B.K. Matilal and A. Chakrabarti (eds.), *Knowing from Words*. Western and Indian Philosophical Analysis of Understanding and Testimony. 1994 ISBN 0-7923-2345-9
231. S.A. Kleiner, *The Logic of Discovery*. A Theory of the Rationality of Scientific Research. 1993 ISBN 0-7923-2371-8
232. R. Festa, *Optimum Inductive Methods*. A Study in Inductive Probability, Bayesian Statistics, and Verisimilitude. 1993 ISBN 0-7923-2460-9
233. P. Humphreys (ed.), *Patrick Suppes: Scientific Philosopher*. Vol. 1: Probability and Probabilistic Causality. 1994 ISBN 0-7923-2552-4
234. P. Humphreys (ed.), *Patrick Suppes: Scientific Philosopher*. Vol. 2: Philosophy of Physics, Theory Structure, and Measurement Theory. 1994 ISBN 0-7923-2553-2
235. P. Humphreys (ed.), *Patrick Suppes: Scientific Philosopher*. Vol. 3: Language, Logic, and Psychology. 1994 ISBN 0-7923-2862-0
Set ISBN (Vols 233–235) 0-7923-2554-0
236. D. Prawitz and D. Westerståhl (eds.), *Logic and Philosophy of Science in Uppsala*. Papers from the 9th International Congress of Logic, Methodology, and Philosophy of Science. 1994 ISBN 0-7923-2702-0
237. L. Haaparanta (ed.), *Mind, Meaning and Mathematics*. Essays on the Philosophical Views of Husserl and Frege. 1994 ISBN 0-7923-2703-9
238. J. Hintikka (ed.), *Aspects of Metaphor*. 1994 ISBN 0-7923-2786-1
239. B. McGuinness and G. Oliveri (eds.), *The Philosophy of Michael Dummett*. With Replies from Michael Dummett. 1994 ISBN 0-7923-2804-3
240. D. Jamieson (ed.), *Language, Mind, and Art*. Essays in Appreciation and Analysis, In Honor of Paul Ziff. 1994 ISBN 0-7923-2810-8
241. G. Preyer, F. Siebelt and A. Ulfig (eds.), *Language, Mind and Epistemology*. On Donald Davidson's Philosophy. 1994 ISBN 0-7923-2811-6
242. P. Ehrlich (ed.), *Real Numbers, Generalizations of the Reals, and Theories of Continua*. 1994 ISBN 0-7923-2689-X
243. G. Debrock and M. Hulswit (eds.), *Living Doubt*. Essays concerning the epistemology of Charles Sanders Peirce. 1994 ISBN 0-7923-2898-1
244. J. Srzednicki, *To Know or Not to Know*. Beyond Realism and Anti-Realism. 1994 ISBN 0-7923-2909-0
245. R. Egidi (ed.), *Wittgenstein: Mind and Language*. 1995 ISBN 0-7923-3171-0

246. A. Hyslop, *Other Minds*. 1995 ISBN 0-7923-3245-8
247. L. Pólos and M. Masuch (eds.), *Applied Logic: How, What and Why*. Logical Approaches to Natural Language. 1995 ISBN 0-7923-3432-9
248. M. Krynicki, M. Mostowski and L.M. Szczerba (eds.), *Quantifiers: Logics, Models and Computation*. Volume One: Surveys. 1995 ISBN 0-7923-3448-5
249. M. Krynicki, M. Mostowski and L.M. Szczerba (eds.), *Quantifiers: Logics, Models and Computation*. Volume Two: Contributions. 1995 ISBN 0-7923-3449-3
Set ISBN (Vols 248 + 249) 0-7923-3450-7
250. R.A. Watson, *Representational Ideas from Plato to Patricia Churchland*. 1995
ISBN 0-7923-3453-1
251. J. Hintikka (ed.), *From Dedekind to Gödel*. Essays on the Development of the Foundations of Mathematics. 1995 ISBN 0-7923-3484-1
252. A. Wiśniewski, *The Posing of Questions*. Logical Foundations of Erotetic Inferences. 1995 ISBN 0-7923-3637-2
253. J. Peregrin, *Doing Worlds with Words*. Formal Semantics without Formal Metaphysics. 1995 ISBN 0-7923-3742-5
254. I.A. Kieseppä, *Truthlikeness for Multidimensional, Quantitative Cognitive Problems*. 1996 ISBN 0-7923-4005-1
255. P. Hugly and C. Sayward: *Intensionality and Truth*. An Essay on the Philosophy of A.N. Prior. 1996 ISBN 0-7923-4119-8
256. L. Hankinson Nelson and J. Nelson (eds.): *Feminism, Science, and the Philosophy of Science*. 1997 ISBN 0-7923-4162-7
257. P.I. Bystrov and V.N. Sadovsky (eds.): *Philosophical Logic and Logical Philosophy*. Essays in Honour of Vladimir A. Smirnov. 1996 ISBN 0-7923-4270-4
258. Å.E. Andersson and N-E. Sahlin (eds.): *The Complexity of Creativity*. 1996
ISBN 0-7923-4346-8
259. M.L. Dalla Chiara, K. Doets, D. Mundici and J. van Benthem (eds.): *Logic and Scientific Methods*. Volume One of the Tenth International Congress of Logic, Methodology and Philosophy of Science, Florence, August 1995. 1997 ISBN 0-7923-4383-2
260. M.L. Dalla Chiara, K. Doets, D. Mundici and J. van Benthem (eds.): *Structures and Norms in Science*. Volume Two of the Tenth International Congress of Logic, Methodology and Philosophy of Science, Florence, August 1995. 1997 ISBN 0-7923-4384-0
Set ISBN (Vols 259 + 260) 0-7923-4385-9
261. A. Chakrabarti: *Denying Existence*. The Logic, Epistemology and Pragmatics of Negative Existentials and Fictional Discourse. 1997 ISBN 0-7923-4388-3
262. A. Biletzki: *Talking Wolves*. Thomas Hobbes on the Language of Politics and the Politics of Language. 1997 ISBN 0-7923-4425-1
263. D. Nute (ed.): *Defeasible Deontic Logic*. 1997 ISBN 0-7923-4630-0
264. U. Meixner: *Axiomatic Formal Ontology*. 1997 ISBN 0-7923-4747-X
265. I. Brinck: *The Indexical 'I'*. The First Person in Thought and Language. 1997
ISBN 0-7923-4741-2
266. G. Hölmström-Hintikka and R. Tuomela (eds.): *Contemporary Action Theory*. Volume 1: Individual Action. 1997 ISBN 0-7923-4753-6; Set: 0-7923-4754-4
267. G. Hölmström-Hintikka and R. Tuomela (eds.): *Contemporary Action Theory*. Volume 2: Social Action. 1997 ISBN 0-7923-4752-8; Set: 0-7923-4754-4
268. B.-C. Park: *Phenomenological Aspects of Wittgenstein's Philosophy*. 1998
ISBN 0-7923-4813-3
269. J. Paśniczek: *The Logic of Intentional Objects*. A Meinongian Version of Classical Logic. 1998 ISBN 0-7923-4880-X
270. P.W. Humphreys and J.H. Fetzer (eds.): *The New Theory of Reference*. Kripke, Marcus, and Its Origins. 1998
ISBN 0-7923-4898-2

271. K. Szaniawski (au.) A. Chmielewski and J. Woleński (eds.): *On Science, Inference, Information and Decision Making*. Selected Essays in the Philosophy of Science. 1998 ISBN 0-7923-4922-9
272. G.H. von Wright: *In the Shadow of Descartes*. Essays in the Philosophy of Mind. 1998 ISBN 0-7923-4992-X
273. K. Kijania-Placek and J. Woleński (eds.): *The Lvov–Warsaw School and Contemporary Philosophy*. 1998 ISBN 0-7923-5105-3

KLUWER ACADEMIC PUBLISHERS – DORDRECHT / BOSTON / LONDON

GPSR Compliance

*The European Union's (EU) General Product Safety Regulation (GPSR) is a set of rules that requires consumer products to be safe and our obligations to ensure this.*

*If you have any concerns about our products, you can contact us on ProductSafety@springernature.com*

In case Publisher is established outside the EU, the EU authorized representative is:

Springer Nature Customer Service Center GmbH
Europaplatz 3
69115 Heidelberg, Germany

**Batch number: 10755284**

Printed by Printforce, the Netherlands